The Gardener's Guide to
Native Plants
of the Southern Great Lakes Region

Rick Gray and Shaun Booth

FIREFLY BOOKS

A Firefly Book

Published by Firefly Books Ltd. 2024

Copyright © 2024 Firefly Books Ltd.
Text copyright © 2024 Rick Gray and Shaun Booth
Photos © 2024 Rick Gray, except as listed on page 352
Cover photos © 2024 Rick Gray

All rights reserved. No part of this publication
may be reproduced, stored in a retrieval system, or
transmitted in any form or by any means, electronic,
mechanical, photocopying, recording or otherwise,
without the prior written permission of the Publisher.

2nd printing, 2024

Library of Congress Control Number: 2023946730

Library and Archives Canada Cataloguing in Publication
Title: The gardener's guide to native plants of the
    Southern Great Lakes region / Rick Gray and Shaun
    Booth.
Other titles: Native plants of the Southern Great
    Lakes region
Names: Gray, Rick (Native plant gardener), author. |
    Booth, Shaun, author.
Description: Includes bibliographical references and
    index.
Identifiers: Canadiana 20230542999 | ISBN
    9780228104605 (softcover)
Subjects: LCSH: Native plant gardening—Ontario,
    Southwestern. | LCSH: Native plant gardening—
    Lake States. | LCSH: Native plant gardens—
    Ontario, Southwestern. | LCSH: Native plant
    gardens—Lake States. | LCSH: Endemic plants—
    Ontario, Southwestern. | LCSH: Endemic plants—
    Lake States.
Classification: LCC SB439.24.G74 G73 2024 | DDC
    639.9/5177—dc23

Published in the United States by
Firefly Books (U.S.) Inc.
P.O. Box 1338, Ellicott Station
Buffalo, New York 14205

Published in Canada by
Firefly Books Ltd.
50 Staples Avenue, Unit 1
Richmond Hill, Ontario L4B 0A7

Cover and interior design: Hartley Millson

Printed in China | E

Canada

We acknowledge the financial support
of the Government of Canada.

Acknowledgments

We would like to thank the following for their
generous support in the making of this book.
Without their help, this would have been an almost
impossible task. First, we'd like to thank the many
amateur and professional botanists who provided
photos. Although we (the authors) have been photo-
graphing native plants for years, it wasn't until we
started putting this book together that we realized
how many images we had not taken. We'd also
like to thank the following professional botanists
and ecologists who looked over the range maps
for their states and provided invaluable feedback:
Scott Namestnik, Botanist, Indiana Department
of Natural Resources; Tyler Bassett, Botanist
and Plant Ecologist, Michigan Natural Features
Inventory; Steve Young, Chief Botanist, New York
Natural Heritage Program; Rick Gardner, Chief
Botanist, Ohio Department of Natural Resources;
and Steve Grund, Botanist, Pennsylvania Natural
Heritage Program. We'd like to offer a special thanks
to William van Hemessen, freelance consulting
Ecologist and Botanist, for the detailed feedback
offered for our range maps in Ontario and parts of
Michigan, and for the encouragement he offered for
this book. His enthusiasm for what we were trying
to accomplish encouraged us to keep going when we
were feeling overwhelmed with what we had taken
on. We'd also like to thank Lorraine Johnson, author
of several great books on native plant gardening,
for her ongoing support in general, and her genu-
inely appreciated advice on the business of getting
published. Finally, we'd like to acknowledge Simone
Hébert Allard, author of a beautiful and comprehensive field guide entitled *Manitoba Butterflies* (Turn-
stone Press). Her work was the inspiration for the
book you hold in your hands.

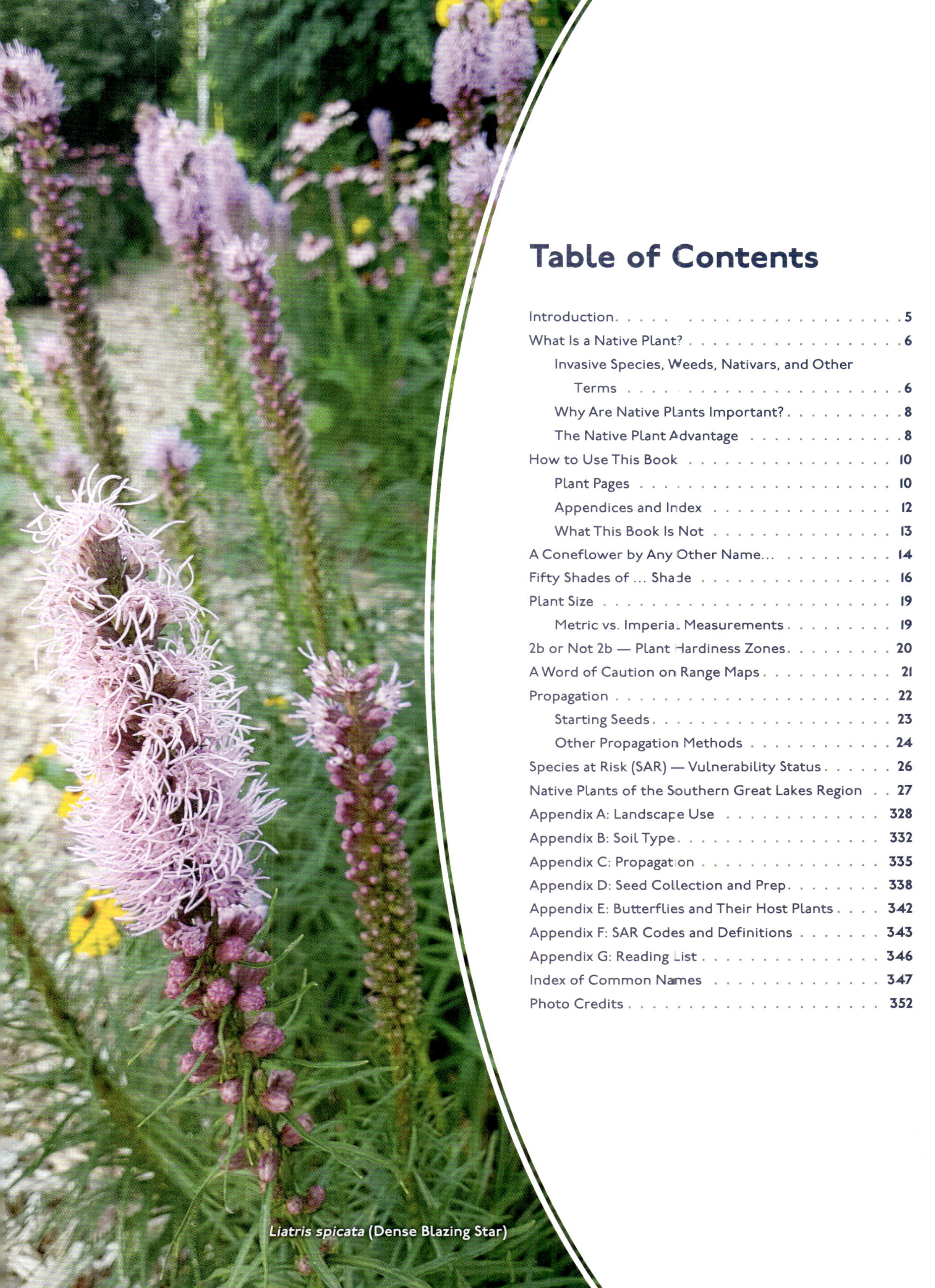

Table of Contents

Liatris spicata (Dense Blazing Star)

A butterfly on *Silphium perfoliatum* (Cup Plant)

Lobelia cardinalis (Cardinal Flower)

Solidago nemoralis (Gray Goldenrod)

Iris versicolor (Blue Flag Iris)

A Great Golden Digger Wasp on
Asclepias incarnata (Swamp Milkweed)

A towering *Lilium canadense* (Canada Lily)

Introduction

Gardening with native plants is, without a doubt, the fastest-growing segment of the horticulture industry in the early 21st century. The publicity surrounding the plight of the Monarch Butterfly and the collapse of bee colonies across the continent, as well as the rapid decline of many of our bird species, has triggered an awareness in folks that not only do we need to do something right away, but each of us actually *can* do something concrete. That something is to start converting our monoculture lawns, and flower beds of exotic plants that do little for our native butterflies and bees, into oases that support native insects which, in turn, support our birds.

But why *another* book on native plant gardening? Don't we have lots already? Many books, indeed, have been written in the last decade or two about which native plants are easy to grow, about how to landscape with native plants, and even about why it's essential to grow native plants in your garden. But none has taken the approach that *The Gardener's Guide to Native Plants* has — to provide you, the gardener, with all the key information you need in order to pick the plants that meet the conditions you have in your garden, tell you how you can propagate those plants, show you what they will look like, tell you how big they will get, tell you which butterflies and moths they are host plants for and, most importantly, identify those that are actually native to where you live.

What Is a Native Plant?

There are many definitions of "native" plants, and these definitions have evolved over time. In North America, the simplest definition is that a native plant is one that was already growing in the region when Europeans arrived.

However, for some, this definition is too simplistic because plants move — either as a response to climate or other environmental influences or through human intervention. For example, some First Nations used fire to modify vegetation to improve hunting and crop production, and they traded seeds from fruit trees and shrubs (*Asimina triloba*, or Pawpaw, is a great example), so when Europeans first arrived, there were plants growing in the Great Lakes Region that may not necessarily have evolved here, even though they may have been here for centuries.

The National Audubon Society defines native plants as those that occur naturally in the region in which they evolved, and the National Wildlife Federation says that a plant is considered native if it has occurred naturally in a particular region, ecosystem, or habitat without human introduction. Depending on how much of a "purist" you want to be, native plants may or may not include those introduced to the region by First Nations.

Is a wildflower the same thing as a native plant? Sometimes. A wildflower is any flowering plant that is not cultivated. Because many plants brought over by early settlers have escaped their gardens, we now have hundreds of "wild" flowers that were not here before settlers arrived and whose origins are Europe, Asia, or even Africa.

Are all the plants listed in this book native to our region? No. But most are. *Agastache foeniculum* (Anise Hyssop), for example, is a western prairie plant. However, it is so often sold as native here that it is almost ubiquitous in the area, so we have included it in the book — more to point out that it is actually *not* native than to promote its use. Another very popular plant that is also not native to most of this region is *Echinacea pallida* (Pale Purple Coneflower). It, too, is often sold as a native — even in some reputable native plant specialty nurseries — and is popular because it is colorful and easy to grow. However, in an effort to limit the number of

Agastache foeniculum (Anise Hyssop) is a western prairie plant that is a favorite of pollinators.

plants in this book to only 150, we decided to leave this one out. They are both beautiful garden plants, though the Hyssop, in particular, is a favorite of pollinators.

Invasive Species, Weeds, Nativars, and Other Terms

We often hear about invasive species, and sometimes (incorrectly) an overly aggressive native plant like *Helianthus tuberosus* (Jerusalem Artichoke) or *Solidago canadensis* (Canada Goldenrod) are called invasive. However, the term "invasive species" is reserved for non-native plants that are so prolific, either because they self-seed, spread vegetatively, or produce toxins to inhibit the growth of any competition, that they spread unchecked and destroy the natural ecological balance of an area. Note that not all non-native species are invasive, but those that are can have devastating impacts on local flora and fauna.

Scilla siberica (Siberian Squill) is an invasive species and spreads quite readily through self-seeding and bulb offshoots.

Weeds, on the other hand, may be non-native *or* native plants. The term weed is an agricultural term and describes any plant that has a negative economic impact on agricultural (food) production. *Erigeron canadensis* (Horseweed or Canada Fleabane) is a prime example of a native plant that is considered a serious weed in agricultural crops, especially since it developed resistance to glyphosate (the major weed-killing ingredient in Roundup) and acetolactate synthase (ALS) herbicides. Essentially, a weed is simply a plant growing where you don't want it. In my all-native garden, I consider squirrel-planted tulips to be weeds.

When many of us start growing native plants, we bring with us a history of growing non-natives. We, therefore, often look to the exotics for unusual colors or forms. This might include a *Lobelia cardinalis* (Cardinal Flower) with a deep burgundy color, an *Asclepias incarnata* (Swamp Milkweed) with a pure white flower, or a *Sanguinaria canadensis* (Bloodroot) with a "double" flower. Most often these plants are natural mutations, and horticulturalists will latch onto these unusual specimens and clone them for resale (some will breed true to the new form, but some are one-off freaks of nature whose seeds, if they even produce any, will revert to the original). Others are the result of selective breeding. And because the horticulture trade is, for the most part, driven by profit, some growers will select for unusual colors or forms to meet the demand for exotic plants.

The term "nativar" was coined to describe a cultivar (itself an abbreviation for **culti**vated **vari**ety) of a native species. For a number of nativars, the jury is still out on whether they are as valuable to pollinators and other insects as a true native, but many of the nativars either do not produce nectar or pollen (the extra petals in double-flowered species

A purple cultivar of *Iris versicolor* (Blue Flag Iris).

are often a mutation where the stamens in a flower are replaced by petals) and are often sterile, while others may be unpalatable to the insects that need them to survive.

Why Are Native Plants Important?

In recent years, one of the more interesting, if not shocking, statistics to appear in native plant gardening circles comes from Dr. Doug Tallamy's work out of the University of Delaware (his book *Bringing Nature Home* is a must-read). Dr. Tallamy's research shows that insects will often not eat non-native plants, so without native plants, insect populations decline. Most of our bird species — even seedeaters like Northern Cardinals and chickadees — supplement their own diet with insects. More importantly, they raise their young on a diet of almost pure insects. (One spring I watched with mixed feelings as the Black Swallowtail butterfly caterpillars I was so proud to have living on my Golden Alexanders got picked off one by one by a Northern Cardinal to feed its babies.) In fact, Tallamy's research indicates that the Carolina Chickadee cannot successfully fledge a nest of babies unless at least 70 percent of the plants within its territory are native species. The huge decline in recent years of bird numbers in North America can likely be ascribed, at least in part, to the disappearance of native plants necessary for the insects they feed on. And without caterpillars feeding on our native plants, we won't have beautiful butterflies, either.

An American Lady caterpillar on *Antennaria parlinii* (Parlin's Pussytoes).

You don't have to give up all your beautiful non-native plants if you don't want to. However, you may find, as many of us have, that the huge increase in biodiversity in your yard once you start growing natives will make you want to keep adding more. The fact that you are reading this book suggests you already know the value of natives and may even have started a native plant garden. This book will help you expand your knowledge and successfully increase your collection of suitable native plants.

The Native Plant Advantage

Not only are native plants important for ecological sustainability, they can make the gardener's life much easier, too. When matched to their preferred growing conditions, those that the plants have evolved with over hundreds or even thousands of years, native plants need very little help in the garden. In fact, thinning some of the more aggressive species (if you choose to grow them) might

Once native plants are established, they'll need very little maintenance, as long as they're well matched to their preferred growing conditions.

be all the work you need to do once the plants are established.

If you follow this guide when choosing your plants, you should not have to water or fertilize your plants — in fact, despite what you may read in some well-meaning books on gardening with native plants, adding compost or artificial fertilizers can be detrimental. Native plants with their typically deep roots are experts at nutrient extraction and can gather all the nutrients they need from the soil profile. Adding extra nutrients may cause native plants to grow tall and spindly, and flop over in the wind, and it can take years to use up the excess nutrients (I learned this the hard way when I first

transitioned to growing natives). The only watering I do now is in the first year after planting. I baby any new plants that I have purchased from the nursery; plants started from seed in the garden get no such special treatment. After that, only a prolonged drought will force me to get the garden hose out — but that is mostly for my benefit because native plants have evolved to simply go dormant under such conditions and will almost always come back the next year as long as they are well matched to their preferred growing conditions. Even weeding your flower beds may eventually become a thing of the past as you allow the natives to fill in and out-compete the non-natives.

How to Use This Book

If I want to buy a car, do I need to know the theory of the internal combustion engine? Or is it sufficient to have a list of options (makes, models, color choices) and a checklist of things I need to know (when to change the oil, what type of fuel it uses, etc.)? As when buying a car, often what we really want when we garden is a simple document that lays out our options. This is true whether we have been gardening for years or if this is our first foray into growing plants. The book you hold in your hands is just such a document. It provides an organized, practical guide to the important information about each plant — information that has been distilled from a multitude of sources — and is presented in a clear, at-a-glance format so you can quickly find the important information.

The plants in this book are listed in alphabetical order by scientific name. We considered many ways to sort and organize them — chronologically by bloom period, sorted by growing conditions or by flower color, by family, by height, or by ecological niche. But each method presented problems. Some of the plants bloom over a period of many months (such as Pearly Everlasting), some spring flowering plants regularly rebloom in the fall (like Wood Poppy). Colors, especially in the pinks/reds/purples/blues, are often quite variable and a challenge to separate out. Some native flowers have two distinct colors (e.g., red and yellow of Wild Columbine), while others come in more than one color choice (e.g., blue or white for Woodland Phlox). Some plants tolerate full shade to full sun (like Harebell) and some can take very dry through very moist conditions (such as Swamp Milkweed). So, in the end, we felt that simply alphabetizing them was the easiest, and once you become more familiar with the Latin names, they will be quick to find in the book. In the meantime, an index will help you locate them by common name, and tables at the back of the book will help you match plants to the conditions in your garden.

Plant Pages

Here's an example of a typical plant page that you'll find in this book:

NAMES For each plant, we provide both its scientific name and common name. We also include alternate common names and the family each plant belongs to.

QUICK GUIDE Each species will have symbols and exposure and moisture bars to show you at-a-glance info about the plant. The symbols on the lefthand side of the page tell you about the plant's key qualities:

 Fragrant

 Fragrant foliage

 Showy fruit

 Cut flower

 Poisonous

 Herbivore resistant

 Salt tolerant

 Attracts pollinators

 Attracts butterflies

 Hosts butterfly or moth larvae

 Attracts hummingbirds

 Attracts berry-eaters

 Attracts seedeaters

 Seeds feed ants

The letter symbols on the righthand side of the page tell you the propagation methods used for this plant. See the section entitled "Propagation" on page 22 for more information.

You will also observe two color-coded bars just underneath the symbols. The top bar represents the exposure requirements for the plant: *dark green*

for full shade, *bright green* for part shade, and *yellow* for full sun (see the section entitled "Fifty Shades of ... Shade" on page 16 for definitions of part shade, etc.). The bottom bar represents moisture requirements for the plants: *dark blue* for plants that tolerate, or even need, constant moisture, *light blue* for moisture-loving/moisture-tolerant plants, *light green* for medium moisture, and *sandy brown* for plants that prefer dry conditions. This color coding will allow you to know, at a glance, what conditions the plant prefers.

<table>
<tr><td>☀</td><td>FULL SHADE</td><td>PART SHADE</td><td>FULL SUN</td></tr>
<tr><td>💧</td><td>WET</td><td>MOIST</td><td>MEDIUM</td><td>DRY</td></tr>
</table>

PLANT DESCRIPTION This section describes the key characteristics of the plant, focusing on the leaves, stems, flowers, and seeds.

IN THE GARDEN This section discusses what the plant is like in the garden and gives suggestions on how best to grow it.

SKILL LEVEL This tells you whether it is an appropriate plant for your skill level as a gardener (most of the plants in the volume we consider suitable for beginners, though there are a few exceptions).

LIFESPAN This tells you the general lifespan of the plant. An annual performs its entire life cycle (seed to flower to seed) in a single growing season. A biennial takes two growing seasons to perform its entire life cycle. A perennial persists for many growing seasons, generally with the top portion of the plant dying back in winter and the plant regrowing on the same root system the following spring.

LIGHT/EXPOSURE This indicates the amount of sun or shade the plant prefers: shade, part shade, or full sun. Keep in mind that very few native plants tolerate constant solid shade, such as is found tight against the north side of a building. Even spring ephemerals found in mature hardwood forests often bloom before the trees leaf out, when they receive the welcoming rays of sunshine through bare branches. See also "Fifty Shades of ... Shade" on page 16.

SOIL TYPE This describes the type of soil the plant prefers as well as its preferred pH. You can also refer to Appendix B on page 332 for a handy table about soil type. This table will allow you to quickly find suitable plants based on the soil type in your garden.

MOISTURE This is the amount of moisture the plant prefers: wet, moist, medium, or dry.

HEIGHT AND SPREAD This gives you an approximate measurement of how tall and wide you can expect your plant to grow. The actual measurements will, of course, depend on the growing conditions in your garden. See "Plant Size" on page 19 for more on this.

BLOOM PERIOD This indicates the months when you can expect the plant to bloom. Note that months shown in parentheses — e.g., (Apr), May, (Jun) — indicate that the flowers may begin opening a week or so before and/or continue blooming a while after the main months of flowering.

COLOR This describes the color(s) of the flowers.

FRAGRANT, SHOWY FRUIT, AND CUT FLOWER Corresponding with the Quick Guide symbols at the beginning of the plant profile, these sections indicate whether the plant is fragrant, has showy fruit, and/or is suitable as a cut flower for bouquets.

PESTS This specifies which pests or diseases, if any, are a concern.

NATURAL HABITAT This describes what conditions the plant would be found in in nature, which can help you get a sense of the conditions where the plant thrives.

WILDLIFE VALUE This section tells you what pollinators, other insects, birds, and mammals benefit from this plant.

BUTTERFLY/MOTH LARVA HOST These sections list which species of butterflies and moths, if any, the plant is a larva (caterpillar) host plant for. Keep in mind that butterflies and moths are just two groups of insects that rely on native plants to complete their life cycles. Native plants are also hosts for a tremendous diversity of beetles, flies, other insects, and, indeed, fungi and countless other organisms.

USDA HARDINESS ZONES For those who like to garden by Plant Hardiness Zone rather than ecological niche, we provide the USDA Hardiness Zone range most commonly published for the plant. Also see the section entitled "2b or Not 2b — Plant Hardiness Zones" on page 20 for more on this subject.

PROPAGATION This details the basic propagation information so you can start more plants for your garden. See "Propagation" on page 22 for more on the various methods.

ADDITIONAL INFO This section provides some interesting facts that didn't fit well in any of the other sections.

SAR STATUS We provide the Species at Risk (SAR) status in Ontario and the states around the southern Great Lakes (Indiana, Michigan, New York, Ohio, and Pennsylvania). For a greater explanation on SAR status, see the section on page 26 entitled "Species at Risk (SAR) — Vulnerability Status."

VISUALS For each plant, we have provided photos of the entire plant in a garden or in its natural setting, a closeup of the flower, a leaf, and of the fruiting body (seed heads, berries, etc.). We also include a range map showing where the plant is considered to be native. See details on page 21 about how to interpret the range maps.

Appendices and Index

At the back of the book, we include some tables for quick reference. The first table suggests possible landscape uses for each plant, indicating what garden type the plant may be suitable for and whether the plant is a foreground, midground, or background plant or suitable as a specimen plant. The second table indicates the ideal soil types for each plant as well as the ideal pH of the soil. The third table encapsulates the propagation techniques for each plant, and the fourth table lists all the butterflies in the book and shows which plants are host plants for which caterpillars. Finally, the last table provides at-a-glance information on seed collection, storage, and sowing.

We also provide a reading list of some of our favorite books on the subject of native plants. The list is by no means comprehensive, but it provides a good starting point.

Finally, we include an index with plant common names. You will observe that a number of names are in **bold** font. These are the most common (at least where we live) of the sometimes multitude of common names. In most cases, we bold only one common name for a plant, but on occasion there may be a couple of very widespread and frequently used names, so we have bolded both (e.g., Wild Leek and Ramps or Jerusalem Artichoke and Sunchoke). We have also tried to make this index as comprehensive as possible, so if your local name for Marsh Marigold is Cow Lily, Publican's Cloak, or Yellow Gowan, you'll still be able to find it in the index.

What This Book Is Not

First and foremost, this book is *not* a book on garden design. It is our belief that garden design is a complete study on its own, that everyone has their own ideas about what looks nice and what doesn't, and that design principles are about color, form, and structure and hold true whether you are planting with native plants, exotic plants, or even fruits and vegetables — or some combination of these. There are lots of excellent books about garden design on the market; simply doing a web search with the keywords "best garden design books" will return lots of choices.

This is also not a book about foraging for wild plants, or even about the food or medicinal uses of our native plants. We originally considered including the edibility of the plants in this book, but in the end felt we didn't have sufficient expertise to provide this information safely. Some of the plants we list are well known as being edible; *Allium tricoccum*, also known as Wild Leeks or Ramps, may be the best known. However, indiscriminate harvesting by foragers has led to the extirpation of this plant in many forests throughout the region. We didn't want to encourage the harvesting of wild plants if the book fell into the wrong hands. In addition, some plants are poisonous — even deadly — unless you know how to properly prepare them. For others, one part of the plant might be safe to eat, and maybe just at a certain time of year, while another part of the same plant might cause gastric upset or even death. And though there is a lot of lore about the medicinal qualities of many of our native plants, only some of it has been verified scientifically. There are lots of books published on edible wild plants and a number on their medicinal uses. If that is what you're looking for, you won't find it here.

Allium tricoccum (Wild Leeks) is a well-known edible plant, but indiscriminate foraging has threatened this plant in many regions — something we wish to discourage.

A Coneflower by Any Other Name...

People sometimes complain "Latin is a dead language, so why do we use it for plants?" The simple answer — tradition. But as you will see in this book, every plant has but one "scientific name" or scientific binomial that is unique to that plant, whereas it may have two or three (and, in some cases, dozens of) common names. For instance, when I talk about Black-eyed Susans, am I referring to *Rudbeckia hirta* (an annual/biennial) or *R. fulgida* (a perennial)? Depending on what part of the country you are in, it could be either. Is Wild Columbine the same plant as Eastern Red Columbine? Yes. It is also known as Cluckies in some parts. And both *Asclepias hirtella* and *A. viridiflora* are called Green Milkweed, though in some locations *A. hirtella* is known as Tall Green Milkweed — which is what we call it in this book. *Caltha palustris* — commonly called Marsh Marigold — has at least 30 distinct common names that we have been able to find, and a whole lot more if you count names that are similar but just spelled differently. And then there is Goat's Rue (not included in this book) — this name is used for both the rare native *Tephrosia virginiana*, which is considered critically imperiled in Ontario, and the non-native *Galega officinalis*, considered an invasive species in Pennsylvania. So knowing the scientific binomial can prevent a lot of confusion.

Why is it called a binomial? Because the name consists of two parts. The first part is the genus (plural genera) — which always starts with a capital letter — and the second part is the species or "specific epithet" — which always starts with a lowercase letter. For example, *Asclepias tuberosa*. *Asclepias* tells us this plant is in the genus that contains the milkweeds, and *tuberosa* tells us which species of milkweed plant it is — in this case, Butterfly Milkweed. Note that, by tradition, foreign words (including Latin) are always written in italics.

"But sometimes I see more than just the two names, and sometimes there is 'var.' or 'subsp.' inserted between the names. What does that mean?" The "var." is the abbreviation for variety (*varietas* in Latin). Botanists will sometimes see that one species has enough genetic variation to separate

Both *Rudbeckia hirta* (left) and *Rudbeckia fulgida* (right) can be called Black-eyed Susan, though the former is an annual/biennial and the latter is a perennial.

Solidago ptarmicoides (Upland White Goldenrod) was called *Aster ptarmicoides* (Upland White Aster) until someone noticed it hybridizing with other species of goldenrods but never asters.

the plants, but they agree the two plants are still the same species, as they can still crossbreed and produce viable offspring. "Subsp." is the abbreviation for subspecies — a similar condition. Both are used, but var. is more common in plants, and subsp. is more common with animals. (We don't make the rules, we just try to follow them as best we can.)

Are the scientific names carved in stone? We wish! As science allows us to better understand the genetics of plants, we have been realizing that early botanists didn't always get it right. Sometimes a plant looks very much like another, so much so that early botanists were convinced they were different species in the same genus. But genetic testing is showing this to not be the case for several species. Sometimes scientists simply move the plant to another genus, as with Barren Strawberry — formerly *Waldsteinia fragarioides*, now *Geum fragarioides*. In some cases, even the species name gets changed, too, as with White Snakeroot — formerly *Eupatorium rugosum*, now *Ageratina altissima*. Sometimes they have to come up with a brand-new genus, as they have done for the asters of North America. The former genus *Aster* is now divided into 11 different genera, three of which are included here — *Doellingeria*, *Eurybia*, and the most numerous of them all, *Symphyotrichum*. Then there is the Upland White Aster, which was for many years known as *Aster ptarmicoides*. One look at it will tell you that it is obviously an aster. Then an observant botanist realized that it hybridized with a couple of species of goldenrod but never with an aster, and they discovered that it was a well-disguised goldenrod, so the name was changed to *Solidago ptarmicoides* or Upland White Goldenrod.

In this book, we have endeavored to use the scientific name that is considered the most current and follow the Database of Vascular Plants of Canada (VASCAN). The primary common name is that which seems to be most prevalent in southern Ontario, where the authors are located, but we also include as many of the other common names as we could verify.

Fifty Shades of … Shade

Most plants use the sun's energy to convert water, carbon dioxide, and minerals into oxygen and energy-rich compounds like sugars (think of that delicious maple syrup produced from the sap of Sugar Maples and other similar trees). This is the process of photosynthesis. Plants that evolved in wide-open spaces, like prairies, typically require lots of light to thrive. Because they have lots of light, and often drier conditions with more desiccating winds than in the forest, they often have small or narrow leaves. If grown in too much shade, these plants tend to stretch to reach the light, becoming "leggy" and weak stemmed. On the other hand, those plants that evolved on the forest floor have adapted to low light levels, often by growing large leaves to catch as much light as possible, but they quickly burn and die in full sun where their leaves dry out and no longer work to convert light to sugars. Then there are plants that evolved where there were at least some trees, but also open spaces. These plants often like some shade through the day but may be able to tolerate either full sun or full shade, or both. Interestingly, many of these plants produce more and/or larger flowers with the more sun they receive.

Full sun, part sun, part shade, dappled shade, full shade? What's the difference, and how do I know what I've got? The answer is not straightforward and will often differ depending on who you ask. Most sources agree that "full sun" means that the plant needs at least six hours of direct sunlight. These plants are often the so-called prairie species.

But is there a difference between part sun and part shade? Yes, but the definitions start to get murky at this point. Some say that part sun is four to six hours of sunlight per day, some say no more than five. Some sources say part shade is four to six hours of sun per day but only in the morning or late in the day. Others say part shade means two to three hours of sun. Some include dappled shade in the part shade definitions, while others keep it separate. (Dappled shade is that found under tree canopies where the tree branches are open enough to allow a fair bit of light to make it past the leaves

Goldenrods and asters bask in the sun. Many species of goldenrod and aster prefer full-sun conditions.

Trillium grandiflorum (Large White Trillium) and *Phlox divaricata* (Wild Blue Phlox) flower on the forest floor. Most flowering plants, even shade-tolerant ones, need some direct sunlight when in bloom.

to the ground. The movement of the tree branches and leaves means that the sun rarely shines directly on a plant for any length of time.) For our purposes, we classify "part shade" as less than six hours of sunlight per day but more than two hours, and we include light, dappled shade in this category, such as you might find under an open-branched tree such as Kentucky Coffeetree. As you can imagine, this is still a wide range, and many part-shade plants will do better at one end of that spectrum or the other. A point to keep in mind: most plants that prefer some shade are just like you and me in that they prefer the shade during the heat of the afternoon.

For our purposes, "full shade" is what you would find on the forest floor. It differs from the full shade (some call it deep shade) at the base of the north side of your house or under large, dense evergreen trees, where the sun never reaches, because most forest floors receive at least a little bit of direct sunlight. Although several ferns will survive in deep shade, providing there is sufficient moisture, most flowering plants need at least some direct sunlight, especially when they are in bloom. Many of our spring ephemerals (ephemeral just means temporary) bloom before the forest canopy leafs out, when the plants receive lots of light through the bare tree branches, and then the plant goes dormant once the forest floor becomes dark.

A Short Note on Boulevard Gardens

Boulevards, those hostile strips of soil between the sidewalk and the street (called verges or curb strips in some places), present many challenges for gardeners. In most municipalities, these often-narrow strips of grass are the responsibility of the homeowner to maintain. However, depending on your local bylaws, you may be restricted in what you can do with them. Always check with your municipality before you put a lot of money and effort into converting them to gardens because in some municipalities you are not allowed to convert them to gardens — they must remain as mowed grass. And keep in mind that boulevards are city property, so they have no protection from damage by city workers doing their jobs. Nor are municipalities required to give any notice of pending work that might damage or destroy your boulevard garden. However, that little patch of lawn in between the sidewalk and road should not be overlooked as a viable site for growing, and saving our wildlife with, native plants.

There are many benefits derived from planting a boulevard garden:

A boulevard garden is a great place to incorporate native plants, promoting wildlife and adding visual interest to the neighborhood.

Year-long natural beauty. Let's face it — lawns are boring. Native plants can produce a stunning display of color and interest in your boulevard all year round. Because boulevard gardens and native plants are a new concept to many people (i.e., neighbors) with more traditional garden views, it may be better to plant the garden more formally to help it be accepted by neighbors who might not be fully on board with a native plant garden instead of grass.

Reduced maintenance. Once a native boulevard garden is established, it will require far less maintenance than the lawn that was there initially (plus, it won't turn an ugly brown color during droughts). Unlike lawn, properly chosen plants won't require supplemental watering or fertilizer to look nice. Added bonus — you get out of having to mow the grass!

Pollinator habitat. A lawn does very little to support beneficial wildlife such as pollinators. By trading your boulevard grass for native plants, you can help restore lost and degraded habitat for pollinators. Even the smallest garden can help!

Environmental education. In urban areas, a lawn seems to be a default landscaping option. By planting a native garden in your boulevard, you can show your neighbors that they, too, can choose a more eco-friendly option than a lawn. You can help educate them further by including habitat signs.

Boulevards require plants that not only tolerate the shade/sun/moisture conditions that may be present but also a certain amount of salt accumulation, dog urine, and even some trampling by pets and neighborhood kids. These plants have to be extra tough. And because of sightline considerations for pedestrians and traffic, your local bylaws will stipulate maximum heights allowed in any plantings. Shrubs are normally not allowed at all, and often plants that get more than 1 m tall are forbidden as well, though many plants can be pruned back in early summer to keep them compact. Despite the challenges, a well-designed boulevard garden can provide long-lasting, low-maintenance beauty to your front yard. You can find a selection of suitable boulevard plants in Appendix A: Landscape Use (see page 328).

Plant Size

Native plants are tough. They will often grow in less than "ideal" conditions where they may survive, reproduce, and carry on as if they were meant to be there. But they may be much smaller than their siblings who landed in a spot with ideal conditions for growth. The sizes described in this book — height and spread — are average measurements as published in seed catalogs, on websites, and in scientific literature. In most instances, a range is given, but bear in mind that native plants don't read the literature and don't always follow the "rules" for height and spread. I have a showstopper of a Canada Lily (*Lilium canadense*) that, despite almost all the references indicating it should be somewhere between 60 and 150 cm (2 to 5 ft for our American readers) with four to 10 (rarely up to 17 to 20) flowers, in my garden it was a towering 242.5 cm (7 ft 11.5 in) with 102 blossoms on eight stems — and this in only its fourth year! The bottom line: the dimensions in the book are for average growth in your average garden with well-matched growing conditions for the plant.

Metric vs. Imperial Measurements

We use metric measurements in this book for two primary reasons: it is the standard for scientific measurements and, like us, it is very Canadian. We considered putting in both sets of units but quickly realized this added a lot of clutter to the description. As a rule of thumb, just remember the following:

1 in = 2.5 cm
1 ft = 30 cm

Therefore, a leaf that is described as 7 cm long is just under 3 in and a plant that gets 90 cm tall is about 3 ft high. Some other conversions I have found handy to memorize (since I'm of an age that I was raised with imperial measurements and often still think in inches and feet):

2 in = 5 cm
3 in = 7.5 cm
4 in = 10 cm
10 in = 25 cm
40 in = 100 cm or 1 m

Technically 1 m = 39.4 in, but with plant heights, that level of precision probably isn't necessary.

My towering *Lilium canadense* (Canada Lily), which in its fourth year measured 242.5 cm tall and had a whopping 102 blossoms on eight stems.

2b or Not 2b — Plant Hardiness Zones

Most gardeners are familiar with the Plant Hardiness Zone (PHZ) maps that are found in many seed catalogs and garden centers and with the paired numbering system (2a, 2b, 3a, 3b, etc.). The earliest PHZs were delineated in the 1920s by the Arnold Arboretum at Harvard University. In the 1960s, the United States Department of Agriculture (USDA) devised their own set of PHZs using different criteria resulting in two different maps. However, the Arnold Arboretum map remained the standard until 1990, when the USDA, in conjunction with the U.S. National Arboretum and using data from thousands of weather stations, created our current maps. These maps are based on the average minimum winter temperatures with each zone marked at 10°F intervals, and the division between a and b at the 5°F interval (e.g., zone 6a has an average winter minimum temperature of –10 to –5°F, 6b is –5 to 0°F) with the idea that plants have a cold threshold they won't survive beyond.

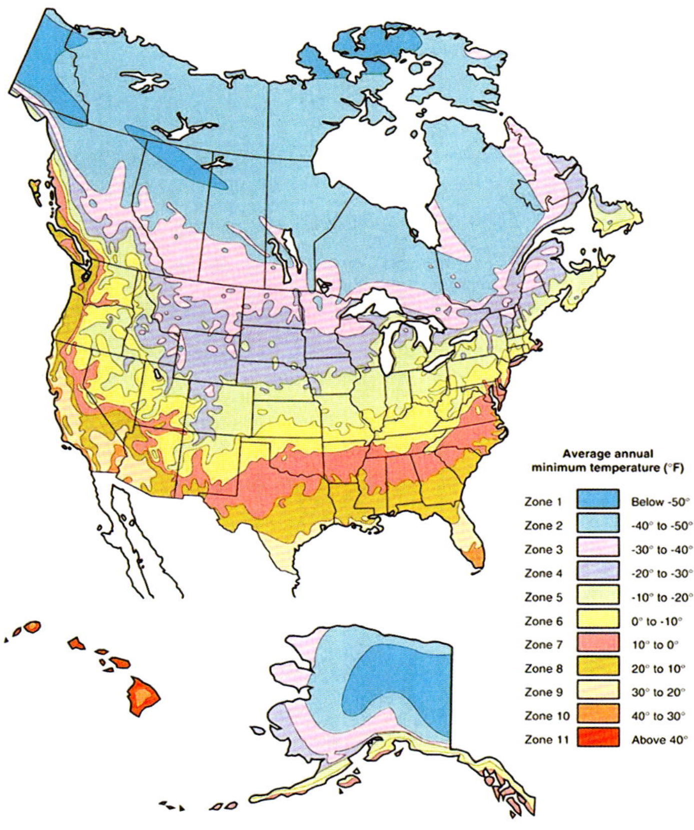

Unfortunately, these zones are a bit too simplified because plants are affected by more than just minimum winter temperatures, especially the farther south you go. Some plants need cold in the winter and thus won't survive in areas that do not receive it. Others are susceptible to heat. Basing the Plant Hardiness Zone on minimum temperature doesn't always work. In response, the American Horticultural Society (AHS) devised their Plant Heat Zones based on the number of days the temperature went above 30°C (86°F), but their methodology created problems as well.

If you've ever looked at the Canadian PHZ map, you've perhaps noticed that it doesn't quite align with the USDA map. This is because the Canadian map is likely more properly called a Plant Suitability Map (though it still goes by the "Plant Hardiness Zone" moniker). It uses a wide range of variables, not just minimum and/or maximum temperature. The variables are as follows: monthly mean of the daily minimum temperatures (°C) of the coldest month; mean frost-free period above 0°C in days; amount of rainfall from June to November, inclusive (mm); monthly mean of the daily maximum temperatures (°C) of the warmest month; winter factor expressed in terms of $(0°C – X_1)$ Rjan where "Rjan" represents the rainfall in January expressed in millimeters; mean maximum snow depth (cm); and maximum wind gust (in km/hr) in 30 years; and it incorporates them into a complex mathematical formula.

Surprisingly, although the American and Canadian calculations are very different, in the southern Great Lakes region, the results are close enough to be compatible for most of the plants. Which is the best (i.e., the most accurate and reliable) system? Being Canadian, we're probably biased. But in this book, we use the USDA system because (a) that's what most people understand, (b) that's what most nurseries use when labeling seed packets or plants, and (c) the U.S. portion of this region is not mapped with the Canadian system, but Ontario has been mapped using the U.S. system.

A Word of Caution on Range Maps

Native plant gardeners, probably more than any other gardener, are interested to know which plants are actually native to where they live. Because of the level of research needed to generate accurate maps, this level of detail is rarely ever included in books on native plant gardening, and instead plants are grouped by region (Northeast, Midwest, etc.) or by state/province (one look at a map of Ontario will tell you how problematic this grouping is). The maps in this book have been generated using all the information we could find online and through scientific journals and other publications. We also had the

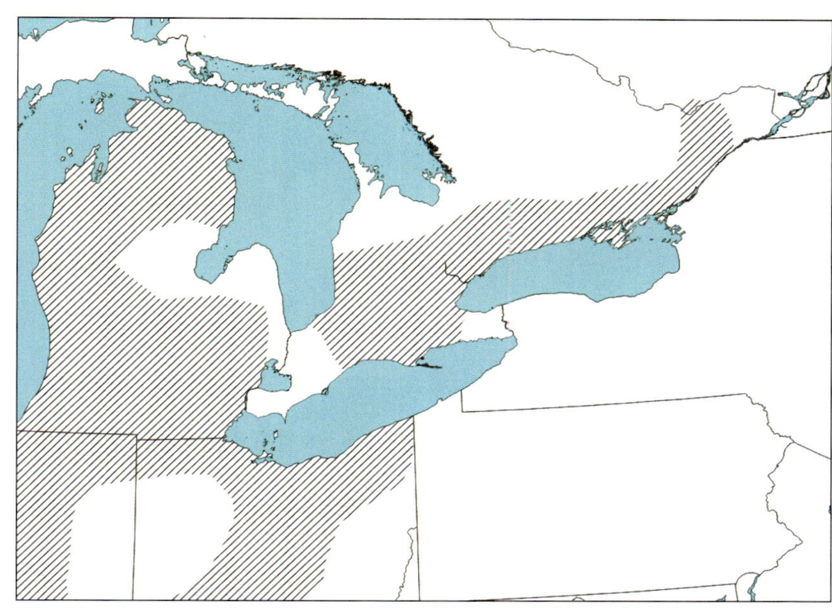

A range map for *Verbena stricta* (Hoary Vervain) showing the accepted native range as the shaded area.

fantastic support of ecologists and state botanists who graciously looked over the maps and pointed out any egregious misinterpretations we might have made. Some species, such as the goldenrods and asters, have been studied extensively in Ontario and elsewhere, and their native range is reasonably well documented. For many plants, however, there is little information available at an appropriate scale. Michigan and New York both publish county presence/absence maps that have been created from historical evidence. Quebec has very little publicly available information that we were able to find, and Ohio, Pennsylvania, and Ontario do not have provincial/state coverage maps. (At the time of writing we have been told that New York and Ohio are working on/updating such maps.) In Ontario, the most detailed maps come from government sources for a select few endangered species. When all else failed, we relied on interpreting the Biota of North America Program (BONAP) U.S. County-Level Species Maps, found at www.bonap.org.

In some cases, multiple sources provided similar ranges, which made mapping easy. However, in a few cases, the range maps were so different that I wondered if they were even talking about the same species. In these latter cases, we took a guess, based on what the majority indicated along with our own personal experience seeing the plants in the wild and with feedback from state and provincial botanists.

Also note that the range maps are very generalized and there may be pockets of a given plant that have grown naturally outside the mapped location but in small enough areas to be overlooked by those mapping them, or in isolated pockets too small to show up on maps at this scale. Also, there will be many areas of unsuitable habitat within the mapped range where the plants never grew.

And though we have tried to be as accurate as possible, these maps are really just a "best guess" and should be used simply to let you know if you might be pushing the limits of that species in your location, not as a definitive guide to where the plants were actually found before Europeans arrived.

Propagation

There are many ways to propagate plants for your garden — these include starting plants from seed, taking cuttings, or dividing rootstocks. Each has its advantages and disadvantages, and some plants may be started by using any of the methods, while others may only be propagated by one or two. In some cases, plants may be technically started by multiple methods, but in practicality, only one is effective. An example is *Helianthus tuberosus* (Jerusalem Artichoke). In theory, it may be started from seed, but the viability of the seeds is sketchy at best, and the plants are so easily started from tubers that it is seldom practical to try starting from seed. These characteristics are noted in the propagation section for each plant, where relevant.

Several authors have developed codes to denote propagation methods, but there is little consistency between them. Those most commonly seen and used in this region are from the website of Prairie Moon Nursery in Minnesota, from William Cullina of the New England Wildflower Society and author of several great books on native plants, and from the Missouri Botanical Garden in St. Louis, Missouri. These codes typically are sets of sequential letters — A through anywhere up to M — and require constant use to remember what each letter represents. For example, Prairie Moon uses H to indicate a seed needs scarification, Cullina uses the letter I, and the Missouri Botanical Garden uses B for the same thing. In this guide, we have created our own codes, which we believe are more intuitive, and these codes are displayed in gray icons on the first page of each plant profile.

PROPAGATION CODE	EXPLANATION
NT	**N**o **T**reatment necessary before sowing.
C(#)	Seeds need **C**old, moist stratification for the number of days **(#)** indicated.
L	Seeds need **L**ight to germinate; do not cover with soil, or cover very lightly.
WC	**W**arm then **C**old: Seeds need a period of warm (summer temperatures) and moist conditions for 60 to 90 days, followed by a period of cold (just above freezing temperatures) and moist conditions for a further 60 to 90 days.
CWC	**C**old, **W**arm, **C**old: Seeds need a cold period (just above freezing) for 60 to 90 days, followed by a warm period (summer temperatures) for another 60 to 90 days, and then a second cold treatment. In nature, these seeds would typically take two years to germinate.
S	These seeds have a tough coat and should be **S**carified before planting.
M	Seeds should be planted fresh or kept **M**oist until planted. These seeds lose viability very quickly if they dry out.
D	Plants may be started by **D**ivision — splitting the roots into sections and replanting.
LYR	Plants may be started by **L**a**y**e**r**ing — a process of covering a section of stem, runner, or another still-attached piece of the plant with soil, where new roots will form and that may then be removed and planted as a new plant.
CR	Plants may be started from **C**uttings from the **R**oots
CS	Plants may be started from **C**uttings from the **S**tems

A handful of *Lupinus perennis* (Wild Lupine) seeds.

Starting Seeds

C(#) **Cold, Moist Stratification** In the southern Great Lakes region, plants have evolved with and adapted to a regular seasonal freeze/thaw cycle. Stratifying seeds is the process of mimicking the cold cycle. If seeds started to germinate as soon as they were shed by the parent plant, many of the tender seedlings would freeze and die during the winter. To protect against this, a lot of our native plants have evolved a strategy to delay germination until the seeds have been subjected to freezing or near freezing temperatures for, typically, anywhere from one to three months. To cold stratify your seeds, unless otherwise noted for an individual plant, place them in a resealable plastic bag with slightly dampened peat moss or dampened sand, and place in the refrigerator for the suggested number of days. For example, for a plant that indicates "C(60)," leave them in the fridge for 60 days. The seeds can then be sown according to their needs, and the resulting germination rates will be much higher than without this process.

L **Light** A number of plants, especially goldenrods, asters, and hyssops, have very tiny seeds. These seeds need light to break dormancy and should not be covered with soil, or with only a very light covering. A rule of thumb to remember (though, as with all things plants, not a rule carved

Some seeds are very small and need light to break dormancy. They should not be covered with soil or covered only very lightly.

in stone) is that the smaller the seed, the more likely it is to need light to germinate.

WC **Warm/Cold** Many of the early spring blooming flowers produce seeds in early summer. These typically need a period of two to three months of warmth (room temperature) before they are cold stratified.

CWC **Cold/Warm/Cold** In nature, a few plants typically take two (or more) years to germinate. These can be tricked into germinating in one year by providing 60 to 90 days (depending on the plant) of cold, moist stratification, followed by an equal period of warmth (room temperature) and a second cold period.

Scarification One relatively common method of seed dispersal is to produce tasty fruit and have your seeds eaten along with the fruit by a passing bird or other animal. The nicks produced from biting the seeds, or the dissolving of a hard protective seed coating by stomach acids, greatly enhance germination rates. We can mimic this process by nicking the seed coat with a sharp knife, or by lightly rubbing the seeds between two pieces of sandpaper, or in a dish with sharp sand. This will allow moisture to more easily reach the embryo inside and trigger germination. This process is often used with large, hard seeds, or with seeds with an otherwise impermeable coating, such as *Liatris*. Note that often these seeds will germinate given sufficient moisture, or even by soaking in hot water for a while, but their germination rate may not be very high.

Moisture Many of our spring blooming plants and plants that grow in damp areas, produce seeds that are termed "hydrophilic," meaning they really like moisture. These seeds quickly lose their viability if they dry out, and therefore should be either planted as soon as possible after they are collected, or stored in damp peat moss or moistened sand until they can be sown.

Other Propagation Methods

Division Dividing the roots of a plant usually works well with any plant that produces suckers or rhizomes, as this is a natural way for these plants to multiply. The simplest process is to take a sharp shovel and slice straight down through the root mass, dividing it into several sections. These can then be transplanted where you want them. The best time to do this is in the spring, once there is little chance of frost, but some plants can be divided this way in the fall, and with a few (such as *Monarda didyma* and *M. fistulosa*), I have done this successfully all summer long.

Layering Layering is an effective technique for getting some plants to start forming roots along their stems, and a quick internet search will provide many different techniques. Not all plants will do this readily, but for those that do — usually plants with soft, flexible stems — the simplest method is to pin the stem to the ground in one or more places, typically at a leaf node, and lightly cover with soil. Nicking or twisting the stem till it cracks at the point to be covered will cause the plant to send repair hormones to the area and will usually speed the formation of roots at that spot. Once new growth is seen at the buried spot, it can then be removed from the parent plant and planted where you want it.

The seed head of *Liatris spicata* (Dense Blazing Star). The seeds of this plant need to be scarified using a knife or sandpaper.

Geum triflorum (Prairie Smoke) can be propagated by root cuttings.

Root Cuttings For those plants that naturally produce root suckers, root cuttings are an effective method of generating new plants. The best time to do this is in late fall or early winter, when the plant is dormant. Take 5 to 10 cm long pieces of pencil-thick root, cut at an angle at the bottom and flat across at the top, and insert at least 5 cm into perlite or other starting mix. In some plants, the roots are very thin and you may not find thick enough sections. In this case, just use twice the length so the plant will have sufficient carbohydrate resources to produce new growth, and lay it horizontally in the planting medium and cover with 1 cm of compost. When the plants show new growth in the spring, they may be potted up and grown out for later planting in the garden.

Stem Cuttings As a kid, I was always fascinated when my grandmother would snip some leaves off her African violets and put the "stem" (petiole) in a glass of water, where it would eventually take root and grow a new plant. A few of our native plants can be started this way, too, though the addition of a bit of rooting hormone often helps. To do this, take a 5 to 12.5 cm piece of stem, remove the lower leaves, and place the stem in a mix of vermiculite and sand. Place in a bright location, out of direct sunlight, for four to eight weeks and then transplant to the garden.

Anaphalis margaritacea (Pearly Everlasting) can be propagated by stem cuttings.

Species at Risk (SAR) — Vulnerability Status

Each political jurisdiction in Canada and the U.S. maintains a list of rare, endangered, and extirpated (wiped out/locally extinct) native plants, both at the federal/national level and at the provincial/state level. Definitions vary widely between jurisdictions, and classification systems vary greatly. In some of these jurisdictions, endangered species acts protect the identified plants, but unfortunately in many of the jurisdictions, the status of a species does not come with any legal protection. In Appendix F: SAR Codes and Definitions (see page 343), you'll find a brief summary for each of the provincial/state Species at Risk classifications that encompass the bulk of the area covered by this book. Also included is a list of the NatureServe Subnational Rank

definitions for species that rank less than a "secure or apparently secure" status. These letters or letter/number combinations at the beginning of each definition are those used in the plant pages under "SAR Status." The NatureServe organization has developed a standardized methodology for determining the risk status of plants, animals, and insects in North America and is used by natural heritage and conservation organizations in their biodiversity protection efforts.

In each plant listing, there is a line for Species at Risk (SAR) status, in which the SAR codes are used to identify the endangered status of that plant. The first letters indicate the province or state, followed by the NatureServe ranking and the official provincial or state status. For example, *Solidago speciosa* (Showy Goldenrod) is shown as

SAR STATUS ON – (not ranked)/E; IN – SU/WL; OH – S2/T; PA – S2/(not listed)

This indicates that in Ontario this plant is not ranked in NatureServe (they do not consider it native to the province), but the province does and classifies it as "Endangered"; it is "Unrankable" in Indiana and the state has placed it on a watch list. In Ohio, NatureServe ranks it as "S2 (Imperiled)," and the state classifies it as "Threatened." Finally, for Pennsylvania NatureServe lists it as "S2 (Imperiled)," but the state does not give it any classification. For Michigan and New York (the remaining two states covered by this book), the plant is not considered endangered and has not been ranked by either the state or by NatureServe.

If you see only "N/A" listed as the SAR status, that is short for "not applicable" and indicates the plant is not considered endangered by NatureServe or any province/state covered in this book.

Solidago speciosa (Showy Goldenrod) in full bloom. Its SAR status will vary depending on the location as well as the authority.

Native Plants
of the Southern Great Lakes Region

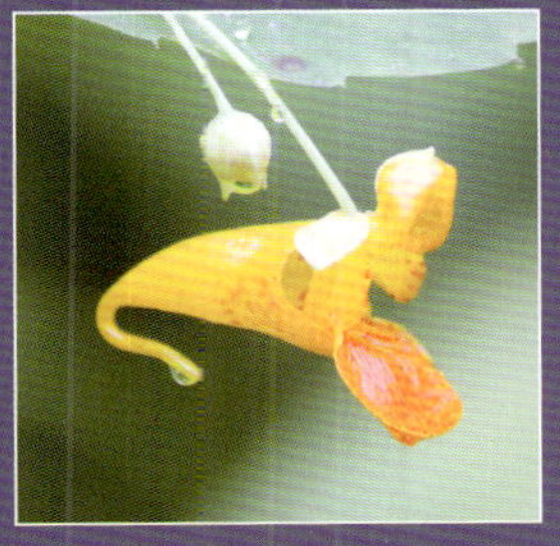

Achillea millefolium
Common Yarrow

QUICK GUIDE

PROPAGATION

| | PART SHADE | FULL SUN |
| | MEDIUM | DRY |

FAMILY *Asteraceae* (Aster Family)

ALTERNATE COMMON NAMES Bloodwort, Devil's Nettle, Milfoil, Nose Bleed, Old Man's Pepper, Sanguinary, Soldier's Woundwort, Stenchgrass, Thousandleaf, Thousand-seal

PLANT DESCRIPTION Common Yarrow features fragrant alternate leaves reaching up to 13 cm long with a lacy, fern-like appearance. The central stem is covered in fine hairs and is unbranched except at the top. The long-lasting flowers are found in compact, flat-topped clusters (10 cm across) at the top of the stems. The seed heads often persist into the winter months.

IN THE GARDEN Despite its delicate appearance, Common Yarrow thrives in the toughest, most inhospitable sites. It is best suited to large gardens or naturalization due to its prolific spread by seeds and rhizomes. The foliage is semi-evergreen.

SKILL LEVEL Beginner

LIFESPAN Perennial

EXPOSURE Full sun to part shade

SOIL TYPE Will tolerate almost any soils if they are not too wet

MOISTURE Dry to medium (drought tolerant)

HEIGHT 100 cm

SPREAD 60–100 cm

BLOOM PERIOD Jun, Jul, Aug, Sep, Oct

COLOR White

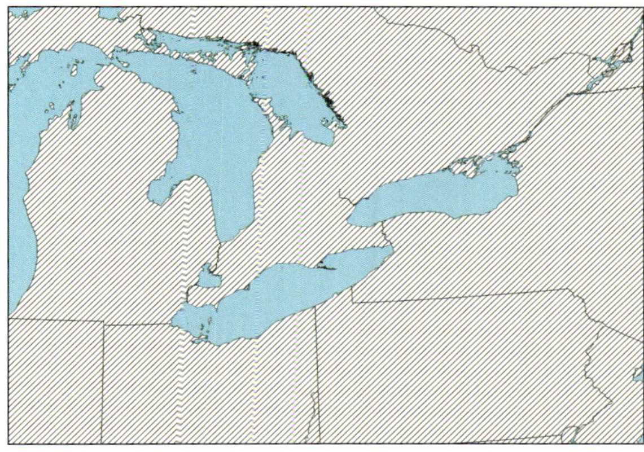

FRAGRANT ✅ (foliage)

SHOWY FRUIT ❌

CUT FLOWER ✅

PESTS Stem rot, powdery mildew, and rust are occasional disease problems

NATURAL HABITAT Sunny meadow-type habitats, open woods, and disturbed sites

WILDLIFE VALUE This is a good plant for pollinators, and the foliage is used by several cavity-nesting birds to line their nests

BUTTERFLY LARVA HOST PLANT FOR Painted Lady (*Vanessa cardui*)

MOTH LARVA HOST PLANT FOR Voluble Dart (*Agrotis volubilis*), Common Pug (*Eupithecia miserulata*), Olive Arches (*Lacinipolia olivacea*), Wavy-lined Emerald (*Synchlora aerata*), Striped Garden Caterpillar (*Trichordestra legitima*), other geometer and noctuid moths

USDA HARDINESS ZONES 3–9

PROPAGATION Seeds require light for germination, so do not cover with soil and, if starting seeds indoors, it germinates best at temperatures between 18 to 24°C (64 to 75°F). Clumps may be divided every two to three years, as needed, to maintain vitality of the planting.

ADDITIONAL INFO Sources are contradictory on the "nativeness" of yarrow. There seems to be little doubt that it was introduced from Europe and Asia, but several sources suggest there was already *Achillea millefolium* on the continent that simply cross-pollinated with the imports. We may never know for sure. Gardeners may notice yellow and pink varieties of yarrow at their local nursery. These are cultivars of the European species of yarrow.

SAR STATUS N/A

Whole plant

Leaves

Seed heads

Actaea pachypoda
White Baneberry

| FULL SHADE | PART SHADE | |
| WET | MOIST | MEDIUM |

Flowers

Young leaves and flowerbuds

FAMILY *Ranunculaceae* (Buttercup Family)

ALTERNATE COMMON NAMES Doll's Eyes, Necklace Weed, Snakeroot, Toadroot

PLANT DESCRIPTION White Baneberry features two to three large compound leaves with each individual leaflet being deeply toothed and hairless. Leaflets are 10 cm long and 6 cm wide. Dense, cylindrical clusters of small, creamy-white flowers rise above the foliage. Each cluster contains 10 to 28 individual flowers, each measuring about 6 mm across. The flower clusters give way to a red raceme (stalk) that bears white berries with a singular black dot on them.

IN THE GARDEN The astilbe-like foliage of White Baneberry adds a delicate look to shade gardens. Creamy-white flowers provide a valuable source of

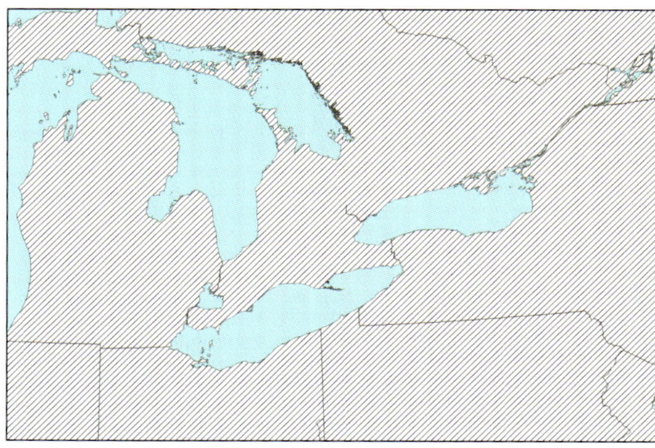

pollen for early-season pollinators. The beautiful, but toxic, white berries and red racemes persist into late fall for stunning visual interest. Slow growing but well worth the wait.

SKILL LEVEL Beginner to intermediate

LIFESPAN Perennial

EXPOSURE Full shade to part shade

SOIL TYPE Acidic (pH<6.8), well-drained, humus-rich soils

MOISTURE Wet to medium

HEIGHT 80 cm

SPREAD 60–90 cm

BLOOM PERIOD May, Jun, Jul

COLOR White

FRAGRANT ✓ (the white flower cluster has a lovely rose-like scent that attracts many insects)

SHOWY FRUIT ✓ (white berries)

CUT FLOWER ✗

PESTS No serious insect or disease problems

NATURAL HABITAT Deciduous woods and thickets

WILDLIFE VALUE A few birds eat the berries

BUTTERFLY LARVA HOST PLANT FOR None

MOTH LARVA HOST PLANT FOR None

USDA HARDINESS ZONES 3–8

PROPAGATION Keep seeds moist — viability is greatly reduced if they dry out. Seeds are best sown outside, 1 to 2 cm deep, as soon as ripe. Most seeds will need two winters to break dormancy or a period of 90 days of moist, cold stratification followed by a warm, moist period and a second cold, moist period. Flowering occurs a year after germination. Root division in early spring or fall is possible, though considered difficult as, once established, Baneberry does not like to be disturbed.

ADDITIONAL INFO Can be grown successfully under Black Walnut (*Juglans nigra*).

SAR STATUS N/A

Actaea racemosa
Black Cohosh

☀	**FULL SHADE**	**PART SHADE**
◌	**WET**	**MOIST**	**MEDIUM**

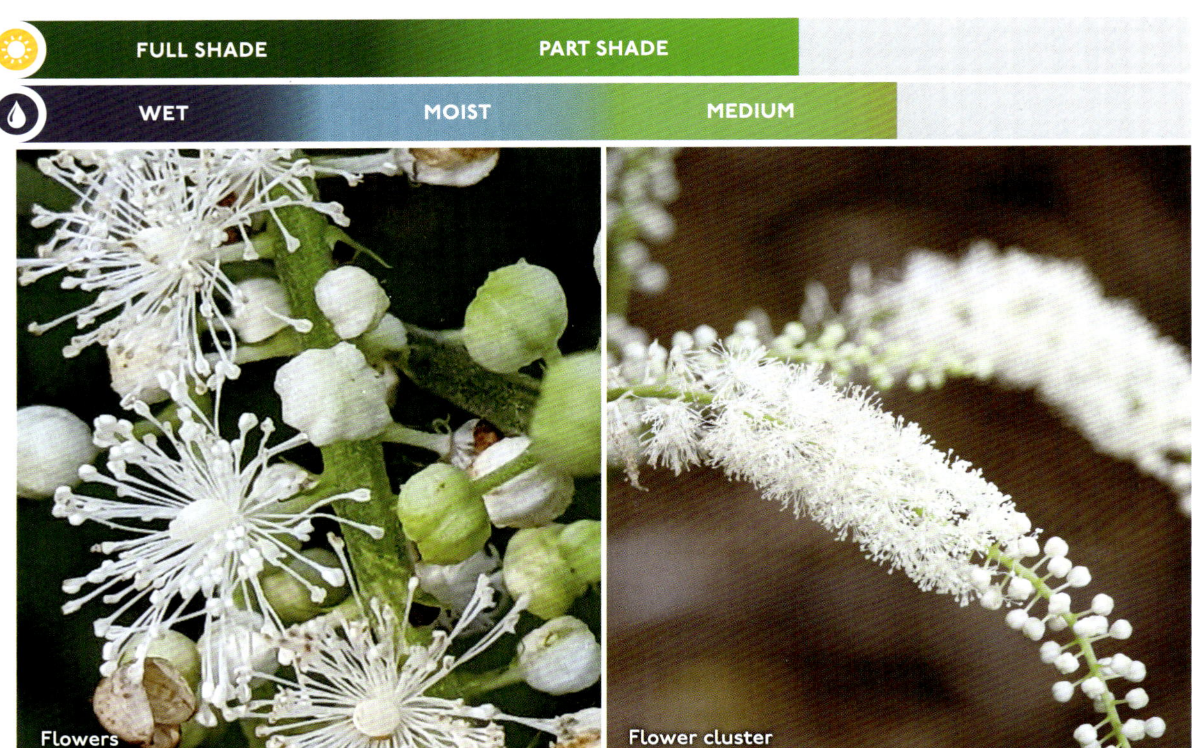

Flowers

Flower cluster

FAMILY *Ranunculaceae* (Buttercup Family)

ALTERNATE COMMON NAMES Black Baneberry, Black Snakeroot, Bugbane, Squaw Root

PLANT DESCRIPTION Black Cohosh features compound, smooth, multi-divided leaves. Each compound leaf is made up of coarsely toothed leaflets. Spires made of many creamy-white flowers rise up above the foliage on light green, hairless stems. Each flower spire can reach 30 to 90 cm in length. Flowers give way to seed pods that split open when ripe to reveal dry, brown seeds.

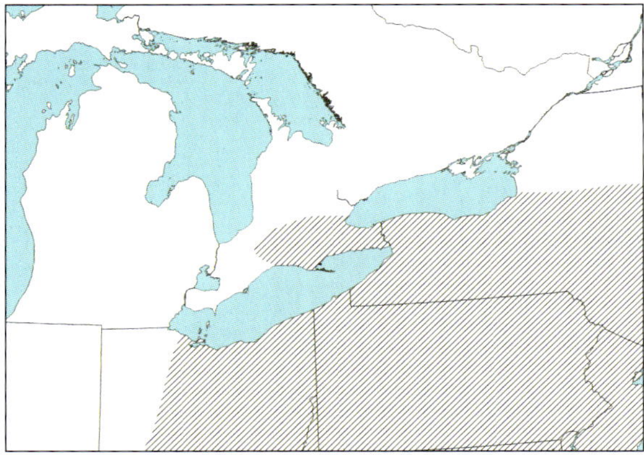

IN THE GARDEN Black Cohosh provides a captivating display of fragrant flowers in the summer shade garden when little else is in bloom. The spires of creamy-white flowers provide a feast for our eyes and pollinators alike. It is also valued for its use as a structural plant and its winter interest. Black Cohosh is a slow grower, so be patient.

Seed pods

Whole plant

Leaves

SKILL LEVEL Beginner

LIFESPAN Perennial

EXPOSURE Part shade to full shade

SOIL TYPE Humus-rich loam

MOISTURE Wet to medium (fairly drought tolerant once established)

HEIGHT 100 cm (in ideal conditions, can grow to 250 cm)

SPREAD 60–120 cm

BLOOM PERIOD Jun, Jul, (Aug)

COLOR White

FRAGRANT ✓ (pungent)

SHOWY FRUIT ✗

CUT FLOWER ✓

PESTS No serious insect or disease problems, though rust and leaf spot are occasional problems

NATURAL HABITAT Mesic to moist deciduous woodlands (where Sugar Maple is often dominant) and the bases of bluffs along rivers

WILDLIFE VALUE Provides both nectar and pollen for various insects

BUTTERFLY LARVA HOST PLANT FOR Spring Azure (*Celastrina ladon*), the sole food source for Appalachian Azure (*Celastrina neglecta major*) — the Appalachian Azure is not known in Ontario, but it can be found south of the Great Lakes

MOTH LARVA HOST PLANT FOR None

USDA HARDINESS ZONES 3–8

PROPAGATION The easiest way to propagate Black Cohosh is by dividing the rhizomes in the spring or fall. You can cut rhizomes into sections, 5 to 7.5 cm in length, making sure there is at least one bud attached to each piece. There can be up to 15 buds on the rhizome of one Black Cohosh plant. Any fibrous roots connected to the rhizome pieces should remain attached.

Because Black Cohosh seeds require a double dormancy (warm, cold, warm, cold) before they will germinate, fall sowing is best. Keep seeds moist. Some seeds may germinate after one winter, but others may take two years to start growing. Plant 6 mm deep in moist shade. Seeds should not be allowed to dry out — if not planting immediately, store in dampened sphagnum or sand.

ADDITIONAL INFO The plant may take several years to reach mature size and once established does not like to be disturbed.

SAR STATUS ON – S2/(not listed); IN – (apparently secure)/WL; MI – SH/(not listed)

Actaea rubra

Red Baneberry

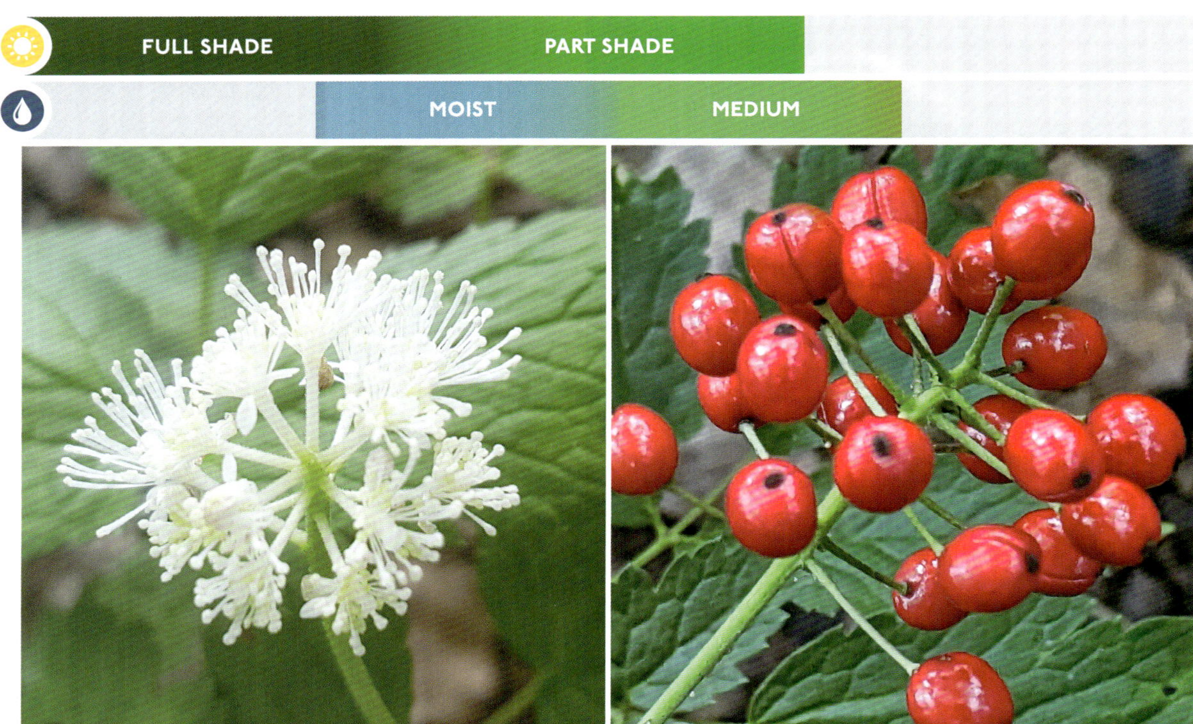

FAMILY *Ranunculaceae* (Buttercup Family)

ALTERNATE COMMON NAMES Cohosh, Necklace Weed, Red Cohosh, Snakeberry

PLANT DESCRIPTION Red Baneberry has smooth, light green stems that are usually unbranched. Each plant has one to four alternate leaves that are twice compound into three to five leaflets per group, with each leaflet measuring just under 9 cm long and wide. Each leaflet has sharply toothed edges, a smooth surface, a rounded base, and a pointed tip. Sometimes the leaflets can feature two to three lobes. Flowers are found in a solitary, rounded cluster at the top of a long flower stem that rises above the leaves. Each plant may have one to a few flower heads and they each measure about 5 cm long. Individual flowers measure about 6 mm wide and are characterized by four to 10 spear-shaped petals and many white stamens that stick out past them. Flowers mature into glossy red (sometimes white), oval berries about 0.8 cm wide. Each berry is borne on a slender green stalk and has a dark spot at its tip.

IN THE GARDEN Red Baneberry provides an attractive, clumping form, fine-textured foliage, and delightful clusters of white flowers in the spring. Come late summer, this plant takes on its signature look as the flowers mature into vibrant red berries that contrast beautifully with its foliage. Herbivores usually ignore this plant.

SKILL LEVEL Intermediate

LIFESPAN Perennial

EXPOSURE Part shade to full shade

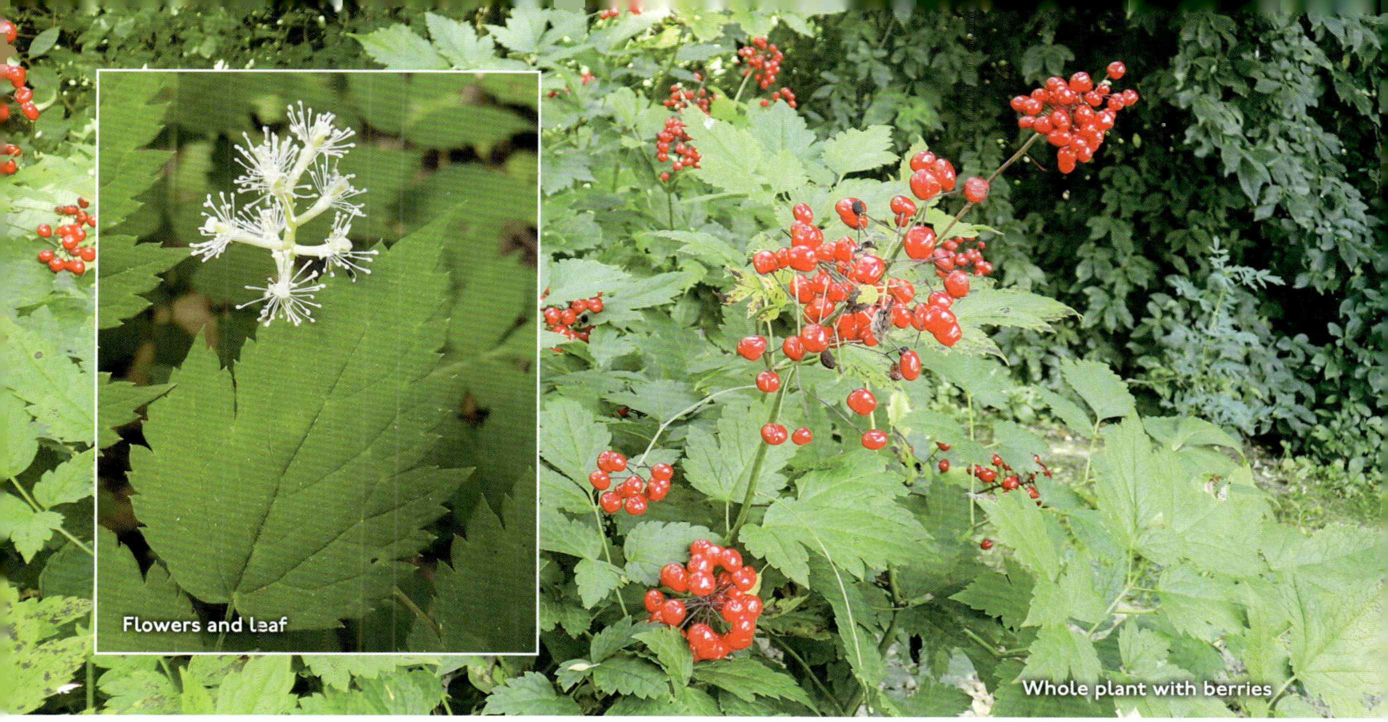

Flowers and leaf

Whole plant with berries

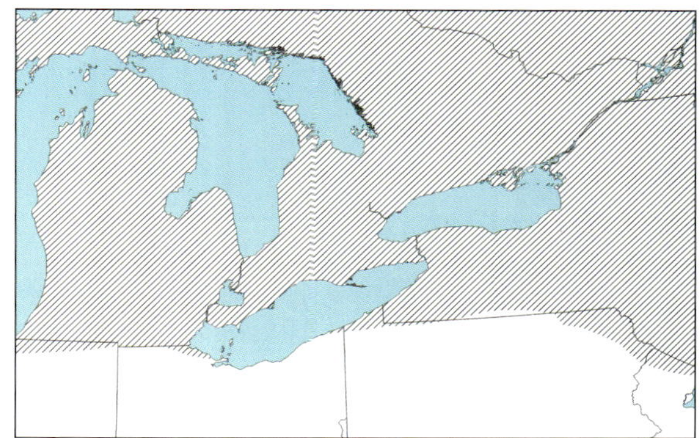

SOIL TYPE Acidic (pH <6.8), well-drained, humus-rich soils, clay, loam, sand

MOISTURE Medium to moist

HEIGHT 80 cm

SPREAD 45–60 cm

BLOOM PERIOD May, Jun, Jul

COLOR White

FRAGRANT ✓ (rose-like fragrance)

SHOWY FRUIT ✓ (red berries)

CUT FLOWER ✗

PESTS No serious insect or disease problems

NATURAL HABITAT Rich, moist deciduous and coniferous woods and thickets

WILDLIFE VALUE The flowers provide pollen for a variety of beetles, flies, and sweat bees, and several birds are known to eat the berries

BUTTERFLY LARVA HOST PLANT FOR None

MOTH LARVA HOST PLANT FOR None

USDA HARDINESS ZONES 3–7

PROPAGATION Red Baneberry is difficult to start from seed. In the wild, it usually takes two years to germinate as it requires a period of cold, moist stratification followed by a warm, moist period and a second cold, moist period. Sow seeds outside, 1 to 2 cm deep, as soon as ripe. Keep seeds moist before planting as they do not like to dry out. Plants flower the year after germination occurs. Red Baneberry may also be divided in early spring or in the fall after the plant has gone dormant.

ADDITIONAL INFO The plants are slow growing and take a few years to grow large enough to flower. Red Baneberry is very similar to White Baneberry (*Actaea pachypoda*), and it doesn't help that both can have white berries! You can tell them apart by the fact that Red Baneberry holds its berries on thin stalks and the berries each have a tiny dot. White Baneberry holds its berries on thicker stalks while each berry has a more prominent dot.

SAR STATUS IN – S1/T; NY – (not ranked)/EV; OH – S2/T; PA – S2/(not classified)

Agastache foeniculum
Anise Hyssop

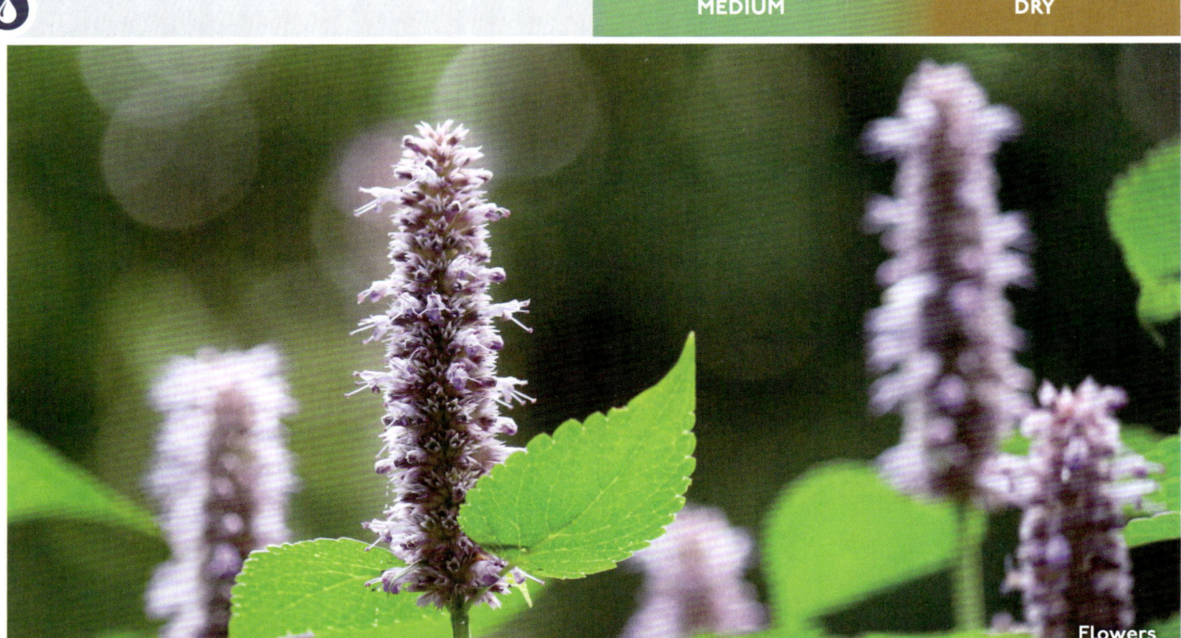

FAMILY *Lamiaceae* (Mint Family)

ALTERNATE COMMON NAMES Blue Giant Hyssop, Fragrant Giant Hyssop, Lavender Giant Hyssop, Licorice Mint

PLANT DESCRIPTION Anise Hyssop features opposite leaves borne on a square stem that may branch towards the top of the plant. Leaves have an anise-like smell and measure 10 cm long and 5 cm across, with a rounded base and pointed tip. The edges are coarsely toothed. Fine hairs give the underside of the leaves a gray-green color. Each stem terminates with a flower spike, up to 15 cm long, made of many small, tightly packed tubular flowers. The dried seed heads persist into the winter months and release small gray seeds.

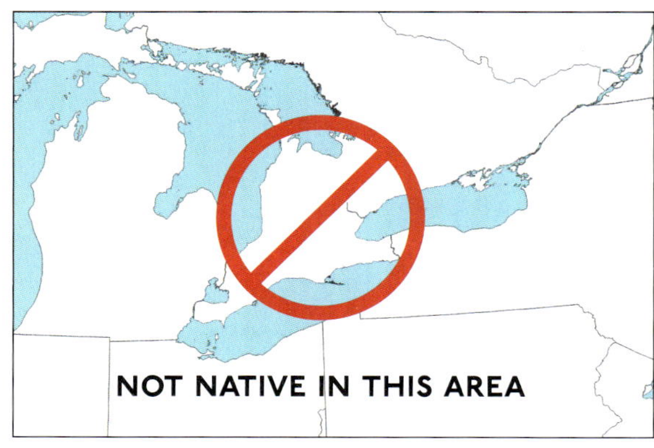

IN THE GARDEN Anise Hyssop is a fragrant, clump-forming plant that sends up spikes of showy lavender-blue flowers. The flowers have an impressively long bloom time and are an excellent nectar

source for a wide range of pollinators, including butterflies and hummingbirds. The dried seed heads persist into winter and provide seasonal interest while feeding songbirds, especially goldfinches.

SKILL LEVEL Beginner

LIFESPAN Short-lived perennial

EXPOSURE Full sun to part shade

SOIL TYPE Sandy, well drained

MOISTURE Dry to medium (moist if well drained)

HEIGHT 100 cm

SPREAD 100 cm

BLOOM PERIOD (Jun), Jul, Aug, (Sep)

COLOR Blue

FRAGRANT ✓ (anise scent)

SHOWY FRUIT ✗

CUT FLOWER ✓

PESTS This plant has no serious pests but may develop root rot in wet soils, and in humid climates, it is susceptible to powdery mildew

NATURAL HABITAT Oak savannas, rocky woodlands, prairies, dry upland forested areas, plains, and fields

WILDLIFE VALUE Bumblebees and many other native bees love it; it attracts hummingbirds and goldfinches flock to the seeds

BUTTERFLY LARVA HOST PLANT FOR None

MOTH LARVA HOST PLANT FOR None

USDA HARDINESS ZONES 4–8

PROPAGATION Seeds need light to germinate, therefore surface sow or cover very lightly. No pretreatment necessary, though 60 days of cold, moist stratification may help germination if starting seeds indoors. Anise Hyssop readily self-seeds. Plants may be divided in spring or fall.

ADDITIONAL INFO *Agastache foeniculum* is native to the Canadian Prairies and American Midwest, west of Lakes Michigan and Superior, but has gained popularity in native plant gardens in the southern Great Lakes region because of its fragrance, attractiveness to bees and finches, and its ease of maintenance in the garden. It is often incorrectly sold as a native wildflower in our region. We have included it in the guide because of its popularity and its benefit to wildlife, but if you are trying to grow a technically native plant garden, you should leave this one out.

A. rugosa (Korean Mint) has often been sold in nurseries as *A. foeniculum*, as the plants are very similar, but Korean Mint has a somewhat stronger aroma/flavor and is more aggressive in self-seeding.

Deadhead spent flowers to promote additional bloom.

SAR STATUS N/A

Agastache nepetoides
Yellow Giant Hyssop

Flowers

Whole plant

FAMILY *Lamiaceae* (Mint Family)

ALTERNATE COMMON NAMES Catnip Giant Hyssop, Giant Yellow Bubble-mint

PLANT DESCRIPTION Yellow Giant Hyssop features opposite leaves found along its strong, square stems. Leaves are toothed and up to 15 cm long, 7 cm wide and taper towards the tips. The central stem splits into secondary stems towards the top. Each stem is topped by a cylindrical cluster of densely packed, tubular flowers. Each flower has four stamens protruding from it. The dried seed heads persist into the winter months and release small, brown seeds.

IN THE GARDEN Yellow Giant Hyssop is a tall, clumping plant with an upright form. Pale yellow flowers are densely packed onto terminal flower spikes and have a long bloom time. An excellent nectar source for a variety of pollinators. Rigid stems and showy seed heads allow Yellow Giant Hyssop to feed birds and provide garden interest well into the winter months. Maintains a clumping habit.

SKILL LEVEL Beginner

LIFESPAN Perennial

EXPOSURE Part shade (will tolerate full sun with sufficient moisture)

SOIL TYPE Sand to clay

MOISTURE Moist to medium

HEIGHT 120–210 cm

SPREAD 30–90 cm

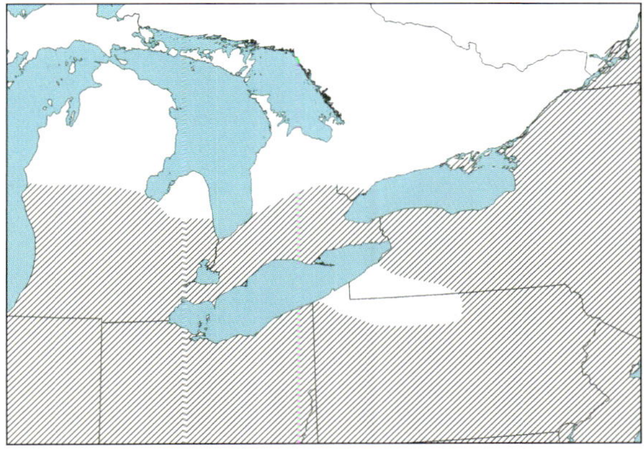

BLOOM PERIOD Jul, Aug, Sep

COLOR Yellow (green)

FRAGRANT ✓ (catnip scent)

SHOWY FRUIT ✗

CUT FLOWER ✓

PESTS No serious insect or disease problems

NATURAL HABITAT Rich, open woodland areas, thickets, and woodland borders

WILDLIFE VALUE Attracts bees, bee flies, and butterflies who feed on the pollen, while the rigid, hollow stems provide nesting opportunities for solitary bees

BUTTERFLY LARVA HOST PLANT FOR Summer Azure (*Celastrina neglecta*)

MOTH LARVA HOST PLANT FOR None

USDA HARDINESS ZONES 2–8

PROPAGATION Surface sow or very lightly cover in the fall, or moist stratify in the fridge for 60 days for spring planting. Seeds require light to germinate. Spreads slowly by rhizomes and can be divided in spring or fall.

ADDITIONAL INFO This plant is walnut (juglone) and deer tolerant.

SAR STATUS NY – S2/T

Leaves

Seed head

Agastache scrophulariifolia
Purple Giant Hyssop

Flowers

Whole plant

FAMILY *Lamiaceae* (Mint Family)

ALTERNATE COMMON NAMES Figwort Giant Hyssop, Lavender Giant Hyssop

PLANT DESCRIPTION Purple Giant Hyssop features opposite, coarsely toothed leaves with a rounded base and pointed tip. The leaves measure about 10 cm long by 5 cm wide and are borne on hairy leaf stems. The stem is rigid and square with scattered hairs. Flowers are densely packed into terminal flower spikes that reach up to 15 cm long. The flowers mature into brown seed heads that contain numerous small, oval seeds. Foliage is fragrant.

IN THE GARDEN The pale purple flowers of Purple Giant Hyssop are not especially striking from a distance; however, they are highly valued for their long bloom time and copious nectar reserves. This is an excellent structural plant as the rigid stems stand upright through the winter months. Expect it to maintain a well-behaved clumping habit with occasional self-seeding in bare soil. This is a fast-growing plant and often blooms in the first year when grown from seed.

SKILL LEVEL Beginner

LIFESPAN Short-lived perennial

EXPOSURE Full sun to part shade

SOIL TYPE Rich, moist soils

MOISTURE Moist to medium

HEIGHT 200 cm

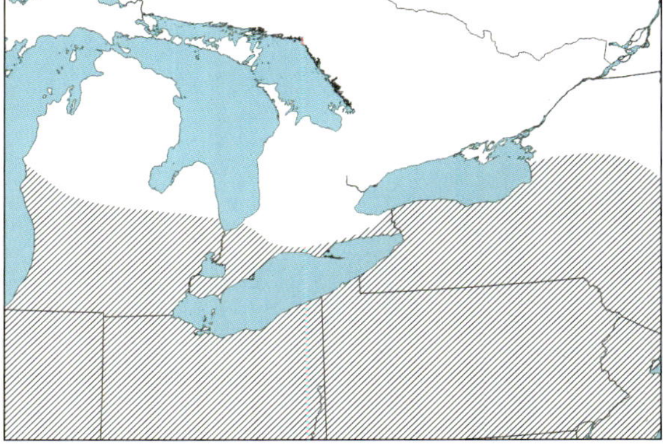

SPREAD 45–60 cm

BLOOM PERIOD Jul, Aug, Sep

COLOR Pale purple

FRAGRANT 

SHOWY FRUIT ✖

CUT FLOWER ✔

PESTS No serious insect or disease problems, though crown/root rot may develop in poorly drained soils

NATURAL HABITAT Rich woodland sites with dappled shade, meadows and thickets, and in recently disturbed, sandier soils where competition is limited

WILDLIFE VALUE The long-blooming, nectar-rich flowers are attractive to butterflies and other pollinators, including hummingbirds, and finches feed on the seeds

BUTTERFLY LARVA HOST PLANT FOR None

MOTH LARVA HOST PLANT FOR None

USDA HARDINESS ZONES 3–8

PROPAGATION Direct sow in the fall, or cold stratify for 60 days for spring planting. Seeds require light for germination. Plants may be divided in the fall or spring.

ADDITIONAL INFO Most sources indicate that this plant is increasingly at risk throughout much of its former range due mainly to habitat loss and competition with non-native plants. It is deer resistant.

SAR STATUS ON – S1/(not classified)

Ageratina altissima
White Snakeroot

FULL SHADE   **PART SHADE**

MOIST

Flowers

FAMILY *Asteraceae* (Aster Family)

ALTERNATE COMMON NAMES Fall Poison

PLANT DESCRIPTION White Snakeroot features an upright form with a firm stem. The coarsely toothed, opposite leaves reach up to 15 cm long and 10 cm wide but become smaller as they rise up the stems. The leaves taper towards their ends. The stems terminate with flat-topped flower clusters made of many small, white flowers with a fuzzy appearance. The flowers give way to small, black seeds with fluffy ends that are carried by the wind.

IN THE GARDEN White Snakeroot is a fast-growing, adaptable plant that lights up shade gardens with clusters of bright white flowers. It spreads quickly by seeds, making it valuable for restoration or larger gardens. The seed heads often remain upright into the winter months. Less vigorous in drier soils.

SKILL LEVEL Beginner

LIFESPAN Perennial

EXPOSURE Part shade to full shade

SOIL TYPE Rich loam or clay-loam

MOISTURE Moist (and well drained)

HEIGHT 50–80 cm (up to 120 cm in ideal conditions)

SPREAD 60–75 cm

BLOOM PERIOD (Jul), Aug, (Sep)

COLOR White

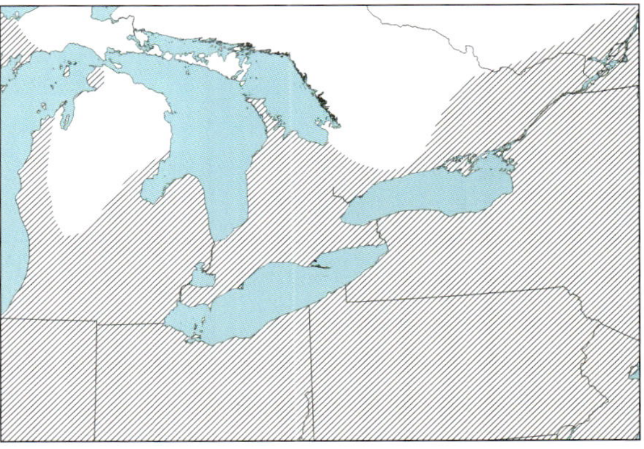

FRAGRANT ❌

SHOWY FRUIT ❌

CUT FLOWER ✅

PESTS Leaf miners and flea beetles may attack the foliage, and the flowers of this plant are a preferred source of nectar for Japanese beetles in my garden

NATURAL HABITAT In partial shade, in thickets and woodland clearings, disturbed sites and along path edges in moist to medium dry soils

WILDLIFE VALUE Nectar attracts butterflies and native bees and is an important source of pollen, and reportedly small birds will eat the seeds in the fall, though I have yet to see this in my garden

BUTTERFLY LARVA HOST PLANT FOR None

MOTH LARVA HOST PLANT FOR Eupatorium Borer Moth (*Carmenta bassiformis*), Burdock Borer Moth (*Papaipema cataphracta*), Ruby Tiger Moth (*Phragmatobia fuliginosa*), Lined Ruby Tiger Moth (*Phragmatobia lineata*), Gracillariid Moth (*Leucospilapteryx venustella*)

USDA HARDINESS ZONES 3–8

PROPAGATION Seeds need light to germinate. Surface sow in the fall and plant thickly as germination is usually low, or cold stratify for 60 days for spring planting. Divide the roots in the fall.

ADDITIONAL INFO In the 1800s, a mysterious "milk sickness" caused the deaths of many people. Entire villages were abandoned because they could not determine the cause of the illness. It turned out that the milk became toxic when cattle had grazed on White Snakeroot.

SAR STATUS N/A

Allium cernuum
Nodding Wild Onion

FAMILY *Liliaceae* (Lily Family)

ALTERNATE COMMON NAMES Lady's Leek, Nodding Pink Onion

PLANT DESCRIPTION Nodding Onion features a tuft of basal leaves originating from a bulb. Its arching, grass-like leaves reach up to 30 cm long and 1 cm wide. The leafless flower stalks rise slightly above the foliage and bend downwards at the top, producing a nodding umbel of flowers (hence the name "Nodding" Onion). All parts have a strong onion smell.

IN THE GARDEN Nodding Onion is a small but showy plant that thrives in tough sites. For best effect, plant it in large groupings. It doesn't like competition from taller plants so plant accordingly.

This is a very well-behaved, clumping plant, but it may self-seed in optimal conditions.

SKILL LEVEL Beginner

LIFESPAN Perennial

EXPOSURE Part shade to full sun

SOIL TYPE Humus-rich, neutral-to-alkaline soils, but will adapt to acidity, and sand to clay

MOISTURE Medium to moist

HEIGHT 40 cm

SPREAD 8–15 cm

BLOOM PERIOD Jun, Jul, Aug

COLOR Pink (white) to light lavender

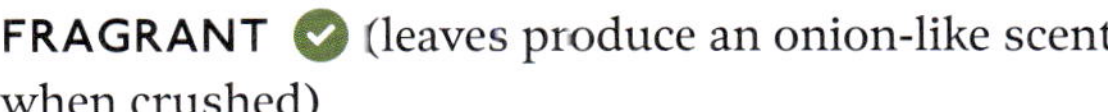

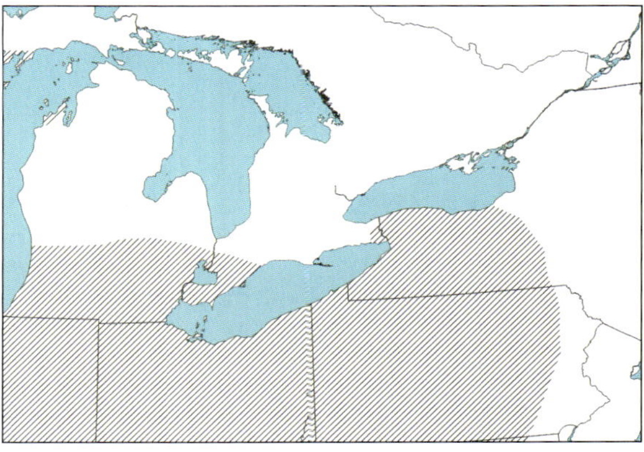

FRAGRANT ✓ (leaves produce an onion-like scent when crushed)

SHOWY FRUIT ✗

CUT FLOWER ✗

PESTS No serious insect or disease problems

NATURAL HABITAT Prairies, rocky outcrops, and the edges of dry, open woodlands

WILDLIFE VALUE Supports a variety of generalist pollinators, including native bees, and the nectar attracts hummingbirds and butterflies

BUTTERFLY LARVA HOST PLANT FOR Hairstreak butterflies (*Satyrium* spp.)

MOTH LARVA HOST PLANT FOR None

USDA HARDINESS ZONES 3–8

PROPAGATION Spreads by seed and bulb off-shoots. Sow seeds in the fall, or provide 60 days of moist, cold stratification if spring planting. Cover lightly with soil/growing medium. Plants benefit from being divided every third year or when eight to 10 bulbs appear in the clump. Plants may be divided any time of the year.

ADDITIONAL INFO Walnut (juglone) tolerant. Nodding Wild Onion is rare in Ontario — it is believed that the only natural populations left are those growing in alvar habitat on Pelee Island.

SAR STATUS NY – S2/T

Allium tricoccum
Wild Leek

Flowers

Flowers

FAMILY *Liliaceae* (Lily Family)

ALTERNATE COMMON NAMES Ramps, Small White Leek

PLANT DESCRIPTION Wild Leek features two broad, glossy leaves reaching about 10 to 22 cm long and up to 9 cm across. They die back once the trees above them have leafed out, giving rise to a solitary, leafless flower stalk. The flower stalk reaches up to 30 cm high and is topped by a 5 cm wide, rounded clump of small flowers. Both flowers and stalk emerge from a green-onion-like bulb. Seed heads often persist into winter months, holding on to their shiny, round, black seeds. All parts have a strong garlic smell.

IN THE GARDEN Wild Leek is a true spring ephemeral as its leaves are some of the first greenery to look forward to in the spring but die back before the heat of summer sets in. Therefore, it is best planted with other shade-tolerant species that will persist later into the season. Clusters of star-like flowers on lanky stems lend a unique look to the garden in the early summer.

SKILL LEVEL Intermediate

LIFESPAN Perennial (spring ephemeral)

EXPOSURE Full shade to part shade

SOIL TYPE Well-drained, rich, loose loam with abundant organic matter

MOISTURE Medium

HEIGHT 30 cm

SPREAD 20–40 cm

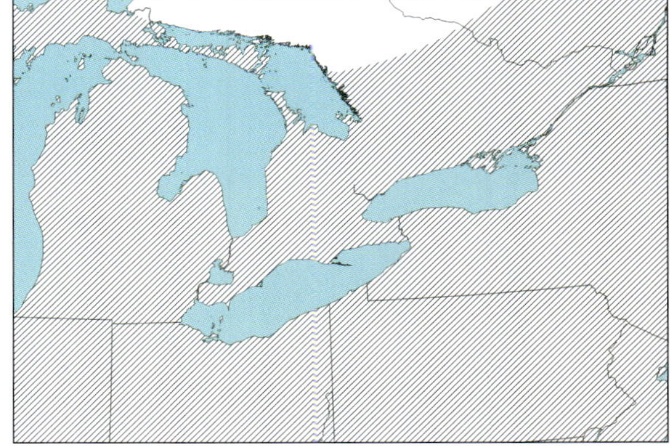

BLOOM PERIOD Jun, Jul (the leaves die off and disappear before the flowers bloom)

COLOR White

FRAGRANT ✓

SHOWY FRUIT ✗

CUT FLOWER ✗

PESTS No serious pests

NATURAL HABITAT Rich, deciduous woods

WILDLIFE VALUE Important nectar source for pollinators

BUTTERFLY LARVA HOST PLANT FOR None

MOTH LARVA HOST PLANT FOR None

USDA HARDINESS ZONES 3–7

PROPAGATION Sow seeds as soon as possible as they do not store well. They require a period of moist warmth followed by 60 days of moist, cold stratification to germinate. Cover with about 2.5 cm of shredded, damp leaves. In the wild, they may take up to two years to break dormancy. Mature bulbs produce offsets, which can be divided in the summer or fall after flowering and planted so that the roots are buried and with just the very tip of the bulb above the surface. These should be mulched with 5 to 7.5 cm of composted hardwood leaves or natural leaf litter. A very slow grower, so be patient.

ADDITIONAL INFO This plant is regularly foraged for its delicious leek-like flavor, but one should use only the leaves as it takes many years for the seeds to grow and many more till the plant can reproduce again. The over-harvesting of Ramps in the wild has led to its disappearance in many woodlots throughout the region.

SAR STATUS N/A

Anaphalis margaritacea
Pearly Everlasting

FAMILY *Asteraceae* (Aster Family)

ALTERNATE COMMON NAMES Western Pearly Everlasting

PLANT DESCRIPTION Pearly Everlasting has many upright stems that are clumped closely together, giving the plant a mounding appearance. The leaves are up to 10 cm long and 2 cm wide and, like the stems, are covered in numerous small hairs, giving the plant a silvery-green appearance. Numerous 1 cm wide, yellow flowers enclosed in white bracts can be found blooming at the end of the stems. The seeds are very small and have tufts of white hairs that carry them off in the wind.

IN THE GARDEN Pearly Everlasting is an easy-to-grow plant that reliably thrives in the driest of sites.

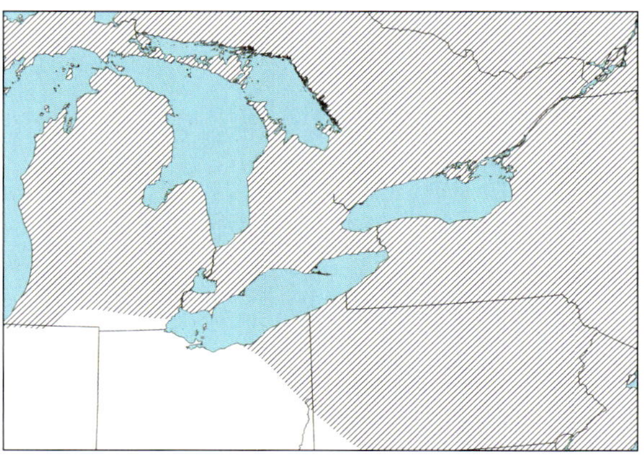

It is highly valued in gardens for its long bloom time. Caterpillars of the American Lady (*Vanessa virginiensis*) and Painted Lady (*Vanessa cardui*)

butterflies feed on the foliage and will weave the leaves together with silk to create small tents. This is nothing to worry about because it is a sign of life in your garden. Your plant will make a full recovery from the hungry caterpillars.

SKILL LEVEL Beginner

LIFESPAN Perennial

EXPOSURE Full sun to part shade

SOIL TYPE Sandy or gravelly soils

MOISTURE Dry to medium

HEIGHT 30–100 cm

SPREAD 30–60 cm

BLOOM PERIOD Jun, Jul, Aug, Sep, Oct

COLOR White

FRAGRANT  (when crushed)

SHOWY FRUIT ❌

CUT FLOWER ✅ (Pearly Everlasting makes excellent dried flowers for arrangements, and it can also be used for fresh cut flowers)

PESTS No serious insect or disease problems, though there is some susceptibility to chewing damage from caterpillars

NATURAL HABITAT Dry prairies, open woods, roadsides, and wastelands

WILDLIFE VALUE Flowers are magnets for pollinators, such as butterflies and bees

BUTTERFLY LARVA HOST PLANT FOR Painted Lady (*Vanessa cardui*), American Lady (*Vanessa virginiensis*)

MOTH LARVA HOST PLANT FOR Everlasting Tebenna Moth (*Tebenna gnaphaliella*)

USDA HARDINESS ZONES 2–7

PROPAGATION No pretreatment of seeds necessary, but surface sow as the tiny seeds require light to germinate. Plants may be divided in the spring or started from stem cuttings. *Anaphalis margaritacea* flowers are dioecious, meaning flowers are either male or female, and only one sex can be found on an individual plant, therefore if you want seeds you will need to have both a male and female plant. Male flowers are ball-like (globular) with many slender, erect yellowish-brown staminate flowers in the yellow center disk. Female flowers are globular to egg-shaped with a yellowish to dark brown bristly ring around the top of the flower head. Both flowers have numerous tiny white bracts in many layers around the center. The bracts on the female flowers do not spread out much until the seed starts forming.

ADDITIONAL INFO Pearly Everlasting flowers, leaves, and stems have been used to produce yellow, gold, green, and brown natural dyes.

SAR STATUS IN – SX/X; OH – (not ranked)/X

Anemonastrum canadense
Canada Anemone

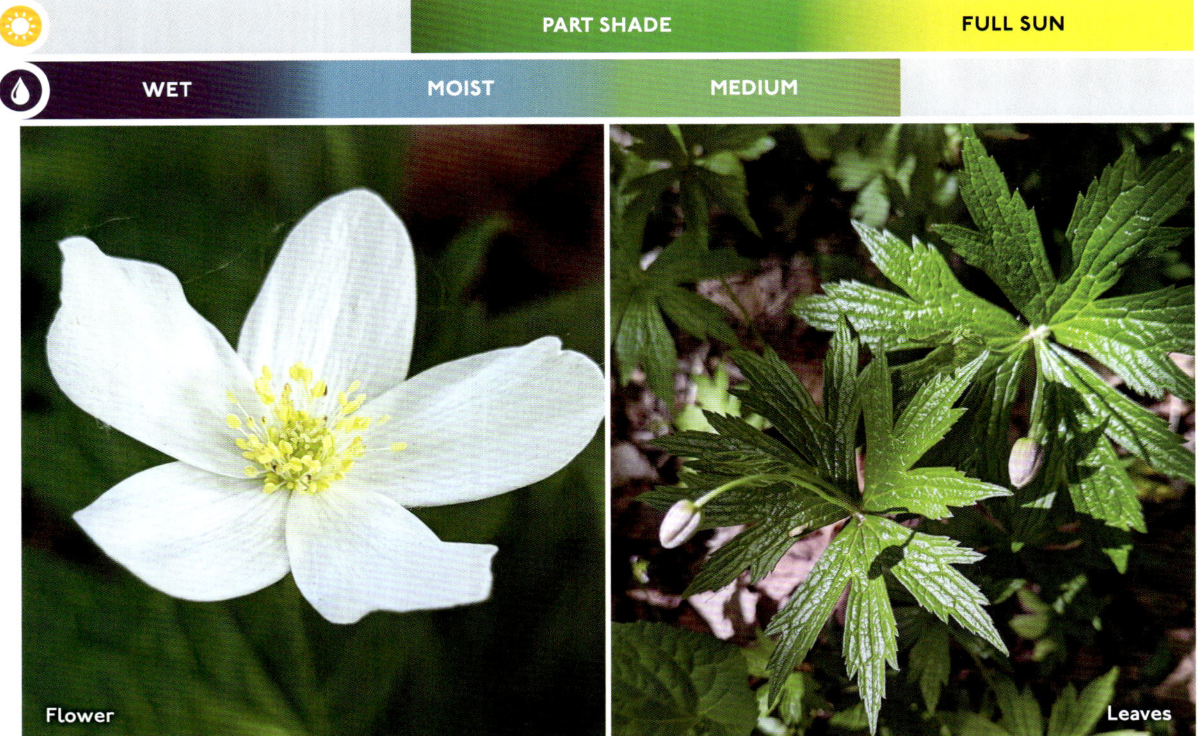

Flower

Leaves

FAMILY *Ranunculaceae* (Buttercup Family)

ALTERNATE COMMON NAMES Meadow Anemone, Round-leaf Thimbleweed, Windflower

PLANT DESCRIPTION Canada Anemone features deeply lobed basal leaves reaching 7 to 15 cm wide and equally as long, with each lobe being further divided into two to three secondary lobes. The leaf stalks reach 10 to 20 cm long and are hairy. The flowering stalks have opposite leaves that look similar to the basal leaves but are attached directly to the stalk. A solitary white flower with five petals and yellow anthers sits atop this stem. Each flower is replaced by spiky clusters of small, dry fruit that drop to the ground soon after ripening.

IN THE GARDEN Canada Anemone is a

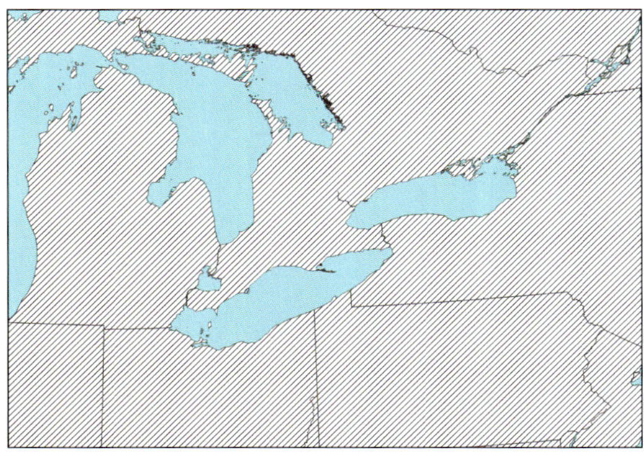

dependable, adaptable, and low-maintenance ground cover. It spreads vigorously by rhizomes to create a lush carpet of leaves. In late spring, the cheery white

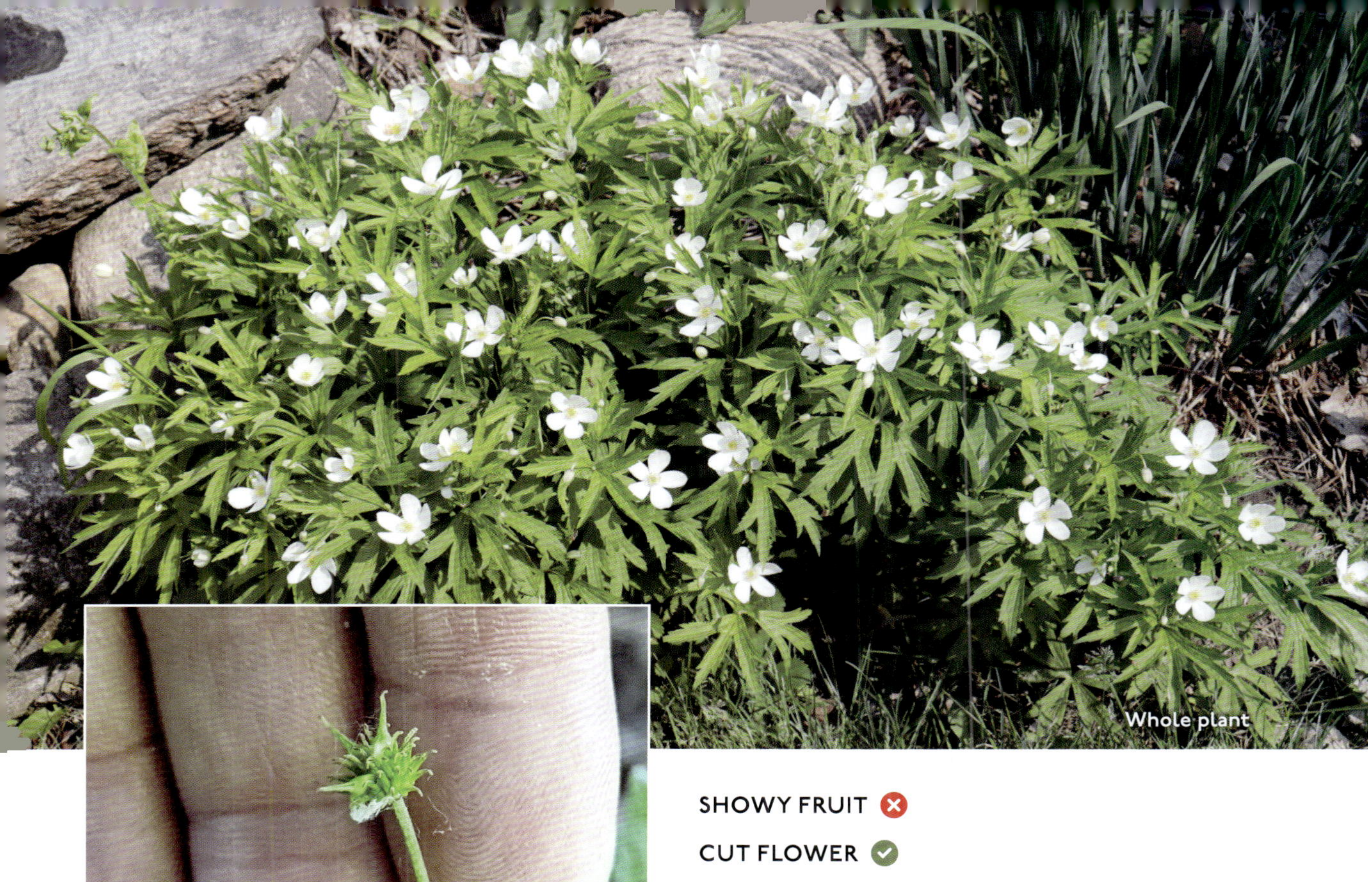

flowers dance in the wind, giving it the nickname of "Windflower." Its vigorous growth can overwhelm smaller plants and gardens, so plant accordingly. Excellent for restoration and erosion control.

SKILL LEVEL Beginner to intermediate

LIFESPAN Perennial

EXPOSURE Full sun to part shade (will tolerate full shade though flowering stems tend to flop in too much shade)

SOIL TYPE Moist, well-drained sandy-loam soil, will tolerate clay

MOISTURE Medium to wet

HEIGHT 60 cm

SPREAD 50–75 cm

BLOOM PERIOD May, Jun, Jul

COLOR White

FRAGRANT ❌

SHOWY FRUIT ❌

CUT FLOWER ✅

PESTS No serious insect or disease problems

NATURAL HABITAT Areas of open, moist ground

WILDLIFE VALUE Attracts a variety of native bees and other pollinators

BUTTERFLY LARVA HOST PLANT FOR None

MOTH LARVA HOST PLANT FOR None

USDA HARDINESS ZONES 3-9

PROPAGATION Seeds can be difficult to germinate and require a double dormancy — a cold, moist resting period followed by a warm, moist period and then another cold, moist period, with the seed germinating in the second year when planted outside. Seeds must be kept moist and not allowed to dry out. Plants may be divided in the fall or by taking root cuttings in the spring (each piece of rhizome should be at least 10 cm long).

ADDITIONAL INFO Canada Anemone is walnut (juglone) tolerant and deer resistant. It relies solely on pollen to attract insects as its flowers do not produce nectar.

SAR STATUS N/A

Anemone cylindrica
Thimbleweed

	PART SHADE	FULL SUN
	MEDIUM	DRY

Flowers

Whole plant

FAMILY *Ranunculaceae* (Buttercup Family)

ALTERNATE COMMON NAMES Candle Anemone, Cottonweed, Long Fruited Anemone, Long Headed Anemone, Long-headed Thimbleweed, Windflower

PLANT DESCRIPTION The erect stalks of Thimbleweed emerge from deeply lobed basal leaves. Leaves are divided into three to five lobes and are up to 10 cm long. Halfway up the stem there is a secondary whorl of smaller leaves. From the secondary whorl of leaves arise the flowering stalks, each topped with a solitary flower. Flowers are greenish white with five petal-like sepals surrounding a small green cone with yellow anthers. Each plant can have two to eight flowering stalks. Seed heads resemble green thimbles and are up to 4 cm long. They are at least twice as long as they are wide. Seed heads turn into fluffy, cotton-like tufts once mature.

IN THE GARDEN Thimbleweed is an adaptable, clump-forming plant with a slender, upright form. It is valued by gardeners for its long bloom time, showy seed heads and ability to support pollinators with its copious pollen reserves. An excellent choice for filling in the transitional bloom time between spring and summer flowers. Note that, in rich soils, Thimbleweed does not tolerate competition from taller plants, so place it accordingly.

SKILL LEVEL Beginner

LIFESPAN Perennial

EXPOSURE Full sun to part shade

SOIL TYPE Prefers sandy, coarse-textured soils but will grow in almost any soil as long as it is dry

MOISTURE Medium to dry (drought tolerant)

HEIGHT 90 cm

SPREAD 15–30 cm

BLOOM PERIOD Jun, Jul

COLOR White, pale green

FRAGRANT ✗

SHOWY FRUIT ✗

CUT FLOWER ✓

PESTS No serious insect or disease problems

NATURAL HABITAT Dry, open woods and prairies

WILDLIFE VALUE Small bees and syrphid flies are attracted to Thimbleweed for its abundant pollen, but mammals usually avoid this plant because the foliage is toxic and causes a burning sensation in the mouth

BUTTERFLY LARVA HOST PLANT FOR None

MOTH LARVA HOST PLANT FOR None

USDA HARDINESS ZONES 3–8

PROPAGATION Easily propagated from seeds sown in fall or early spring. Thirty days of cold, moist stratification is said to improve germination, but this is usually not necessary as this plant readily self-seeds. Surface sow or very lightly cover

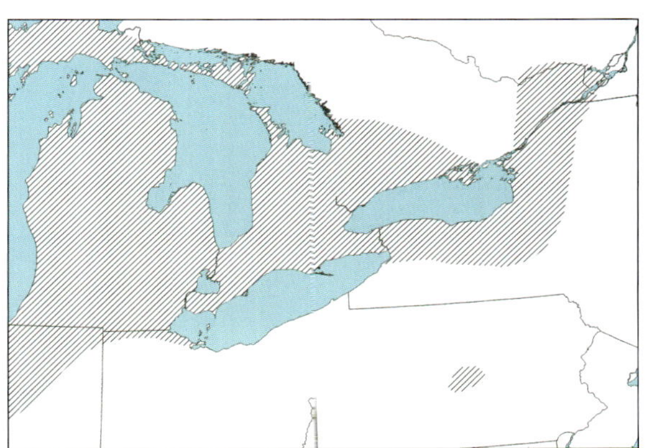

as the seeds need light to break dormancy. Germinates best in cool soil. Plants can be divided in early spring. Thimbleweed can take from two to five years to flower.

ADDITIONAL INFO *Anemone cylindrica* can be distinguished from other anemones by the cylindrical seed heads, which are at least twice as long as they are across. Although often confused with Tall Thimbleweed (*A. virginiana*), this plant's leaflets are wedge-shaped at the base with the lobes fanning out, whereas the outer lobes of *A. virginiana*'s leaflets are rounded with teeth along the tip. Also, Thimbleweed is shorter, usually less than 60 cm tall, whereas Tall Thimbleweed can reach up to 120 cm. Another short anemone, Canada Anemone (*Anemonastrum canadensis*), can be differentiated in that the leaves are sessile, whereas Thimbleweed's leaves have petioles.

SAR STATUS OH – S3/T; PA – S1/E

53

Anemone quinquefolia
Wood Anemone

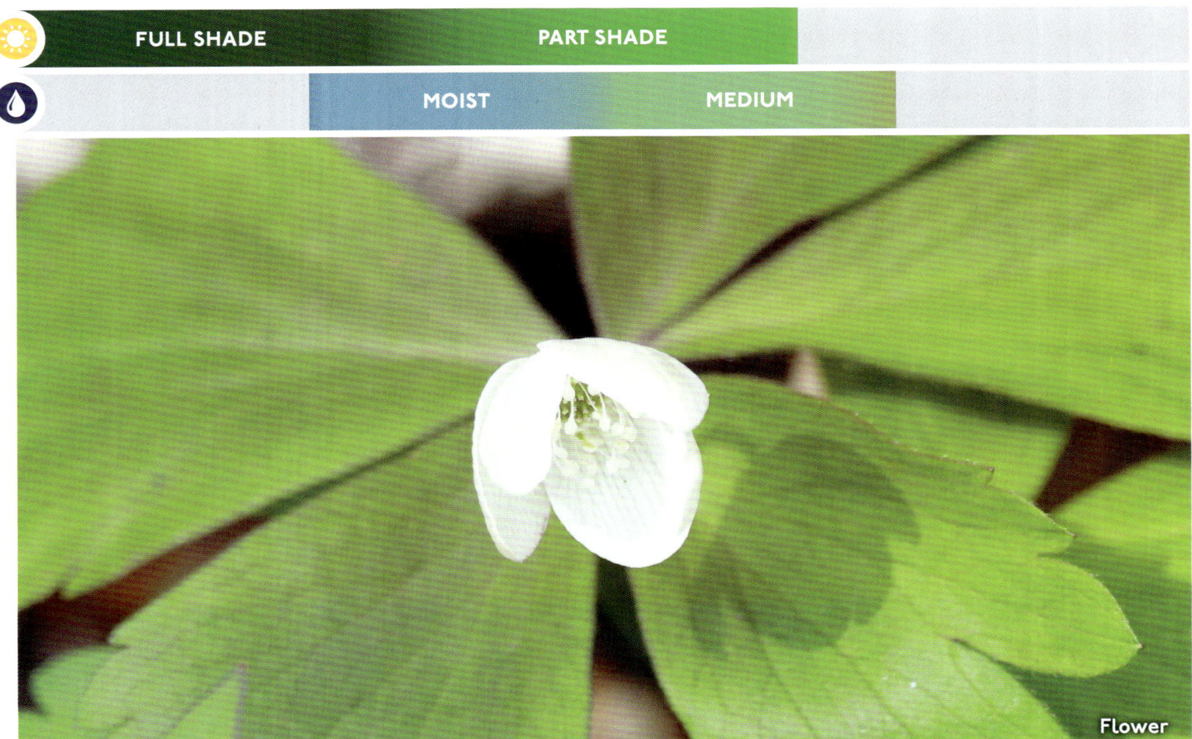

FAMILY *Ranunculaceae* (Buttercup Family)

ALTERNATE COMMON NAMES Five Leaved Anemone, Herb Trinity, Mayflower, Nightcaps, Nimbleweed, Wood Flower, Wood Windflower

PLANT DESCRIPTION Wood Anemone features a finely hairy stem with a group of three compound leaves and one basal leaf. Each leaflet is up to 4 cm long, deeply divided into two or three parts, coarsely toothed, finely hairy, and wedge shaped at the base. Some leaves are so deeply cleft that they give the impression of having four to five leaflets. Leaflets are attached by very short leaf stalks. A flower stalk arises from the center of the whorled leaves and terminates with a single 2.5 cm wide flower. Flowers are characterized by four to nine

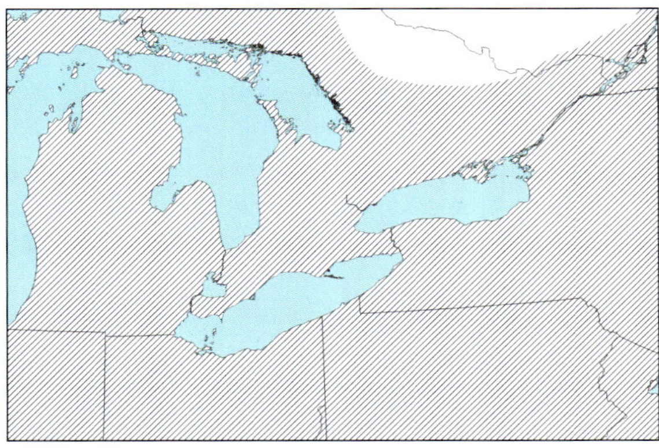

sepals (usually five) surrounding a cluster of many white-tipped stamens. Flowers give way to a round seed head containing tiny, hairy, beaked, oval seeds.

IN THE GARDEN Wood Anemone is a true spring ephemeral, meaning it emerges early in the spring to take advantage of the extra sunlight coming down through the leafless trees above it. The delicate-looking flowers bloom for a short time, then the whole plant goes dormant by midsummer. Because of this, it is recommended that it be planted with long-lasting plants that can fill in the gap it leaves behind. The short and sweet springtime display of Wood Anemone encourages us to stop and truly appreciate the cycles of nature.

SKILL LEVEL Beginner

LIFESPAN Perennial (spring ephemeral)

EXPOSURE Full shade to part shade

SOIL TYPE Rich, well drained

MOISTURE Moist to medium

HEIGHT 10–20 cm

SPREAD 10–20 cm

BLOOM PERIOD Apr, May, Jun

COLOR White (pink)

FRAGRANT ✖

SHOWY FRUIT ✖

CUT FLOWER ✖

PESTS Pest free and deer resistant

NATURAL HABITAT Deciduous or mixed evergreen-deciduous forests and forest edges, riverbanks, and fields

WILDLIFE VALUE A variety of native bees collect the pollen

BUTTERFLY LARVA HOST PLANT FOR None

MOTH LARVA HOST PLANT FOR None

USDA HARDINESS ZONES 3–8

PROPAGATION Very little information is available on starting seeds for *Anemone quinquefolia*, except that it is difficult to germinate, and artificial stratification may not work. It apparently requires a cold period, followed by a warm, and then another cold (i.e., two successive winters) before seeds germinate. Fortunately, this plant can be propagated by dividing the root as it spreads naturally by rhizomes into thick mats.

ADDITIONAL INFO Contact with the sap of this slow-growing plant may irritate skin.

SAR STATUS N/A

Anemone virginiana
Tall Thimbleweed

FAMILY *Ranunculaceae* (Buttercup Family)

ALTERNATE COMMON NAMES Large Anemone, Tall Anemone, Thimbleweed, Virginia Anemone

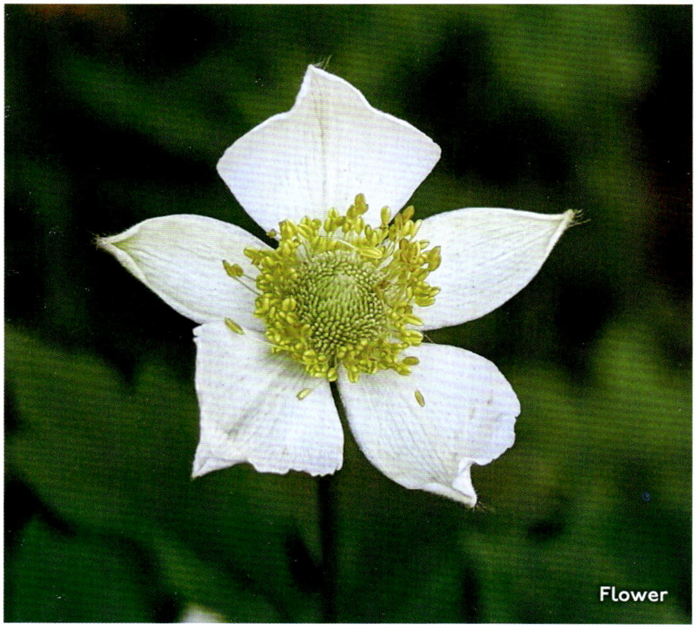

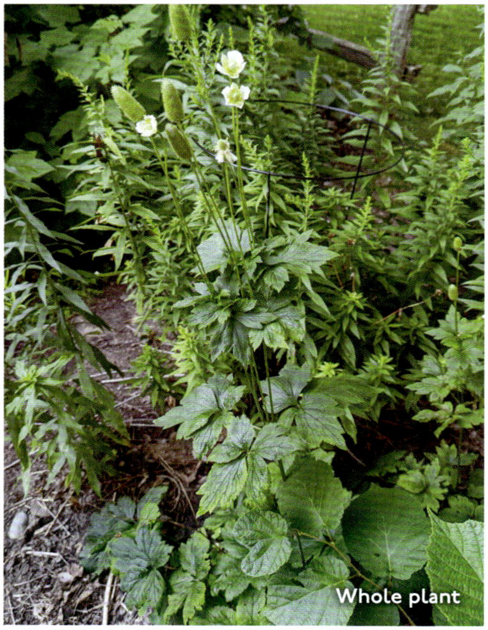

PLANT DESCRIPTION Tall Thimbleweed is characterized by upright, rigid stems that are mostly unbranching and sparsely hairy. Leaves occur in one or two whorls along the stem, in groups of two to three. They are about 12.5 cm long and attached to the stem with long leaf stalks. Each leaf is divided into two to three deep lobes that are further divided into two to three shallower lobes. Each flower stalk has a solitary flower about 2.5 cm across with five petal-like sepals and many green stamens (the part with the pollen). Once the flowers fade, they give way to green, cylindrical seed heads that measure up to 2.5 cm long and just over 1 cm wide. Seeds ripen into fluffy, cotton-like seed heads.

Tall Thimbleweed is similar to Thimbleweed (*Anemone cylindrica*) but can be differentiated by the leaves. Thimbleweed has wedge-shaped leaflets, while Tall Thimbleweed has rounded leaflets. In addition, Tall Thimbleweed is a taller plant but has shorter cones (<2.5 cm) while Thimbleweed is a shorter plant but has longer cones (up to 4 cm long).

IN THE GARDEN Tall Thimbleweed is valued in gardens for its charming anemone-like flowers and upright form. The flowers don't contain nectar but still support pollinators with copious pollen reserves. The rigid stems and fluffy seed heads make this a great choice for adding late-season interest in your garden. Tall Thimbleweed is a clumping plant and is

not aggressive like many other anemones are. Herbivore resistant and useful for shallow, rocky soils.

SKILL LEVEL Beginner

LIFESPAN Perennial

EXPOSURE Full sun to part shade (but will tolerate full shade)

SOIL TYPE Prefers acidic-to-neutral soils and has best growth where there is plenty of organic matter in the soil

MOISTURE Dry to (moderately) moist (drought tolerant once established)

HEIGHT 100 cm

SPREAD 30–60 cm

BLOOM PERIOD May, Jun, Jul, Aug

COLOR White

FRAGRANT ❌

SHOWY FRUIT ❌

CUT FLOWER ✅

PESTS No serious insect or disease problems

NATURAL HABITAT Moist meadows and prairies, savannas, moist open woodlands and along the edges of woods and thickets

WILDLIFE VALUE Native bees and other pollinators visit the flowers, usually seeking pollen rather than nectar

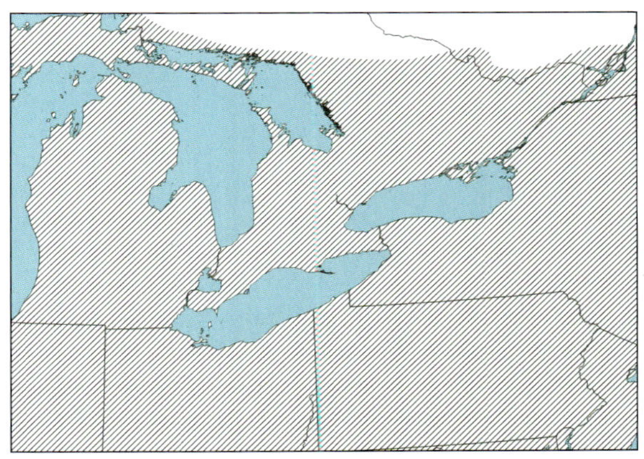

BUTTERFLY LARVA HOST PLANT FOR None

MOTH LARVA HOST PLANT FOR None

USDA HARDINESS ZONES 3–8

PROPAGATION No treatment of seeds required, though germination is said to increase in cooler soils. Older plants can be divided.

ADDITIONAL INFO Walnut (juglone) tolerant. Tall Anemone is purported to produce allelopathic chemicals that inhibits or stunts the growth of neighboring plants. It is usually found near wooded areas, while *Anemonastrum canadensis* prefers sun. Flowering may stop if plants become too crowded. The remedy for this situation is to divide plants in the fall.

SAR STATUS N/A

Antennaria neglecta
Field Pussytoes

Flowers

Flowering stalks

FAMILY *Asteraceae* (Aster Family)

ALTERNATE COMMON NAMES Cat's Foot, Early Everlasting, Prairie Everlasting, Prairie Pussytoes

PLANT DESCRIPTION Field Pussytoes features basal leaves up to 5 cm long and 1 cm across. They have smooth margins, hairy undersides, and pointy tips. Hairy flowering stalks rise from the center of the basal leaves with small alternate leaves growing along them. Stalks terminate with flower clusters that resemble a cat's paw (hence the common name) and are made up of about 20 small, white flower heads. Flowers are replaced by tiny seeds with cotton-like tufts that allow them to blow away in the wind.

IN THE GARDEN If you need a low-growing, drought-tolerant, dependable groundcover for

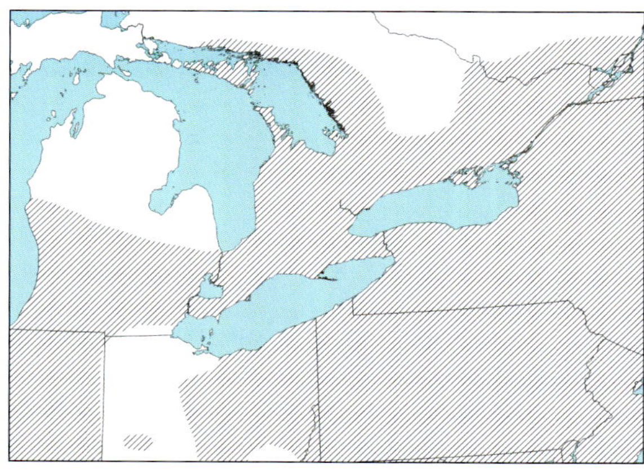

a sunny area then look no further than Field Pussytoes. The early spring flowers are an important food source for pollinators and turn into fluffy,

cotton-like seed heads in early summer. They don't like being shaded by taller plants, so plant accordingly. Suitable for growing in flowering lawns.

SKILL LEVEL Beginner

LIFESPAN Perennial

EXPOSURE Full sun

SOIL TYPE Lean, gritty to rocky, well-drained soils (it does not do well in fertile, humus-rich soils, particularly if drainage is poor)

MOISTURE Dry to medium

HEIGHT 10–15 cm

SPREAD 15–20 cm

BLOOM PERIOD May, Jun, Jul

COLOR White (red, green, brown)

FRAGRANT ❌

SHOWY FRUIT ❌

CUT FLOWER ❌

PESTS No serious insect or disease problems

NATURAL HABITAT Open sandy to rocky meadows and fields, alvars, and forests

WILDLIFE VALUE Parts of the plant are poisonous so deer, rabbits, and other small animals normally do not eat them, though some upland gamebirds are known to feed on the foliage and/or seed heads

BUTTERFLY LARVA HOST PLANT FOR Painted Lady (*Vanessa cardui*), American Lady (*Vanessa virginiensis*) — the caterpillars build little protective "tents" of silk between pairs of leaves

MOTH LARVA HOST PLANT FOR Everlasting Tebenna Moth (*Tebenna gnaphaliella*)

USDA HARDINESS ZONES 3–8

PROPAGATION Does not require any pretreatment (though 60 days of cold, moist stratification is said to enhance germination if starting seeds indoors). Keep seeds moist or plant immediately after collecting. Some seeds will germinate within a month, the rest will germinate the following spring. The seeds need light to germinate, so do not cover them. Mature plants may also be divided.

ADDITIONAL INFO This plant forms dense colonies. It differs from *Antennaria parlinii*, which has broad obovate (egg-shaped) basal leaves, 2 to 6 cm long and 1.5 to 5 cm wide, that taper to a thin stalk at the base and have three to five prominent veins.

SAR STATUS N/A

Antennaria parlinii

Parlin's Pussytoes

FAMILY *Asteraceae* (Aster Family)

ALTERNATE COMMON NAMES Ladies' Tobacco, Smooth Pussytoes

PLANT DESCRIPTION Parlin's Pussytoes feature both basal and alternate leaves. Basal leaves are up to 9.5 cm long and 4.5 cm wide, toothless, and rounded at the tip. The leaves have three to five prominent veins and fine hairs that give the leaves a gray-green appearance, with the undersides of the leaves taking on a more silvery-gray look. Widely spaced alternate leaves are found along the flowering stalk and are much smaller than the basal leaves but are still hairy with smooth margins. From the basal leaf clumps emerge hairy flowering stalks, each topped by a cluster of small flowers that give the appearance of a cat's paw. Flowers give way to tiny, brown seeds topped with cotton-like tufts of white hair that allow them to be carried by the wind.

IN THE GARDEN Parlin's Pussytoes are an adaptable, low-growing groundcover that thrives in tough conditions. It spreads at a slow-to-moderate rate by rhizomes, carpeting the ground with attractive velvety leaves.

SKILL LEVEL Beginner

LIFESPAN Perennial

EXPOSURE Full sun to full shade (but prefers full sun)

SOIL TYPE Lean, gritty to rocky or sandy clay, well-drained soils

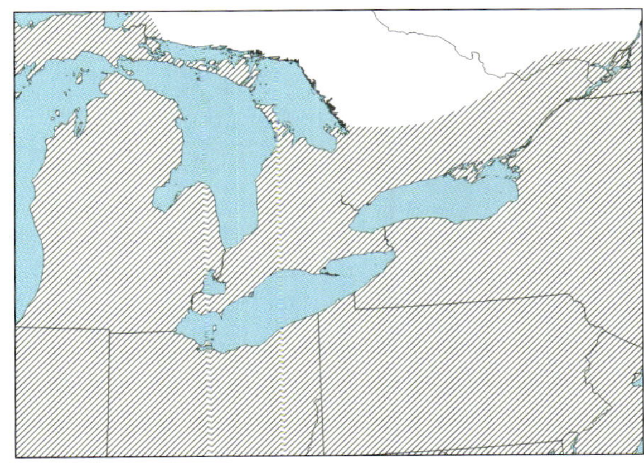
Seed heads

Whole plant

MOISTURE Dry to medium

HEIGHT 15–25 cm

SPREAD 20 cm

BLOOM PERIOD May

COLOR White

FRAGRANT ✗

SHOWY FRUIT ✗

CUT FLOWER ✓

PESTS No serious insect or disease problems

NATURAL HABITAT Prairies, dry meadows, sloped open woodlands, and disturbed sites like eroded banks or abandoned fields

WILDLIFE VALUE The leaves are a preferred food for deer, quail, and rabbits

BUTTERFLY LARVA HOST PLANT FOR Painted Lady (*Vanessa cardui*), American Lady (*Vanessa virginiensis*)

MOTH LARVA HOST PLANT FOR Everlasting Tebenna Moth (*Tebenna gnaphaliella*)

USDA HARDINESS ZONES 5–9

PROPAGATION There is little information available on starting Parlin's Pussytoes from seed other than that it may be difficult and is slow to germinate. The very small seeds require light to germinate, therefore surface sow. New plants may be propagated by dividing clumps in the spring, or from cuttings.

ADDITIONAL INFO One of the few native plants that does well in dry, shady locations, Parlin's Pussytoes does not do well in fertile, humus-rich soils, particularly if drainage is poor.

SAR STATUS N/A

Aquilegia canadensis
Wild Columbine

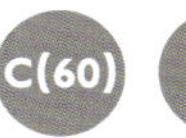

Flowers

A hummingbird feeds on the nectar

FAMILY *Ranunculaceae* (Buttercup Family)

ALTERNATE COMMON NAMES Canada Columbine, Cluckies, Common American Columbine, Eastern Red Columbine, Jack-in-trousers, Rock Lily, Wild Red Columbine

PLANT DESCRIPTION Wild Columbine features light-green-to-blue-green compound leaves that occur as basal foliage around the base of the plant and as alternating leaves up the flower stems. Each compound leaf is made up of three leaflets, each of which is lobed and measures 6 cm long and wide. Nodding flowers, up to 5 cm long, are borne on thin, branching stalks that rise above the basal leaves. Flowers are defined by five rolled-up, yellow petals that taper upwards, ending in nectar-rich spurs. Dangling yellow stamens (the part that carries the

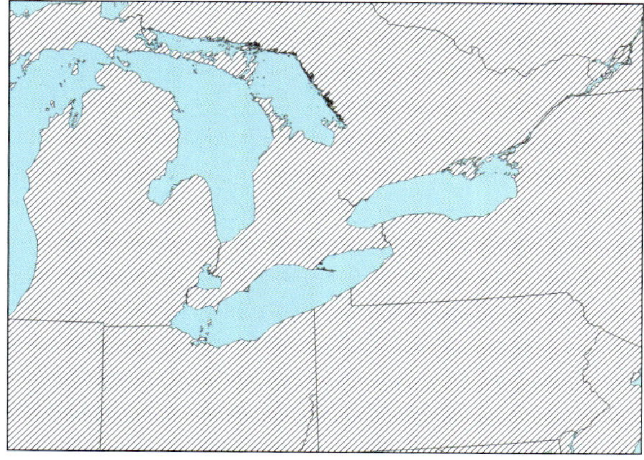

pollen) protrude from the bottom of the flowers. The flowers give way to erect green seed pods that turn brown as they dry and then split open to release shiny black seeds.

IN THE GARDEN The graceful, nodding flowers of Wild Columbine are a unique and welcome addition to native plant gardens. This plant is valued by gardeners for a stunning floral display, adaptability, and ease of growth. This drought-tolerant beauty is a jack-of-all-trades in the garden. It is extremely adaptable to light and moisture conditions as long as drainage is good.

SKILL LEVEL Beginner

LIFESPAN Short-lived perennial, but self-seeds readily

EXPOSURE Full sun to full shade (but does best in part shade)

SOIL TYPE Sandy, well-drained soils, not too rich

MOISTURE Dry to moist

HEIGHT 30–90 cm

SPREAD 30–60 cm

BLOOM PERIOD May, Jun, Jul

COLOR Red or pink and yellow

FRAGRANT ❌

SHOWY FRUIT ❌

CUT FLOWER ✅

PESTS Leaf miners

NATURAL HABITAT Woodlands and rocky slopes, slopes of deep ravines, steep streams and riverbanks, and old fields

WILDLIFE VALUE Blooms attract hummingbirds, bees, butterflies, and sphinx moths while seeds are consumed by finches and buntings

BUTTERFLY LARVA HOST PLANT FOR Columbine Duskywing (*Erynnis lucilius*)

MOTH LARVA HOST PLANT FOR Pink-patched Looper (*Eosphoropteryx thyatyroides*), Columbine Borer Moth (*Papaipema leucostigma*), Purple-lined Sallow (*Pyrrhia exprimens*)

USDA HARDINESS ZONES 3–8

PROPAGATION Most easily propagated by seed sown on the surface in the fall (seeds need light to germinate, so do not cover). If starting indoors, it does best with at least 60 days of cold, moist stratification. Division of mature plants when not flowering is difficult but possible with care. Young seedlings, however, transplant easily when less than 15 cm tall and can be a great source of new plants as Wild Columbine readily self-seeds.

ADDITIONAL INFO Wild Columbine is a short-lived perennial and will persist by self-seeding into bare soil.

SAR STATUS N/A

Aralia nudicaulis
Wild Sarsaparilla

FAMILY *Araliaceae* (Ginseng Family)

ALTERNATE COMMON NAMES American Sarsaparilla, False Spikenard, Rabbit Root, Small Spikenard, Sweet-root, Virginia Sarsaparilla, Wild Licorice

PLANT DESCRIPTION Wild Sarsaparilla produces a single compound leaf on a smooth leaf stalk. It splits into three parts, each made up of three to seven leaflets. These leaflets each measure 7.5 to 12.5 cm long and 5 cm wide. They are widest just above the middle, taper to a pointed tip, and have finely toothed edges. The smooth, leafless flower stem rises directly from the base of the stem and splits into three stems, each of which terminate with a globe-shaped flower cluster. Each cluster measures 2.5 to 5 cm across and is made up of

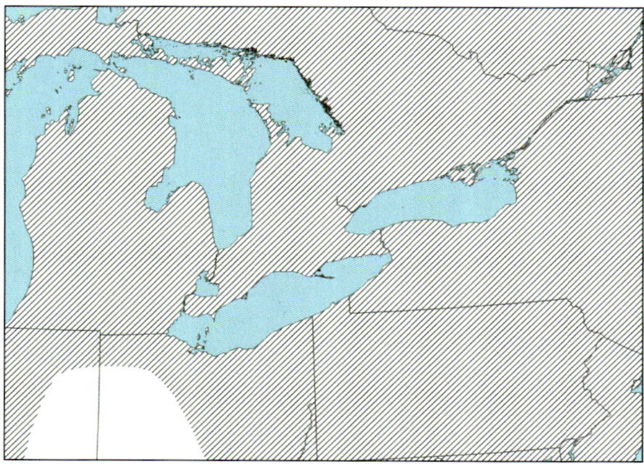

20 to 40 smaller, stalked flowers that each measure less than 1 cm across. Flowers give way to green berries that ripen to dark purple.

IN THE GARDEN Wild Sarsaparilla adds a unique look to woodland gardens. As it emerges in early spring, the leaves are a shiny copper color then change to a medium green by summer. Once the flowers fade, they give way to dark purple berries that are relished by birds. The leaves remain lush and green well into late summer, at which point they take on stunning shades of yellow.

SKILL LEVEL Beginner

LIFESPAN Perennial

EXPOSURE Part shade to full shade (will tolerate full sun)

SOIL TYPE Sandy soil to deep loam, moderate to rich in nutrients

MOISTURE Dry to moist

HEIGHT 30–60 cm

SPREAD 60 cm

BLOOM PERIOD May, Jun

COLOR Greenish white

FRAGRANT ❌

SHOWY FRUIT ✅

CUT FLOWER ✅

PESTS No serious pests, although the caterpillars of Aralia Shoot Borer Moth (*Papaipema araliae*) are known to bore through the stems

NATURAL HABITAT Shaded mixed woods, in rich soil with lots of decaying plant material

WILDLIFE VALUE Native bees as well as some species of flies and beetles visit the flowers, and the berries of *Aralia* spp. are eaten by sparrows, thrushes, and small mammals

BUTTERFLY LARVA HOST PLANT FOR None

MOTH LARVA HOST PLANT FOR None

USDA HARDINESS ZONES 4–8

PROPAGATION Seeds benefit from scarification, either by a sulfuric acid bath or rubbing lightly with sandpaper, followed by three to five months of cold, moist stratification. Keep seeds moist or plant as soon as harvested. Some success has been had with putting the seeds through a warm/cold/warm, moist stratification at one month for each period. Seedlings mature very slowly. Wild Sarsaparilla can also be propagated by dividing the rootstock while the plant is dormant.

ADDITIONAL INFO Wild Sarsaparilla was used to flavor root beer when the actual plant (Sarsaparilla) was not available.

SAR STATUS N/A

Aralia racemosa
Spikenard

Flowers

Berries

FAMILY *Araliaceae* (Ginseng Family)

ALTERNATE COMMON NAMES American Spikenard, Hungry Root, Indian Root, Life of Man, Old Man's Root, Pettymorrel, Small Spikenard, Spiceberry, Spignet

PLANT DESCRIPTION The smooth, maroon-colored stems of Spikenard hold a few large (up to 60 cm long) compound leaves that are made up of many smaller leaflets, each measuring about 14 cm long. Leaflets are heart shaped at the base, sharply toothed, and abruptly taper to a pointed tip. Spikenard often grows wider than it is tall. Large, tapered flower clusters are made up of many tiny, stalked flowers. Individual flowers are under 1 cm across with five triangular petals. Flower stalks are covered

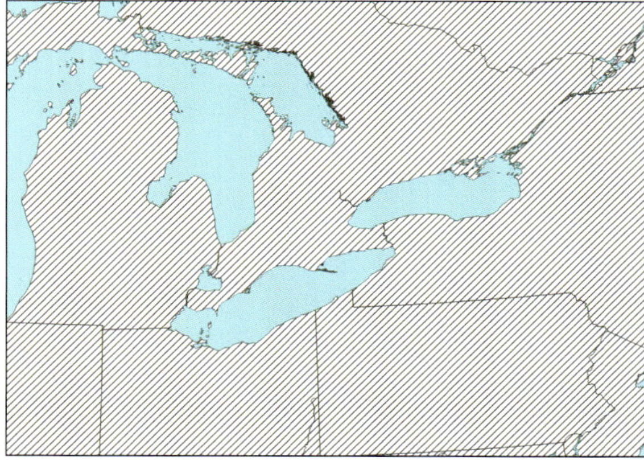

in many fine hairs. Flowers give way to dense, hanging clusters of dark-red-to-purple fruit, 0.5 cm in diameter.

IN THE GARDEN Spikenard is a large, spreading perennial that gives the appearance of a shrub but dies back to the ground every year. It is valued in gardens for its lush leaves, ability to tolerate deep shade, and attractive broad form. The summer flowers bloom at a time when not much else is blooming in shade gardens. The dark red berries look like little jewels but don't stick around long as they are a favorite food of many songbirds. Excellent as a specimen or in small groups.

SKILL LEVEL Beginner

LIFESPAN Perennial

EXPOSURE Part shade to full shade

SOIL TYPE Fertile, humus-rich loams, but tolerates a wide range of soils including rocky and clay ones

MOISTURE Wet to medium

HEIGHT 90–150 cm

SPREAD 120 cm

BLOOM PERIOD (Jun), Jul, (Aug)

COLOR White

FRAGRANT ✓

SHOWY FRUIT ✓

CUT FLOWER ✓

PESTS No serious insect or disease problems

NATURAL HABITAT Rich, usually moist beech-maple and hemlock-hardwood forests, especially along edges and clearings and in cedar swamps

WILDLIFE VALUE The berries of *Aralia* spp. are eaten by some woodland songbirds and some small mammals

BUTTERFLY LARVA HOST PLANT FOR None

MOTH LARVA HOST PLANT FOR None

USDA HARDINESS ZONES 3–8

PROPAGATION Sow seeds immediately or keep moist. Germination is enhanced with scarification, either in a mild sulfuric acid or by rubbing with sandpaper. Cold, moist stratification may help germination, but artificial stratification seems to be hit-and-miss with these seeds. The easiest propagation method is to divide old rootstocks when the plants go dormant in the fall, though plants may be slow to bloom after being disturbed. Plants may also be started from root cuttings.

ADDITIONAL INFO Plants will slowly spread over time by self-seeding and creeping rhizomes to form thickets.

SAR STATUS N/A

Arisaema triphyllum
Jack-in-the-pulpit

	FULL SHADE	PART SHADE	
	WET	MOIST	MEDIUM

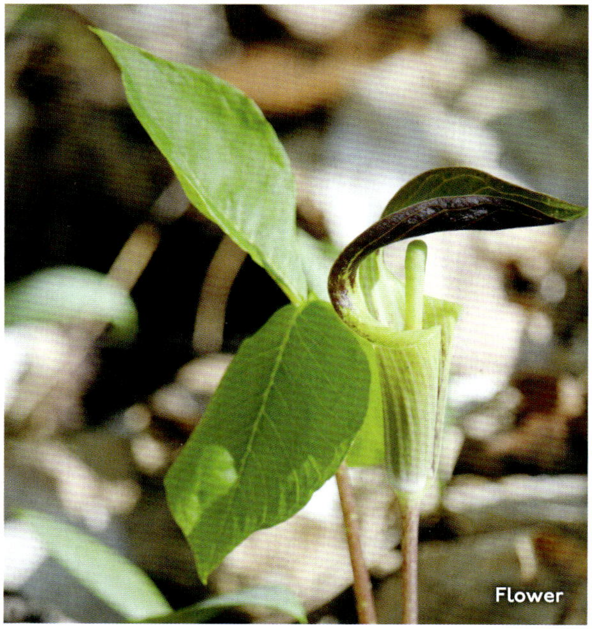

Flower

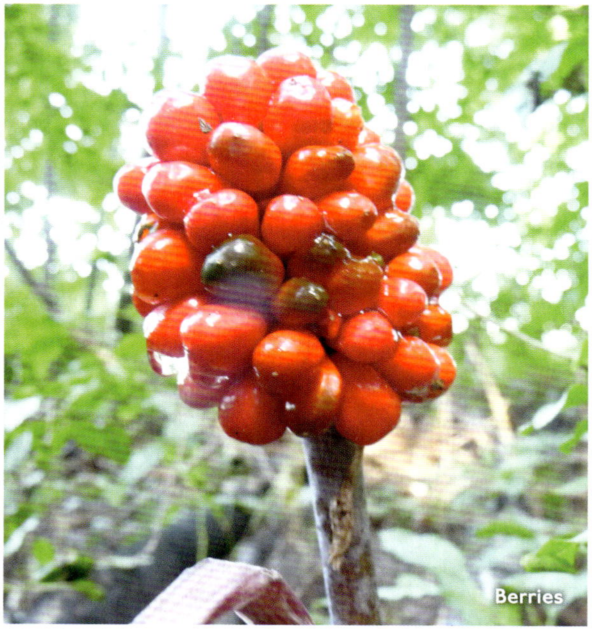

Berries

FAMILY *Araceae* (Arum Family)

ALTERNATE COMMON NAMES Indian Turnip, Small Jack-in-the-pulpit

PLANT DESCRIPTION From Jack-in-the-pulpit's corm (a bulb-like tuber) emerge one to two leaves and a single flowering stalk. The leaves are divided into three leaflets that each measure up to 18 cm long and 7.5 cm wide. Leaflets are oval-shaped and glossy and have smooth margins. A single flower occurs on a separate stalk. It consists of the spadix (a spike of tiny flowers) that is enclosed by the spathe (a leaf-like sheath), which forms a hood over the top. The spadix and spathe are nicknamed "Jack" and the "pulpit," respectively. The spathe is usually green with variably colored stripes that

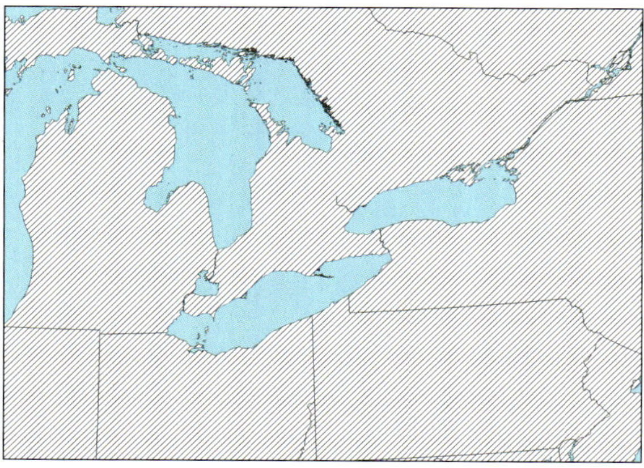

can be maroon, dark purple, or dark green. Tightly packed clusters of smooth, green berries ripen into red to reddish-orange berries in mid to late summer.

IN THE GARDEN Jack-in-the-pulpit is an intriguing spring wildflower with a unique bloom. It is valued by gardeners for its lush appearance and tropical feel. Most plants will disappear by mid to late summer, leaving behind their red berries, which add interest and wildlife value to shade gardens. Due to its ephemeral lifecycle, it is best planted with other shade-tolerant species that can fill in the gap when it goes dormant.

SKILL LEVEL Beginner

LIFESPAN Perennial (spring ephemeral)

EXPOSURE Full shade to part shade

SOIL TYPE Rich in organic matter, does poorly in heavy clay soils

MOISTURE Wet to medium

HEIGHT 75 cm

SPREAD 30–45 cm

BLOOM PERIOD Apr, May, Jun

COLOR Green, purple

FRAGRANT ❌

SHOWY FRUIT ✅

CUT FLOWER ❌

PESTS No serious insect or disease problems

NATURAL HABITAT Forests, woodlands, swamps, and marshy areas

WILDLIFE VALUE Upland gamebirds, wood thrushes, and some small mammals eat the berries, which ripen in mid to late summer

BUTTERFLY LARVA HOST PLANT FOR None

MOTH LARVA HOST PLANT FOR None

USDA HARDINESS ZONES 3–9

PROPAGATION Seeds require a double dormancy of cold, warm, cold, warm of 60 to 90 days for each part of the cycle to germinate. May be directly sown outdoors, 20 mm deep, but it may take two or more years to germinate. Plants will take five years to flower when started from seed. May also be propagated by root division and cormlets can be separated from the parent corm in fall.

ADDITIONAL INFO Jack-in-the-pulpit leaves are sometimes confused with Trillium leaves, but are actually easily distinguished in that the leaves of Jacks tend to form a T shape, whereas Trillium leaves are spread equally around the point of origin and resemble the Mercedes-Benz emblem or the peace sign of the 1960s.

Jack-in-the-pulpit has both male and female plants, and they can change sex from one year to the next depending, apparently, on the previous year's reproductive success. Another adaptation for reproductive success is the presence of a small hole at the base of the smaller male flower that allows pollinators to exit more easily, laden with pollen. Female plants do not have this hole so, with only one way out, pollinators are more likely to pollinate the female flower.

SAR STATUS N/A

Asarum canadense
Wild Ginger

QUICK GUIDE

PROPAGATION

FAMILY *Aristolochiaceae* (Birthwort Family)

ALTERNATE COMMON NAMES Asarabaca, Canada Ginger, Catfoot, Ginger Root, Heart Snakeroot, Indian Ginger, Woodland Ginger

PLANT DESCRIPTION Wild Ginger is a stemless plant with a pair of basal leaves measuring up to 15 to 20 cm long and wide. Leaves are heart shaped and have smooth margins, and the undersides are covered in fine hairs. They are borne on stalks that are equally as hairy. Each plant has a single, maroon-colored, cup-shaped flower that rests on the ground, underneath the foliage. The flower measures up to 5 cm across from tip to tip and is characterized by three triangular petal-like sepals. Flowers give way to seed pods that look very similar to the flowers and contain small, brown seeds. All parts of the plant smell like ginger.

IN THE GARDEN Wild Ginger is a classic native groundcover for shade. It spreads slowly, but surely, by rhizomes to form dense colonies. Even once the spring flowers fade, the leaves remain lush well into fall. The hairy leaves and strong ginger smell mean herbivores leave this plant alone.

SKILL LEVEL Beginner

LIFESPAN Perennial

EXPOSURE Part shade to full shade

SOIL TYPE Moist, rich soils with a pH of 6 to 7 is best, sandy to clay, will tolerate dry soils once established

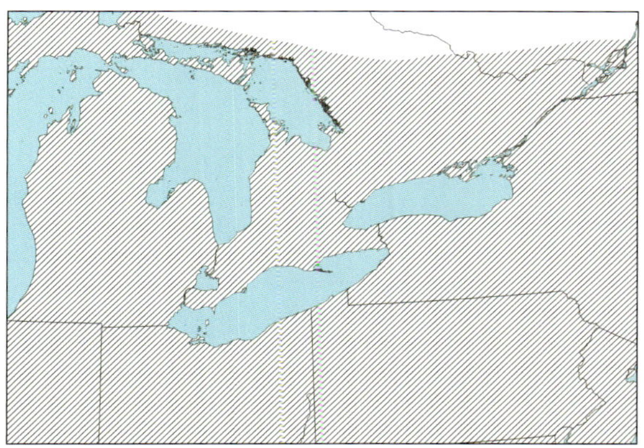

MOISTURE Moist to medium

HEIGHT 10–15 cm

SPREAD 30 cm

BLOOM PERIOD Apr, May

COLOR Brown

FRAGRANT ✘

SHOWY FRUIT ✘

CUT FLOWER ✘

PESTS Wild Ginger has few pests, though slugs and snails can sometimes cause problems

NATURAL HABITAT The understory of deciduous, and occasionally coniferous, forests — with light shade and moist to slightly dry conditions

WILDLIFE VALUE Flies and beetles are likely the main pollinators, and Wild Ginger is apparently unpalatable to mammals, which rarely eat it

BUTTERFLY LARVA HOST PLANT FOR Pipevine Swallowtail (*Battus philenor*)

MOTH LARVA HOST PLANT FOR None

USDA HARDINESS ZONES 3–7

PROPAGATION Seeds should not be allowed to dry out and are difficult and/or slow to germinate and grow to maturity. They require a two-to-three-month warm period followed by a two-to-three-month cold period. If sown outdoors, they require a full year to germinate. You can also propagate by root division in the fall or by stem cuttings in the summer.

ADDITIONAL INFO Wild Ginger seeds are naturally spread by ants. Each seed has an oil-rich outgrowth attached to it, called an elaiosome, which is a favorite food of ants. After eating these attachments, they discard the seed, and it later germinates.

SAR STATUS N/A

Asclepias exaltata
Poke Milkweed

FULL SHADE	PART SHADE

	MEDIUM

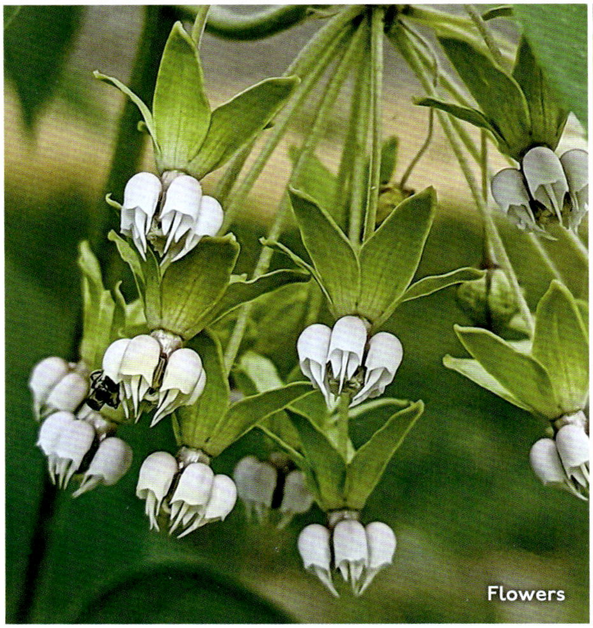

Flowers

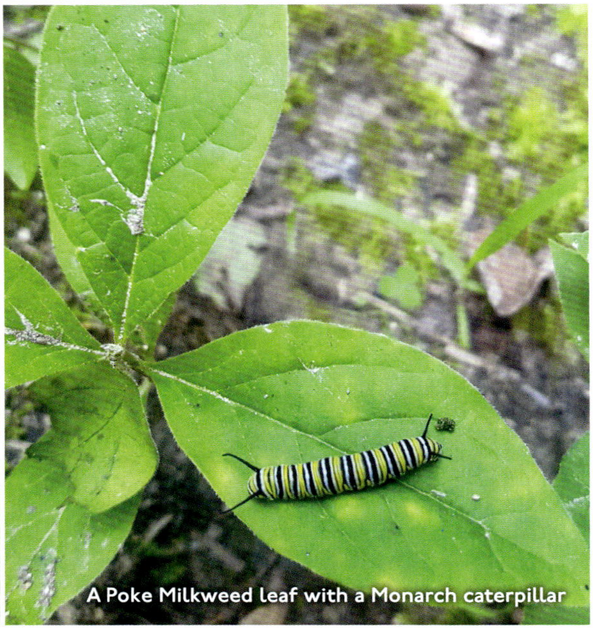

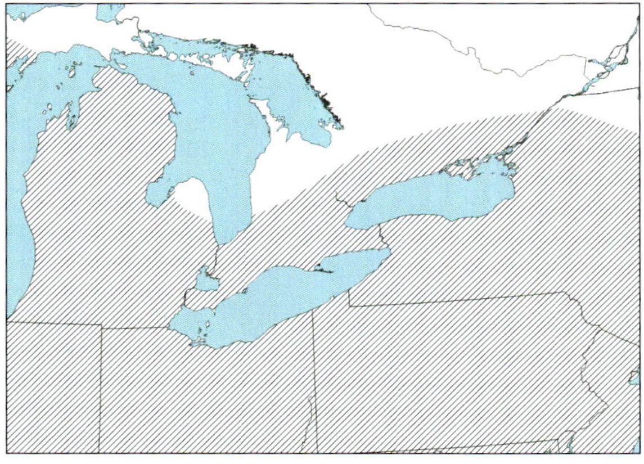

A Poke Milkweed leaf with a Monarch caterpillar

FAMILY *Asclepiadaceae* (Milkweed Family)

ALTERNATE COMMON NAMES Tall Milkweed, White Woodland Milkweed

PLANT DESCRIPTION Poke Milkweed is characterized by smooth, round, unbranched stems that taper towards the tip. Large, broad leaves are found in an opposite arrangement and measure 7.5 to 20 cm long and 2.5 to 7.5 cm across (possibly larger in deep shade). They are roughly egg shaped and taper to a point at both ends of the stalk. Leaf margins are smooth as is the leaf surface. Leaves contain milky sap. Stems terminate with one or more droopy, open, globular flower clusters that are stemmed. Individual flowers are 1 cm across and just over 1 cm long. Each flower is exaracterized by five cylindrical hoods that surround a central column and five greenish petals that flare back, away from the crown. Flowers are replaced by long, erect, downy seed pods that taper to a tip. Once they mature, they release flat, brown seeds with a tuft of white hair that allows them to be carried off by the wind.

IN THE GARDEN Poke Milkweed is one of our most shade-tolerant milkweed species and is highly beneficial for wildlife and gardeners alike. The large, oval leaves and tall stature of Poke Milkweed make it an excellent structural plant in shady gardens. The droopy flower heads are highly fragrant and provide early summer flowers for pollinators. Poke Milkweed is not an aggressive spreader and is very suitable for a garden setting. Tolerates dry shade.

SKILL LEVEL Beginner

LIFESPAN Perennial

EXPOSURE Part shade to full shade (one of the most shade-tolerant milkweeds, but will also grow in full sun)

SOIL TYPE Sand, loam, does best with moderate moisture and rich organic content

MOISTURE Medium

HEIGHT 180 cm (can exceed 180 cm in ideal conditions)

SPREAD 45 cm

BLOOM PERIOD Jun, Jul

COLOR White, pink, green

FRAGRANT ✅ (strikingly aromatic when in bloom)

SHOWY FRUIT ❌

CUT FLOWER ❌

PESTS Milkweed Tussock Moth (*Euchaetes egle*) caterpillar

NATURAL HABITAT Open forests and forest edges where oaks, maples, and similar trees are dominant

WILDLIFE VALUE Poke Milkweed flowers attract bumblebees and butterflies. Other insects feed on the foliage, flower tissues, seed pods, pith of the stems, or plant juices. Mammals usually avoid eating any milkweeds because their foliage is bitter tasting, but of all the milkweeds, this one is the most likely to be eaten — by rabbits in particular. Birds such as the Baltimore Oriole (*Icterus galbula*) and the Orchard Oriole (*Icterus spurius*) use the strong plant fibers from last year's milkweed stems to construct their nests, and other birds use leftover silky seed tufts to line their nests

BUTTERFLY LARVA HOST PLANT FOR Monarch Butterfly (*Danaus plexippus*)

MOTH LARVA HOST PLANT FOR Unexpected Cycnia (*Cycnia inopinatus*), Delicate Cycnia (*Cycnia tenera*), Milkweed Tussock Moth (*Euchaetes egle*), Stalk Borer Moth (*Papaipema nebris*), Isabella Tiger Moth (*Pyrrharctia Isabella*), Striped Garden Caterpillar (*Trichordestra legitima*)

USDA HARDINESS ZONES 3–7

PROPAGATION Thirty to 60 days of cold, moist stratification will help germination if starting seeds indoors, but direct sowing in the fall or early spring is easiest. This plant has a deep taproot and does not transplant easily once established.

ADDITIONAL INFO This species has been known to hybridize with Common Milkweed (*Asclepias syriaca*) when the two plants grow near each other.

SAR STATUS N/A

Asclepias incarnata
Swamp Milkweed

Flowers

Leaves

FAMILY *Asclepiadaceae* (Milkweed Family)

ALTERNATE COMMON NAMES Pink Milkweed, Red Milkweed, Rose Milkweed, Wetland Milkweed

PLANT DESCRIPTION Swamp Milkweed is an upright plant with smooth stems. Lance-shaped leaves are arranged in an opposite pattern along the stem and are attached by short stalks. Leaves measure about 15 cm long and 4 cm wide. They are toothless and taper towards their tips. Multiple, 8 cm wide, convex flower clusters can be found at the top of the plant. Individual flowers measure about 6 mm across and have five sectioned crowns. Each flower has five cylindrical hoods, each with a curved horn protruding from it and arching towards the center. Five downward-flared petals are found underneath. Both stems and leaves contain a white milky sap. Slender, pointed seed pods contain flat, brown seeds, each with a tuft of white hair that allows them to be carried off by the wind.

IN THE GARDEN Swamp Milkweed is a well-behaved, clump-forming milkweed. It is valued by gardeners for its showy pink flowers and ability to support a wide diversity of beneficial wildlife, including caterpillars of the Monarch Butterfly. Despite its affinity for moist areas, it is fairly drought tolerant and will thrive in drier areas once established. The rigid stems stand tall through the winter months.

SKILL LEVEL Beginner

LIFESPAN Perennial

EXPOSURE Full sun to part shade (not shade tolerant)

Seed pods

Whole plant

A butterfly visitor

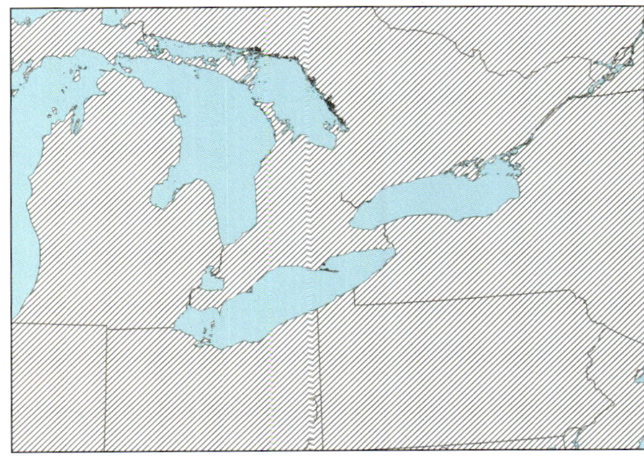

SOIL TYPE One of the few ornamentals that thrives in mucky clay soils; it prefers neutral to slightly acidic soil but will tolerate heavy clay and a pH of up to 8

MOISTURE Moist to wet, but it will also grow in drier areas, such as prairies, fields, and roadsides

HEIGHT 100 cm

SPREAD 60–90 cm

BLOOM PERIOD Jun, Jul

COLOR Pink to mauve and white

FRAGRANT 

SHOWY FRUIT ❌

CUT FLOWER ✅

PESTS Swamp Milkweed will inevitably have aphids, especially the Oleander Aphid (*Aphis nerii*) — also known as the Orange Milkweed Aphid. The insects are not a problem unless the plant looks sick, at which point an effective treatment is to spray the plant and aphids with soapy water. Several weevil larvae prey on developing seeds, so look for signs of damage including small entry holes in the pod exuding latex. And, of course, Monarch caterpillars *love* Swamp Milkweed

NATURAL HABITAT Wet meadows and prairie, along waterways, swamps, marshes, and other wet areas

WILDLIFE VALUE Nectar source for butterflies and hummingbirds

BUTTERFLY LARVA HOST PLANT FOR Monarch Butterfly (*Danaus plexippus*)

MOTH LARVA HOST PLANT FOR Unexpected Cycnia (*Cycnia inopinatus*), Delicate Cycnia (*Cycnia tenera*), Milkweed Tussock Moth (*Euchaetes egle*), Isabella Tiger Moth (*Pyrrharctia Isabella*), Striped Garden Caterpillar (*Trichordestra legitima*)

USDA HARDINESS ZONES 3–9

PROPAGATION No pretreatment of seeds is necessary, though cold, moist stratification for at least four to 12 weeks is believed to enhance germination. Alternatively, seeds can be soaked in hot water for 12 hours, and the soaking process repeated two additional times for an expected seed germination of 50 percent. Fall sowing with light soil cover provides good germination. Plants can be divided in late spring.

ADDITIONAL INFO The juice of this wetland milkweed is less milky than that of other species. It tends to bloom twice in a growing season when in gardens. Rare occasional white specimens are found in the wild and these have led to cultivars such as "Ice Ballet."

SAR STATUS N/A

Asclepias purpurascens
Purple Milkweed

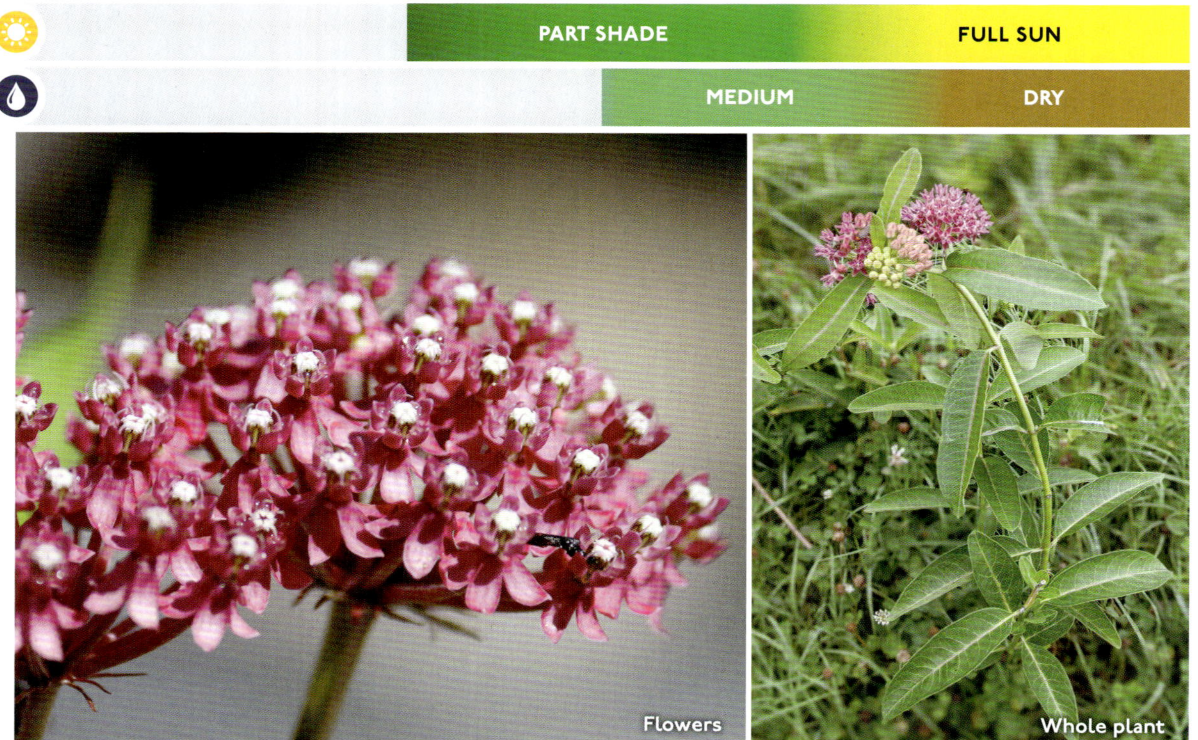

FAMILY *Asclepiadaceae* (Milkweed Family)

ALTERNATE COMMON NAMES None

PLANT DESCRIPTION Purple Milkweed features upright stems with short-stalked, opposite leaves up to 20 cm long. Leaves are oblong to oval with fine hairs underneath and are toothless with wavy margins. Leaves may curl upwards from the middle vein. Stems and leaves contain a milky white sap. Multiple rounded flower clusters, up to 8 cm wide, are held above the leaves at the top of the plant. Individual flowers measure slightly less than 1 cm across and 1.2 cm long and are characterized by five hoods that are more than twice as tall as the center column. Each hood has a short, curved horn that arches towards the center. Each flower has five petals that bend down and back then up at the tips.

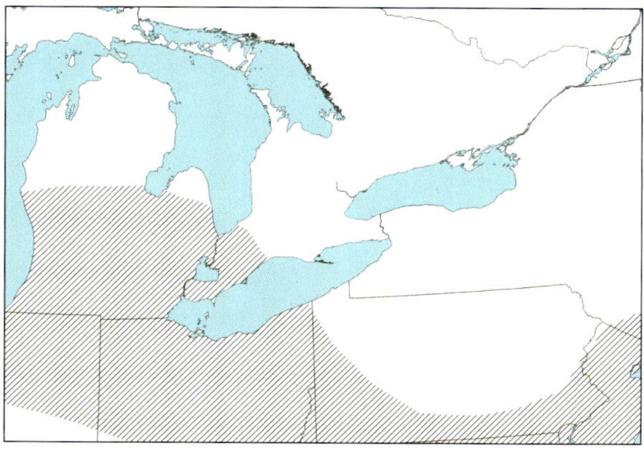

Flowers give way to smooth seed pods that reach up to 15 cm long. Each pod splits open to release numerous flat, brown seeds with tufts of white silk that allow them to be carried by the wind.

Purple Milkweed can be differentiated from Common Milkweed (*Asclepias syriaca*) by its longer, stiffer flower stems at the top of the plant compared to the droopy flower heads of Common Milkweed. The flowers are also a rich purple color compared to the soft pink to white of Common Milkweed.

IN THE GARDEN Purple Milkweed is valued by gardeners for its showy, deep purple flowers and high wildlife value. This milkweed can spread quickly by rhizomes so make sure to give it some space.

SKILL LEVEL Beginner

LIFESPAN Perennial

EXPOSURE Full sun to part shade (prefers morning sun and afternoon shade)

SOIL TYPE Well-drained soil, enriched with a light amount of composted cow manure (high-nutrient requirement), and it might possibly favor calcium-rich sites

MOISTURE Dry to medium

HEIGHT 90 cm

SPREAD 30–90 cm

BLOOM PERIOD Jun, Jul

COLOR Red, purple

FRAGRANT ✅

SHOWY FRUIT ❌

CUT FLOWER ✅

PESTS No serious insect or disease problems

NATURAL HABITAT Fields and other open areas

WILDLIFE VALUE Purple Milkweed attracts a variety of butterflies who sip on the nectar, and bees (bumblebees in particular) who drink nectar and collect pollen. Said to be a favored nectar plant of the Great Spangled Fritillary (*Speyeria cybele*)

BUTTERFLY LARVA HOST PLANT FOR Monarch Butterfly (*Danaus plexippus*)

MOTH LARVA HOST PLANT FOR Unexpected Cycnia (*Cycnia inopinatus*), Delicate Cycnia (*Cycnia tenera*), Milkweed Tussock Moth (*Euchaetes egle*), Stalk Borer Moth (*Papaipema nebris*), Isabella Tiger Moth (*Pyrrharctia Isabella*), Striped Garden Caterpillar (*Trichordestra legitima*)

USDA HARDINESS ZONES 3–8

PROPAGATION Sowing seeds directly outdoors in the fall or early spring is easiest, though some seeds may not sprout for two seasons. Seeds started indoors require at least 30 days of cold, moist stratification. As this plant spreads by rhizome, it may be divided by digging new shoots, though it does not like to be moved once established, so division should occur when the shoots are very young.

ADDITIONAL INFO This plant is considered by many gardeners to be too vigorous and weedy for small gardens, though it is not as aggressive as Common Milkweed.

SAR STATUS ON – S1/(not listed); IN – S3/(not listed); MI – S2/T; NY – S2/T

Asclepias syriaca
Common Milkweed

FAMILY *Asclepiadaceae* (Milkweed Family)

ALTERNATE COMMON NAMES Kansas Milkweed, Milkplant, Silkweed, Silky Swallow-wort

PLANT DESCRIPTION Common Milkweed is characterized by stout, unbranched, slightly hairy stems. Leaves occur in an opposite pattern, are up to 20 cm long and 9 cm wide, are broadly oblong, and have smooth margins. The lower leaf surface is densely covered in short hairs but is hairless above. Both stems and leaves exude a milky sap when cut open. Flowers are found in globular clusters up to 10 cm across and emerge from the point where the upper leaves attach to the stem. Individual flowers are about 6 mm across and are characterized by five raised hoods, each with a curved horn protruding from it. Each flower has five petals that bend

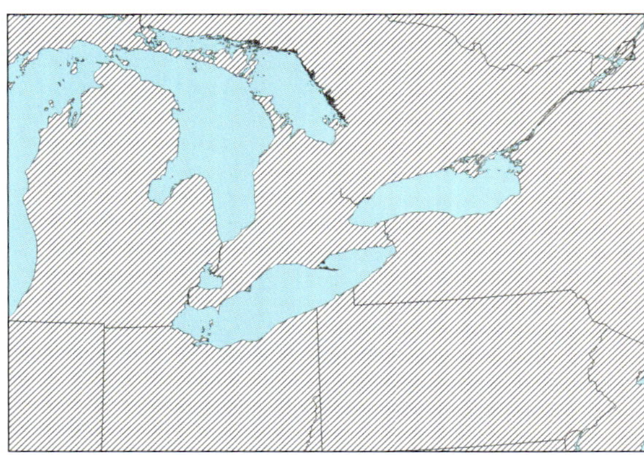

backwards towards the center of the flower cluster. Flowers give way to 10 cm long plump seed pods with soft spines and woolly hairs. Once mature, the seed pods split open to reveal many flat, brown

seeds with silky hairs that allow them to be carried by the wind.

IN THE GARDEN Common Milkweed is a very familiar milkweed species with a wide distribution. The conspicuous midsummer flowers are highly fragrant and fill the air with an appealing vanilla-like fragrance. This milkweed is very easy to grow but is known for its vegetative spread, so it may not be suitable for small or formal gardens. Excellent for naturalization projects.

SKILL LEVEL Beginner

LIFESPAN Perennial

EXPOSURE Full sun to part shade

SOIL TYPE Rich sandy or gravelly loam soils that are well drained

MOISTURE Dry to moist

HEIGHT 120 cm (can reach over 200 cm in ideal conditions, though not common)

SPREAD 20–30 cm

BLOOM PERIOD Jun, Jul, Aug

COLOR Pink (white, purple)

FRAGRANT ✅

SHOWY FRUIT ❌

CUT FLOWER ✅

PESTS No serious insect or disease problems

NATURAL HABITAT Old fields, roadsides, and wastelands

WILDLIFE VALUE Source of nectar for humming-birds and many butterflies and bees

BUTTERFLY LARVA HOST PLANT FOR Monarch Butterfly (*Danaus plexippus*)

MOTH LARVA HOST PLANT FOR Unexpected Cycnia (*Cycnia inopinatus*), Delicate Cycnia (*Cycnia tenera*), Milkweed Tussock Moth (*Euchaetes egle*), Stalk Borer Moth (*Papaipema nebris*), Isabella Tiger Moth (*Pyrrharctia Isabella*), Virginian Tiger Moth (*Spilosoma virginica*), Striped Garden Caterpillar (*Trichordestra legitima*)

USDA HARDINESS ZONES 3–8

PROPAGATION No treatment needed if sowing seeds outdoors in the fall or early spring. Some sources suggest 30 days of cold, moist stratification will aid germination if starting indoors. Surface sow. *Asclepias syriaca* is said to be tricky to start in that in some years, the seeds are dormant and require stratification, while in other years, they do not. They can be slow to germinate. Seedlings do not flower until the second year and usually only one or two flowers in each cluster mature into a seed pod. This plant spreads by underground rhizomes, and when a young shoot pops up, it can be dug up (keeping as much of the root as possible) and transplanted. When mature, Common Milkweed does not transplant well.

ADDITIONAL INFO The hypoallergenic downy floss (the silky hairs that surround the seeds), which is five to six times more buoyant than cork and six times lighter and warmer than wool, has had a variety of uses in history. It has been used as stuffing for pillows and mattresses, in life preservers, and some companies are now using it as an animal-cruelty-free insulation in jackets (instead of down from geese).

SAR STATUS N/A

Asclepias tuberosa
Butterfly Milkweed

FULL SUN

MEDIUM **DRY**

Flowers

Whole plant

FAMILY *Asclepiadaceae* (Milkweed Family)

ALTERNATE COMMON NAMES Butterflyweed, Chigger Flower, Orange Milkweed, Pleurisy Root

PLANT DESCRIPTION Butterfly Milkweed is characterized by rigid, hairy stems with lance-shaped, alternate leaves attached with little-to-no leaf stalk. Leaves measure about 5 to 15 cm long and 2.5 cm wide and are toothless, glabrous on top, and sparsely hairy underneath and end with a pointed tip. Only the foliage exudes a milky sap. Stems are mostly unbranching, except for at the top where several flat-topped flower clusters, up to 8 cm across, can be found. Each cluster is made up of up to 25 individual flowers measuring about 1 cm across. Flowers are characterized by five hoods with a curved horn emerging from each one and arching towards the central crown. Each flower has five backwards-flared petals. Flowers give way to narrow, smooth, 15 cm long seed pods. Each pod contains numerous flat, brown seeds with tufts of white silk that allow them to be carried by the wind.

IN THE GARDEN Butterfly Milkweed is valued in gardens for its cheerful orange flowers, long bloom time, and high drought tolerance. It maintains a clumping form and is not an aggressive spreader, which makes it suitable for small or formal gardens. The deep taproot makes it hard to transplant, so choose its location wisely. Stems remain upright well into the winter months.

SKILL LEVEL Beginner

LIFESPAN Perennial

EXPOSURE Full sun (not shade tolerant)

SOIL TYPE Prefers sandy or rocky soil that is well drained

MOISTURE Dry to medium

HEIGHT 80 cm

SPREAD 45 cm

BLOOM PERIOD Jun, Jul, Aug

COLOR Orange

FRAGRANT ✗

SHOWY FRUIT ✗

CUT FLOWER ✓

PESTS No serious insect or disease problems, though crown rot can be a problem in wet, poorly drained soils, and it is susceptible to rust and leaf spot

NATURAL HABITAT Prairies, open woods, or along roadsides

WILDLIFE VALUE Nectar source for native bees, butterflies, and hummingbirds

BUTTERFLY LARVA HOST PLANT FOR Queen Butterfly (*Danaus gilippus*), Monarch Butterfly (*Danaus plexippus*), Gray Hairstreak (*Strymon melinus*)

MOTH LARVA HOST PLANT FOR Unexpected Cycnia (*Cycnia inopinatus*), Delicate Cycnia (*Cycnia tenera*), Milkweed Tussock Moth (*Euchaetes egle*), Stalk Borer Moth (*Papaipema nebris*), Isabella Tiger

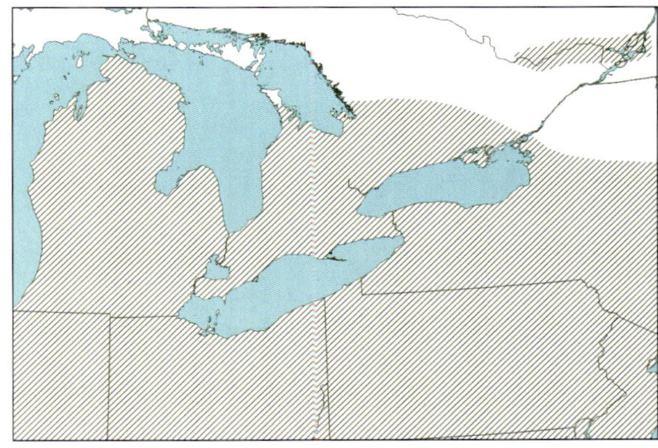

Moth (*Pyrrharctia Isabelia*), Striped Garden Caterpillar (*Trichordestra legitima*)

USDA HARDINESS ZONES 3–9

PROPAGATION Seeds sown in the spring require 30 days of cold, moist stratification. Plants are easily grown from seed but are somewhat slow to establish and may take two to three years to produce flowers. Butterfly Milkweed does not transplant well due to its deep taproot, and it is probably best left undisturbed once established. The quickest method of propagation is root cuttings. In the fall, cut the taproot into 5 cm sections and plant each section vertically, keeping the area moist.

ADDITIONAL INFO Unlike many of the other milkweeds, this species does not have milky-sapped stems. *Asclepias tuberosa* will host Monarch Butterfly caterpillars, but if other milkweeds are present this one is often ignored.

SAR STATUS N/A

Asclepias verticillata
Whorled Milkweed

Flowers

Flowers with seed pods

FAMILY *Asclepiadaceae* (Milkweed Family)

ALTERNATE COMMON NAMES Horsetail Milkweed

PLANT DESCRIPTION Whorled Milkweed features an erect, unbranched stem with lines of fine hairs along it. Whorls of four to six narrow, lance-shaped leaves occur along the stem. Leaves are 7.5 cm long and 3 mm wide with a prominent middle vein. New leaves point upwards while older leaves are droopy. Stems exude a milky sap. Umbrella-like flower clusters are found on the upper part of the plant. They measure up to 8 cm across and arise from the point where the leaf attaches to the stem. Each cluster is made up of 10 to 20 individual flowers measuring about 6 mm long. Flowers are characterized by five short hoods around a central column, each with an arching horn protruding from it that reaches for the center of the flower. Five flared petals bend backwards from each flower. Flowers give way to thin seed pods, 7 to 10 cm long, each containing numerous flat, brown seeds that are carried off in the wind by tufts of white silky hairs.

IN THE GARDEN Whorled Milkweed is valued for its elegant creamy-white flowers and grass-like leaves. It is easy to grow and thrives in poor soil and drought conditions. It is a small milkweed but spreads to form colonies so it may not be suitable for formal gardens. Whorled Milkweed blooms a little bit later than other milkweeds, so it is a good choice for extending the bloom season of milkweeds in your garden.

SKILL LEVEL Beginner

LIFESPAN Perennial

EXPOSURE Full sun to part shade

SOIL TYPE Well-drained sand or loam, but will tolerate clay if dry

MOISTURE Dry to medium

HEIGHT 60–90 cm

SPREAD 30–60 cm

BLOOM PERIOD Jul, Aug, Sep

COLOR White

FRAGRANT  ✓

SHOWY FRUIT ✗

CUT FLOWER ✓

PESTS No serious insect or disease problems, but may occasionally be bothered by aphids

NATURAL HABITAT Sandy or rocky soils of prairies, badlands, floodplains, and open woods. It is often one of the first species to colonize disturbed roadsides, ditches, and railroad rights-of-way

WILDLIFE VALUE The nectar of the flowers attracts many kinds of insects, including a variety of native bees, wasps, flies, butterflies, skippers, and beetles. Because this species is one of the last milkweeds to die off in the fall, it is a common late-season host plant for Monarch Butterfly (*Danaus plexippus*) larvae

BUTTERFLY LARVA HOST PLANT FOR Monarch Butterfly (*Danaus plexippus*)

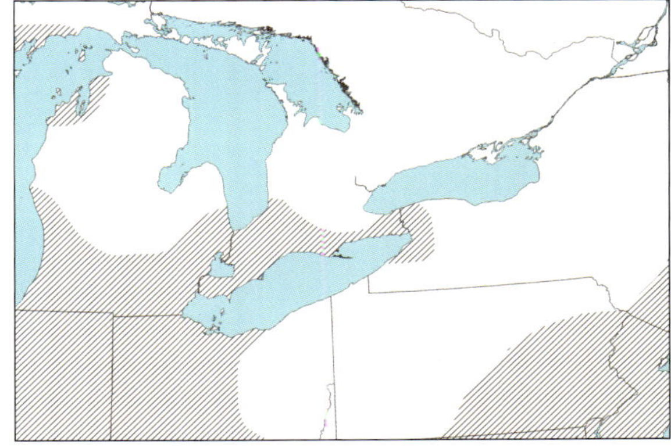

MOTH LARVA HOST PLANT FOR Unexpected Cycnia (*Cycnia inopinatus*), Delicate Cycnia (*Cycnia tenera*), Milkweed Tussock Moth (*Euchaetes egle*), Stalk Borer Moth (*Papaipema nebris*), Isabella Tiger Moth (*Pyrrharctia Isabella*), Striped Garden Caterpillar (*Trichordestra legitima*)

USDA HARDINESS ZONES 3–9

PROPAGATION As with many of the milkweeds, no pretreatment is necessary if sowing outdoors in the fall or early spring. Thirty days of cold, moist stratification can improve germination if starting seeds indoors. Some seeds will take two seasons to germinate. Whorled Milkweed sends out lots of runners and new plants pop up, up to 1 m from the parent. These new plants can be carefully divided, ensuring lots of roots are attached, and planted.

ADDITIONAL INFO This is reportedly among the most toxic of the milkweeds and is said to be very poisonous to livestock and horses.

SAR STATUS NY – S2/R; PA – S2/T(P)

Asclepias viridiflora
Green Milkweed

Flowers

Leaves

FAMILY *Asclepiadaceae* (Milkweed Family)

ALTERNATE COMMON NAMES Green Comet Milkweed, Green-flowered Milkweed, Short Green Milkweed, Wand Milkweed

PLANT DESCRIPTION Green Milkweed features a light green and slightly hairy stem with variably shaped leaves from oblong to lance shaped. Lance-shaped leaves measure up to 15 cm long and 2.5 cm wide. They are often folded lengthwise, toothless, and wavy around the edges. Oblong leaves are shorter, up to 6 cm wide, wavy, and toothless and have a pointed tip. Leaves are found in an opposite, occasionally alternate, pattern. Plants from dry sites typically have narrow leaves and plants from moist sites have more oblong leaves. Each plant has a few half-dome-shaped flower

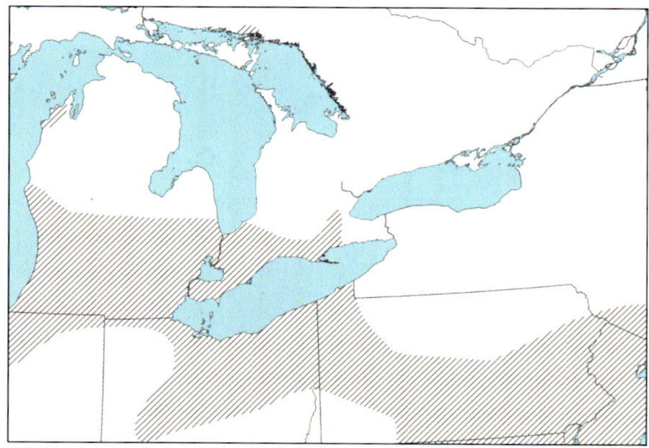

clusters found near the top that each measure about 5 cm across. They emerge from where the leaves attach to the stem. Twenty to 80 individual flowers each measure 8 mm long and 3 mm across. They

are characterized by a five-parted crown with five hoods and five petals that flare backwards. Flowers are without horns. Flowers give way to smooth seed pods that measure about 10 cm long and are pointed at both ends.

IN THE GARDEN Green Milkweed is a tough plant, thriving in dry, nutrient-poor soils. Its interesting flowers look like small comets with a trailing tail and come together to form attractive, domed clusters. Herbivores rarely bother this plant.

SKILL LEVEL Intermediate to advanced

LIFESPAN Perennial

EXPOSURE Full sun to part shade

SOIL TYPE Poor soil is preferred, containing gravelly or sandy material (requires very good drainage)

MOISTURE Dry

HEIGHT 30–60 cm

SPREAD 30–45 cm

BLOOM PERIOD Jun, Jul (blooms last about three weeks)

COLOR Green (sometimes with a purplish tint)

FRAGRANT ✅

SHOWY FRUIT ❌

CUT FLOWER ✅

PESTS No serious insect or disease problems

NATURAL HABITAT Usually found in prairie habitats, sandy and gravelly soils, and old fields

WILDLIFE VALUE Monarch Butterfly (*Danaus plexippus*) caterpillars occasionally use Green Milkweed as a host plant; the flowers attract ants, bumblebees, and other long-tongued bees, which are the most common pollinators

BUTTERFLY LARVA HOST PLANT FOR Monarch Butterfly (*Danaus plexippus*)

MOTH LARVA HOST PLANT FOR Unexpected Cycnia (*Cycnia inopinatus*), Delicate Cycnia (*Cycnia tenera*), Milkweed Tussock Moth (*Euchaetes egle*), Stalk Borer Moth (*Papaipema nebris*), Isabella Tiger Moth (*Pyrrharctia Isabella*), Striped Garden Caterpillar (*Trichordestra legitima*)

USDA HARDINESS ZONES 3–8

PROPAGATION No pretreatment is needed if sowing outdoors in the fall, but seeds require at least 30 days of cold, moist stratification if starting indoors. Scarification of seeds (lightly rubbing with sandpaper or placing seeds in a bowl of sand and gently stirring them) can enhance germination. Some seeds will not germinate till the second season.

ADDITIONAL INFO Plants are unlikely to produce flowers for the first three to four years.

SAR STATUS ON – S2/(not listed); NY – S2/T

Caltha palustris
Marsh Marigold

Flowers

Leaf

FAMILY *Ranunculaceae* (Buttercup Family)

ALTERNATE COMMON NAMES Boots, Brave Bassinets, Bull Flower, Cow Lily, Cowslip, Crazy Beth, Crowfoot, Drunkards, Goldes, Gools, Horse Blob, King's Cup, Mare Blob, Marybuds, May Blob, Meadow-bright, Meadow Buttercup, Meadow Cowslip, Meadow Gowan, Mireblob, Publicans-and-sinners, Publican's Cloak, Soldier's Buttons, Water Boots, Water Buttercup, Water Cowslip, Water Dragon, Water Goggles, Water Gowan, Yellow Gowan, Yellow Marsh Marigold (and a whole lot more!)

PLANT DESCRIPTION Marsh Marigold features hairless, hollow, regularly branching stems. The leaves are mostly basal, although a few alternate leaves can be found along the stems. Leaves measure 10 cm long and across (alternate leaves are smaller), are round to kidney shaped with a deeply heart-shaped base, and have scalloped to toothless edges and a succulent look to them. Small clusters of two to five yellow flowers are found on upper stems. Each flower measures 2 to 4 cm across and features five to nine rounded, waxy, petal-like sepals surrounding a ring of numerous yellow stamens (the part that holds the pollen).

Flowers give way to clusters of flattened, curved capsules (resembling a jester's hat), with each capsule measuring about 1 cm long. They start out erect and green then curve outwards and become a light brown color with age. They split open at the top to release greenish-brown seeds.

Seed head | A grouping of Marsh Marigold

IN THE GARDEN In early spring, the cheerful yellow flowers of Marsh Marigold glow in enthusiasm for the warmer weather ahead. This plant is easily grown in wet, mucky soil, in the shallow water of a pond or a rain garden. Give it rich soil and never allow it to dry out. It is a well-behaved plant with a low-mounding habit. It spreads by seeds and rhizomes to form colonies but is not an aggressive spreader. Although its name suggests that it is related to marigolds, it belongs to the buttercup family.

SKILL LEVEL Beginner

LIFESPAN Perennial

EXPOSURE Full sun to part shade (it will tolerate full shade but is less likely to produce flowers)

SOIL TYPE Muddy, humus-rich soil

MOISTURE Wet

HEIGHT 30–45 cm

SPREAD 30–45 cm

BLOOM PERIOD Apr, May, Jun

COLOR Yellow

FRAGRANT ❌

SHOWY FRUIT ❌

CUT FLOWER ❌

PESTS No serious insect or disease problems, though it may be susceptible to powdery mildew and rust

NATURAL HABITAT Marshes, swamps, wet meadows, and stream margins

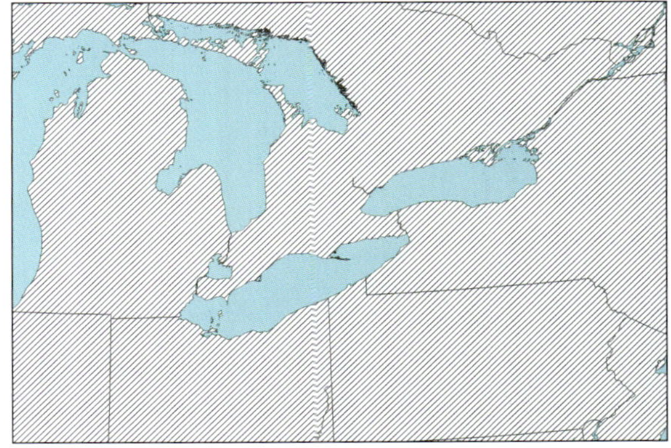

WILDLIFE VALUE The nectar and pollen of the flowers attract primarily flies and bees

BUTTERFLY LARVA HOST PLANT FOR None

MOTH LARVA HOST PLANT FOR None

USDA HARDINESS ZONES 3–7

PROPAGATION Seeds should be sown as soon as they're ripe and cannot dry out before sowing. Seedlings do not flower until the third year following germination. Plants also reproduce easily by division in early spring as the plants are emerging.

ADDITIONAL INFO In researching this plant, we discovered 30 different common names for this one flower — that does not even include all the variations of spellings of each name. *This* is why we like using the scientific binomial when identifying a species!

SAR STATUS N/A

Campanula rotundifolia
Harebell

Flowers

FAMILY *Campanulaceae* (Bellflower Family)

ALTERNATE COMMON NAMES Bluebell, Bluebell Bellflower, Bluebell of Scotland, Lady's Thimble, Round-leaved Bellflower, Witches' Thimble

PLANT DESCRIPTION Harebell features multiple erect, hairless, slender stems. They are unbranched except for some side stems near the top. It has both basal and alternate leaves. Basal leaves are heart shaped, up to 2.5 cm long, hairless, and borne on long stalks. The leaf edges are toothless to scalloped. Alternate leaves are found along the stems and measure up to 7.5 cm long. They are narrowly linear and stalkless, angle upwards, and have smooth margins. Each stem terminates with a solitary, nodding, bell-shaped flower. Each flower is 2 cm long and features five pointed lobes that flare slightly outwards. Flowers give way to brown seed capsules that hang the same way the flowers do. Each capsule contains numerous minute seeds, less than 1 mm long.

IN THE GARDEN Harebell is a charming plant with delicate blue flowers, but don't let its appearance fool you — it is one tough plant! It is valued by gardeners for its ability to tolerate tough sites and for its long bloom time. It doesn't compete well with taller plants, so keep this in mind when choosing its location.

SKILL LEVEL Beginner

LIFESPAN Perennial

EXPOSURE Full sun to full shade (seems to do best with full morning sun and afternoon part shade)

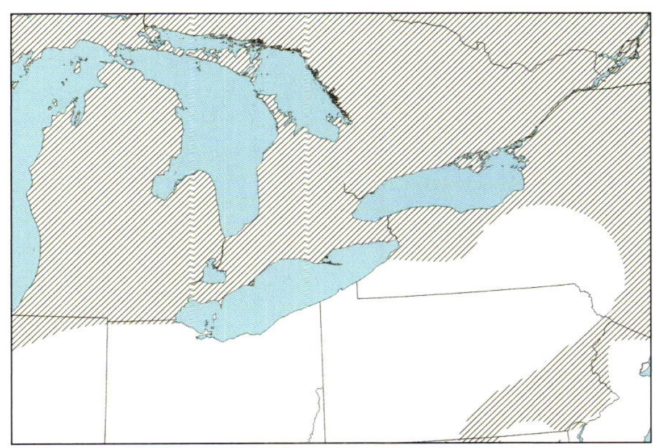

SOIL TYPE Sandy, well-drained soils

MOISTURE Dry

HEIGHT 30–50 cm

SPREAD 15–30 cm

BLOOM PERIOD Jun, Jul, Aug, Sep

COLOR Blue

FRAGRANT ❌

SHOWY FRUIT ❌

CUT FLOWER ✅

PESTS No serious insect or disease problems, though slugs and snails are occasional visitors, and watch out for aphids

NATURAL HABITAT Open fields, rocky areas, meadows, and shorelines

WILDLIFE VALUE Attracts hummingbirds, insect pollinators, such as butterflies and bees, and other beneficial insects

BUTTERFLY LARVA HOST PLANT FOR None

MOTH LARVA HOST PLANT FOR None

USDA HARDINESS ZONES 3–6

PROPAGATION Harebells readily self-sow and no treatment is necessary if seeds are collected and spread on the garden as soon as they turn brown. If starting indoors, they should have 30 days of cold, moist stratification. They require light to germinate, so do not cover with soil. Harebells can also be multiplied by dividing the roots, by taking cuttings and putting them in a glass of water, or by layering — the process of tacking a stem to the ground and covering with an inch or two of soil, at which location the plant will produce roots. This new plant can then be separated and planted.

ADDITIONAL INFO Deadhead spent flowers to encourage additional bloom. Plants are often short lived, but they will easily remain in the garden by self-seeding. Will spread in the garden by creeping roots. Walnut (juglone) tolerant.

SAR STATUS OH – S2/T; NY – (not ranked)/EV

Campanulastrum americanum
Tall Bellflower

FAMILY *Campanulaceae* (Bellflower Family)

ALTERNATE COMMON NAMES American Bellflower

PLANT DESCRIPTION Tall Bellflower features erect, hairy, mostly unbranching stems with slight grooves along them. The leaves are found in an alternate pattern and measure about 7.6 to 15 cm long and 1.2 to 5 cm across, becoming smaller as they ascend the stem. They are lance to egg shaped, taper to a sharp tip, are hairy along major veins on the underside with a rough upper surface, and have serrated margins. The leaf base narrows to hairy leaf stalks. Stems terminate with a flower spike measuring 15 to 60 cm long with shorter flower spikes emerging from leaf axils (where the leaf meets the stem). Individual flowers are 2.5 cm

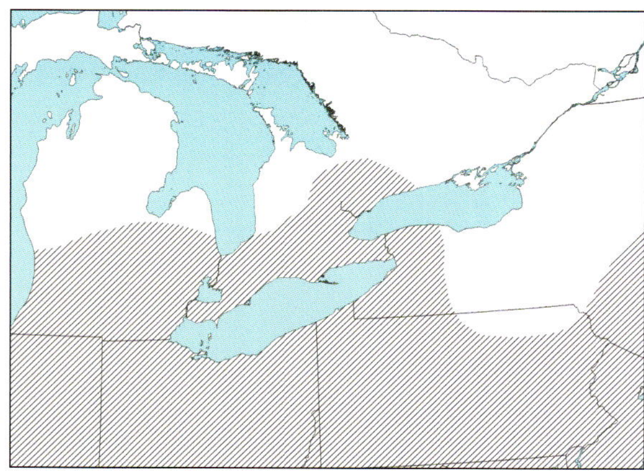

across with five blue petals and a creamy-white center ring. The petals have wavy edges and pointed tips. Each flower has a style (reproductive organ) protruding from the center of the flower. Flowers give way to three-sectioned seed capsules up to 1.2 cm long that release numerous tiny, brown seeds when mature.

Not to be confused with the common exotic garden weed Creeping Bellflower (*Campanula rapunculoides*), which has more bell-shaped flowers compared to the saucer-shaped flowers of Tall Bellflower. Creeping Bellflower is also much shorter and more aggressive.

IN THE GARDEN Tall Bellflower adds a strong vertical presence to gardens with delightful spires of violet-blue, star-shaped flowers. If its seeds are started in fall, then it acts as an annual. If its seeds are started in spring, then it acts as a biennial. Due to its short-lived nature, it will persist in your garden via self-seeding.

SKILL LEVEL Beginner

LIFESPAN Annual or biennial

EXPOSURE Full sun to part shade

SOIL TYPE Rich loam, clay, sand, circumneutral (pH 6.8–7.2)

MOISTURE Moist to medium (plants need regular and even moisture)

HEIGHT 150–200 cm

SPREAD 30–60 cm

BLOOM PERIOD Jun, Jul, Aug

COLOR Blue

FRAGRANT ❌

SHOWY FRUIT ❌

CUT FLOWER ✅

PESTS No serious insect or disease problems, though slugs and snails are occasional visitors, and watch out for aphids

NATURAL HABITAT Marshy ground, streambanks, openings in deciduous forests, and disturbed areas, such as trails, edges of fields, and along railroads

WILDLIFE VALUE A number of bees, including bumblebees and leaf-cutting bees, butterflies, and skippers seek nectar and or pollen, and deer occasionally eat the flowers and foliage

BUTTERFLY LARVA HOST PLANT FOR None

Seed capsules

Leaves

MOTH LARVA HOST PLANT FOR None

USDA HARDINESS ZONES 4–7

PROPAGATION No treatment needed as seeds germinate easily, but they require light to break dormancy, so do not cover the seeds.

ADDITIONAL INFO Deadhead spent flowers to encourage additional bloom.

SAR STATUS NY – S1/E

Capnoides sempervirens
Pale Corydalis

Flowers

FAMILY *Fumariaceae* (Fumitory Family)

ALTERNATE COMMON NAMES Colic Weed, Harlequin Corydalis, Harlequin Flower, Pale Fumewort, Pink and Yellow Corydalis, Pink Corydalis, Rock Harlequin, Tall Corydalis

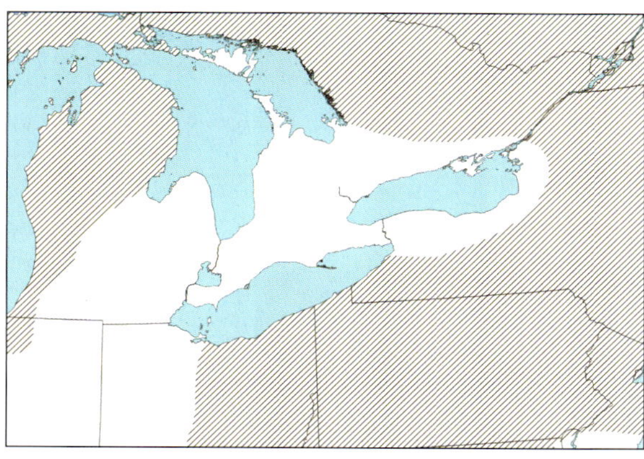

PLANT DESCRIPTION Pale Corydalis is a biennial, producing a basal rosette of leaves during its first year of growth. These leaves are compound in groups of three to five, with each individual leaflet being deeply cleft into two to three parts. These are further divided into two to three narrow, rounded segments. Leaves have a blue-green waxy appearance. During its second year of growth, it sends up many branching, erect, hairless, blue-green stems from its basal rosette. Lower leaves are stalked while upper leaves become increasingly stalkless as they ascend. Stems terminate with clusters of dangling, tubular flowers that each measure 1.5 cm long and are somewhat flattened with varying shades of pink tubes and yellow lips at the end of the tubes. Each

Leaves

Whole plant

Seed pods

flower has a pair of teardrop-shaped sepals clasping it. Flowers give way to long, narrow seed pods that split open when ripe to reveal tiny, black seeds.

IN THE GARDEN Pale Corydalis is a unique looking plant with lacy foliage and captivating pink and yellow flowers. It is valued by gardeners for its ease of growth in tough, dry, rocky sites and for its long bloom time. As a biennial, it will only live for two years, so expect it to persist in your garden via self-seeding. It is an excellent addition to just about any garden but is especially effective in the rock garden.

SKILL LEVEL Beginner

LIFESPAN Biennial

EXPOSURE Full sun to part shade

SOIL TYPE Poor, dry, gravelly soil

MOISTURE Dry

HEIGHT 80 cm

SPREAD 30 cm

BLOOM PERIOD May, Jun, Jul, Aug, Sep

COLOR Pink and yellow

FRAGRANT ❌

SHOWY FRUIT ❌

CUT FLOWER ✅

PESTS May be affected by the *Macrosiphum corydalis* aphid

NATURAL HABITAT Dry woods, rocky ledges, and recent clearings

WILDLIFE VALUE Long-tongued bees, particularly bumblebees, are pollinators, and the seeds have an elaiosome (food package) that ants take back to their nest, and thus help to distribute the seeds

BUTTERFLY LARVA HOST PLANT FOR None

MOTH LARVA HOST PLANT FOR None

USDA HARDINESS ZONES 2–10

PROPAGATION Easy to start from seed; no treatment is necessary if the seeds are collected and sown early in the summer. These will produce a rosette by fall and bloom the next year. Seeds collected in the fall or kept over winter should be cold, moist stratified for 30 days. These spring-sown seeds will flower the following year.

ADDITIONAL INFO Formerly known as *Corydalis sempervirens*, it was recently changed to *Capnoides sempervirens* and is the only species in the genus *Capnoides*. This tough little flower will withstand harsh conditions, but it does not fare well with a lot of competition.

SAR STATUS IN – S1/E; OH – S1/E

Cardamine concatenata
Cut-leaved Toothwort

QUICK GUIDE

PROPAGATION

FULL SHADE PART SHADE

MOIST MEDIUM

Flowers

Whole plant

FAMILY *Brassicaceae* (Mustard Family)

ALTERNATE COMMON NAMES Crinkleroot, Crow's Foot, Crow's Toes, Pepper Root, Pepperwort, Purple-flowered Toothwort, Toothache Root, Toothwort, Wild Horseradish

PLANT DESCRIPTION Cut-leaved Toothwort features basal leaves that are divided into three to five narrow, coarsely toothed lobes that radiate from a central point on the leaf stalk (palmately lobed). Leaves are hairless and measure 7.6 cm in length and width. Erect, hairless flowering stems rise separately from the basal leaves and feature a whorl of three leaves (similar to the basal leaves) on the upper stem. These stems terminate with clusters of droopy, four-petaled flowers measuring up to 2 cm in width when fully open. They are backed by four

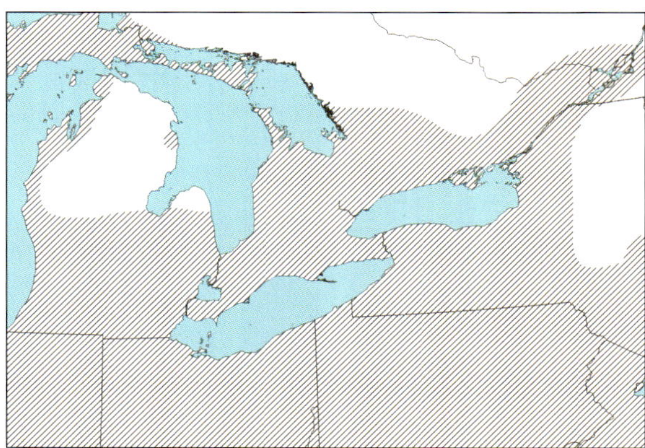

green sepals with pale edges. Flower color varies from white to pink. Flowers give way to long, slender seed pods that split open when ripe to reveal dry seeds.

IN THE GARDEN Cut-leaved Toothwort is a charming woodland plant whose early spring flowers are a delight to gardeners and pollinators alike. It is a true spring ephemeral, meaning it will bloom before the trees above it have leafed out, then retreat back into the ground by midsummer, anticipating the spring. Due to its ephemeral habit, it is best planted with something that can fill in the gap it leaves behind.

SKILL LEVEL Intermediate

LIFESPAN Perennial (spring ephemeral)

EXPOSURE Part (dappled) shade to full shade

SOIL TYPE Fertile, humus-rich, well-drained forest soils

MOISTURE Moist to medium (but it tolerates seasonal flooding)

HEIGHT 10–20 cm

SPREAD 10–20 cm

BLOOM PERIOD May

COLOR White tinged with pink

FRAGRANT ❌

SHOWY FRUIT ❌

CUT FLOWER ✅

PESTS No serious insect or disease problems

NATURAL HABITAT Rich woods, wooded bottomlands, rocky banks and bluffs, and limestone outcrops

WILDLIFE VALUE Nectar from the flowers attracts both long-tongued and short-tongued bees

BUTTERFLY LARVA HOST PLANT FOR Mustard White (*Pieris oleraceae*), West Virginia White (*Pieris virginiensis*)

MOTH LARVA HOST PLANT FOR None

USDA HARDINESS ZONES 3–8

PROPAGATION Cut-leaved Toothwort is difficult to grow from seed, but the rhizome can be easily divided when the plant is dormant and planted a little below the surface of the ground in an area with sparse groundcover that receives shade during the summer. Care must be taken as the rhizomes are easily broken. Seeds should not be allowed to dry out and should be sown on a moist, shaded seedbed immediately after collection. Expect the seedlings to flower in three to four years.

ADDITIONAL INFO The presence of this species in a woodland indicates that its soil has not been subjected to much human disturbance. This is one of the spring wildflowers that quickly declines when Garlic Mustard (*Alliaria petiolata*) invades.

SAR STATUS N/A

Cardamine diphylla
Two-leaved Toothwort

FAMILY *Brassicaceae* (Mustard Family)

ALTERNATE COMMON NAMES Broad-leaved Toothwort, Crinkleroot, Pepper Root, Pepperwort, Twin-leaved Toothwort

PLANT DESCRIPTION Two-leaved Toothwort produces a solitary, smooth stem bearing two opposite (or nearly opposite) leaves about midway. These leaves are deeply lobed as to appear compound and are coarsely toothed. Early spring leaves emerge with a dark red tinge that quickly fades to green as they mature. Flowers are found in a cluster at the top of the stem. They have four petals and measure 1 cm to just under 2 cm wide. Flowers produce long, thin seed pods that split open when ripe to reveal small, brown seeds.

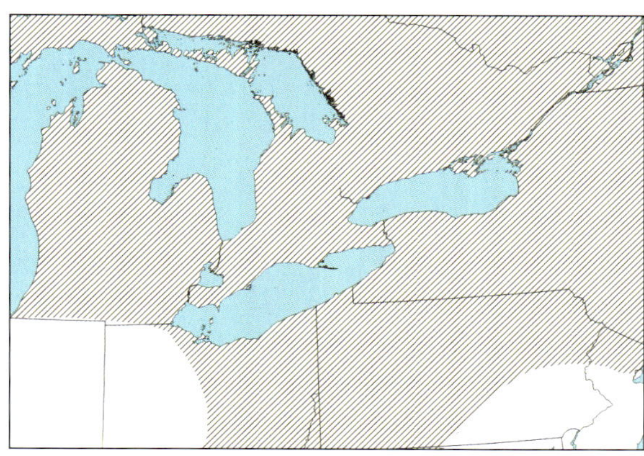

IN THE GARDEN Two-leaved Toothwort is a delightful spring wildflower that provides some of the earliest greenery in the spring. Its lush foliage

provides the perfect green backdrop to allow its white flowers to really stand out. It is ephemeral, meaning it starts growing in early spring before the trees above it have leafed out then goes dormant by midsummer. In the right conditions it will spread via rhizomes to form colonies.

SKILL LEVEL Intermediate

LIFESPAN Perennial (spring ephemeral)

EXPOSURE Full shade to part shade (needs early spring sun to flower, such as under the canopy of deciduous trees)

SOIL TYPE Fertile, humus-rich, sandy to loamy soils

MOISTURE Moist

HEIGHT 20–35 cm

SPREAD 30 cm

BLOOM PERIOD Apr, May

COLOR White (to light pink)

FRAGRANT ❌

SHOWY FRUIT ❌

CUT FLOWER ✅

PESTS No serious insect or disease problems, although occasionally flea beetles may be a problem

NATURAL HABITAT Prefers damp areas often flooded in spring in deciduous and mixed forests, especially in rather moist or seepy areas, and in cedar swamps and damp meadows

WILDLIFE VALUE Attracts a variety of bees and butterflies

BUTTERFLY LARVA HOST PLANT FOR West Virginia White Butterfly (*Pieris virginiensis*)

MOTH LARVA HOST PLANT FOR None

USDA HARDINESS ZONES 3–8

PROPAGATION The seeds have low viability in storage, therefore the seed should be sown immediately after collection. The seedlings may take up to three to four years before they bloom. As with *Cardamine concatenata*, it is easier to multiply this plant by rootstock division when the plant is dormant. However, the rhizomes are easily damaged or broken so great care must be used.

ADDITIONAL INFO Note that this species rarely produces viable seed.

SAR STATUS IN – S3/WL

Caulophyllum giganteum
Giant Blue Cohosh

FAMILY *Berberidaceae* (Barberry Family)

ALTERNATE COMMON NAMES Early Blue Cohosh, Papoose Root, Purple-flowered Blue Cohosh, Squaw Root

PLANT DESCRIPTION Giant Blue Cohosh features an erect, smooth, light-green-to-purple central stem. Leaves are found at the top of the stem and are compound into groups of three. Each leaflet is toothless and smooth and measures about 6 cm long and wide. When the stems and leaves first emerge in the spring, they are dark purple but fade to green as they mature. Flowers are found in clusters of four to 18 flowers at the top of the stem. Each flower is characterized by five to six dark purplish-red sepals surrounding a yellow center. These flowers turn into bright blue, hard, berry-like seeds.

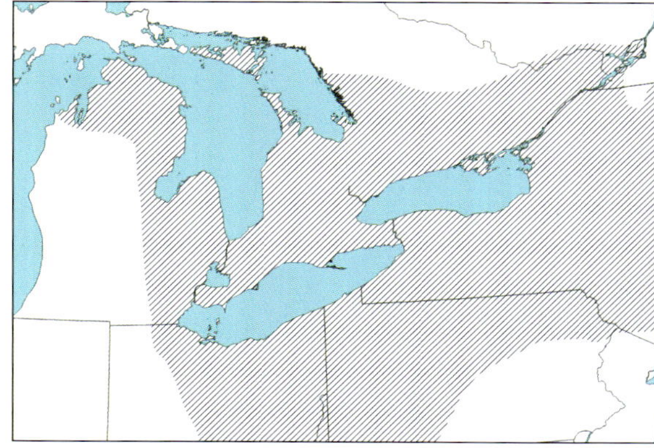

IN THE GARDEN The ethereal purple and yellow flowers of Giant Blue Cohosh look like they come from another planet! In early spring, they rise up from the ground clasped by deep purple, unfurled

leaves. Once the leaves unfurl, they turn green and form attractive clumps of lacy foliage. Flowers become bright blue seeds that often persist into the winter months. A unique and desirable, although slow-growing, addition to shade gardens.

SKILL LEVEL Intermediate to advanced

LIFESPAN Perennial

EXPOSURE Full shade to part shade (prefers dappled shade)

SOIL TYPE Fertile, well-drained but moisture-retentive, humus-rich soil

MOISTURE Moist to medium

HEIGHT 30–60 cm

SPREAD 30–60 cm

BLOOM PERIOD Apr, May

COLOR Dark purple or brown, occasionally yellowish green

FRAGRANT ❌

SHOWY FRUIT ✅

CUT FLOWER ✅ (berries)

PESTS No serious insect or disease problems

NATURAL HABITAT Rich deciduous forests

WILDLIFE VALUE Self-pollination is common, but the flowers are occasionally visited by sweat bees and various flies, while woodland birds and mice are the primary dispersers of seeds after eating the blue fruit

BUTTERFLY LARVA HOST PLANT FOR None

MOTH LARVA HOST PLANT FOR Black-patched Clepsis (*Clepsis melaleucanus*)

USDA HARDINESS ZONES 3–8

PROPAGATION *Caulophyllum* seeds are notoriously hard to germinate, sometimes remaining dormant for three to four years, though more typically they will germinate in two to three years. The seeds are hydrophilic and should not be allowed to dry out before planting. They also need multiple cycles of cold and warm to break dormancy, though nicking (scarifying) the seeds may hasten germination. They can then take several more years before they flower. The rootstock can be divided in the fall or before growth starts in the spring, but established plants are generally best left undisturbed. Can spread very slowly by rhizomes over time to form colonies.

ADDITIONAL INFO This species has unusual blue fruits that are technically not fruits but are actually exposed seeds that have ripened too quickly and have caused the ovary wall to break and shrivel away. In this way, they are unique among all flowering plants. Plants can live for more than 50 years.

Giant Blue Cohosh (*Caulophyllum giganteum*) is very similar to Blue Cohosh (*C. thalictroides*). You can tell the two apart by noting how *C. giganteum* has dark purple flowers that bloom earlier in the season (up to two weeks earlier) and before its leaves have unfurled. On the other hand, *C. thalictroides* has light greenish flowers that bloom after its leaves unfurl in the spring.

SAR STATUS IN – S1/E; PA – S3/(not listed)

Caulophyllum thalictroides
Blue Cohosh

FULL SHADE	PART SHADE

MOIST	MEDIUM

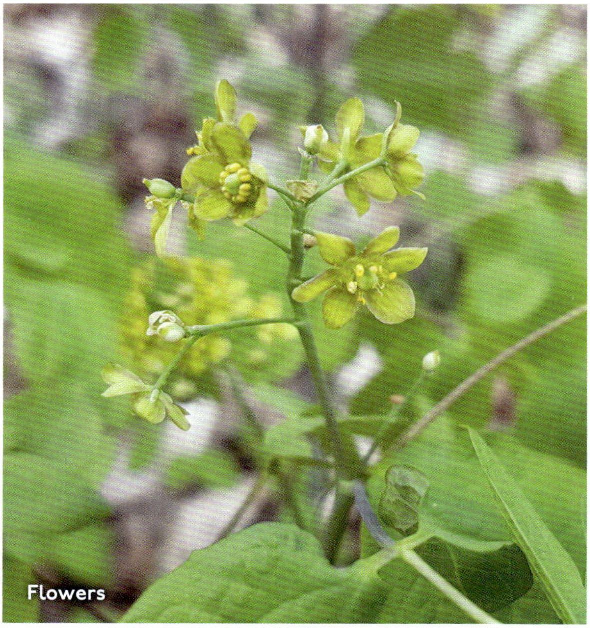

Flowers

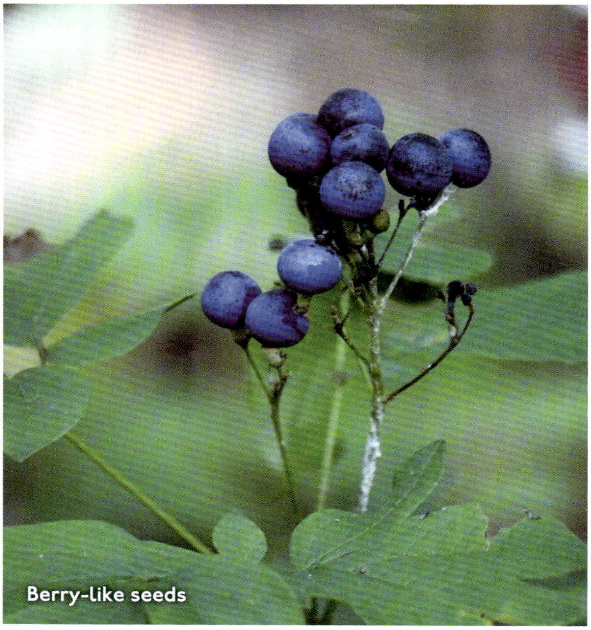

Berry-like seeds

FAMILY *Berberidaceae* (Barberry Family)

ALTERNATE COMMON NAMES Blue Ginseng, Common Blue Cohosh, False Cohosh, Green Vivian, Papoose Root, Squaw Root

PLANT DESCRIPTION Blue Cohosh features an erect, smooth central stem with a light green to pale purple color. Leaves (and flower clusters) are found in a whorl at the top of the stem and are compound in groups of three. Leaflets are toothless and smooth and measure just over 6 cm long and wide with two to five pointed lobes. Non-flowering plants have a single compound leaf; flowering plants have two. The central stem terminates with a 7.6 cm long flower cluster made up of five to 30 flowers. Each flower measures 8 mm across and features six yellowish-green petal-like sepals and six yellow stamens surrounding a green center. Flowers give way to hard, berry-like seeds that measure just under 1 cm across. They start out green and change to blue as they mature.

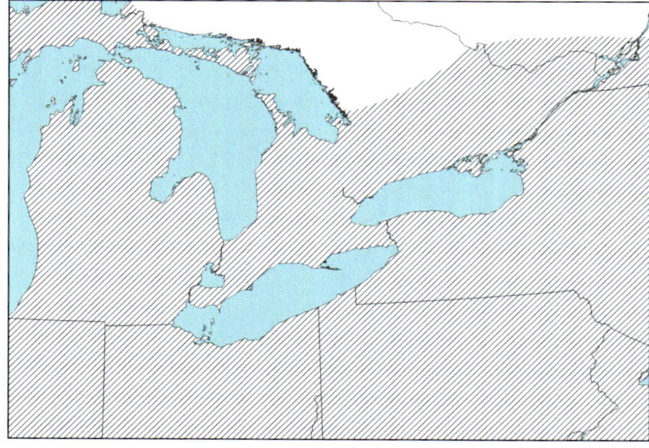

Leaves

Whole plant

IN THE GARDEN Blue Cohosh pushes up through its bed of leaves in early spring, adding a unique look to gardens with its lacy leaves and yellow, star-like flowers. Flowers eventually turn into blue berry-like seeds that will persist into winter, adding strong winter interest to your garden. A desirable but slow-growing addition to shade gardens.

SKILL LEVEL Intermediate to advanced

LIFESPAN Perennial

EXPOSURE Full shade to part (dappled) shade

SOIL TYPE Fertile, humus-rich, well-drained soil

MOISTURE Consistently moist to medium

HEIGHT 45–60 cm

SPREAD 30 cm

BLOOM PERIOD May

COLOR Yellow, green, brown

FRAGRANT ❌

SHOWY FRUIT ✅

CUT FLOWER ✅ (berries)

PESTS No serious insect or disease problems

NATURAL HABITAT Rich, mesic woods and meadows

WILDLIFE VALUE Early pollinators

BUTTERFLY LARVA HOST PLANT FOR None

MOTH LARVA HOST PLANT FOR Black-patched Clepsis (*Clepsis melaleucanus*)

USDA HARDINESS ZONES 3–7

PROPAGATION *Caulophyllum* seeds are notoriously hard to germinate, sometimes remaining dormant for three to four years, though more typically they will germinate in two to three years. The seeds are hydrophilic and should not be allowed to dry out before planting. They also need multiple cycles of cold and warm to break dormancy, though nicking (scarifying) the seeds may hasten germination. They can then take several more years before they flower. The rootstock can be divided in the fall or before growth starts in the spring, but established plants are generally best left undisturbed. Can spread very slowly by rhizomes over time to form colonies.

ADDITIONAL INFO Blue Cohosh has unusual blue fruits that are technically not fruits but are actually exposed seeds that have ripened too quickly and have caused the ovary wall to break and shrivel away. In this way, they are unique among all flowering plants. Plants can live for more than 50 years.

Caulophyllum giganteum (Giant Blue Cohosh) generally has dark purple/red flowers that open with or before the leaves do, whereas *C. thalictroides* (Blue Cohosh) has lighter greenish flowers that bloom after the leaves have opened.

SAR STATUS N/A

Chelone glabra
White Turtlehead

Flowers

Whole plant

FAMILY *Scrophulariaceae* (Figwort Family)

ALTERNATE COMMON NAMES Balmony, Bitter-weed, Fishhead, Fishmouth, Shellflower, Snakehead, Snakemouth, Turtlebloom

PLANT DESCRIPTION White Turtlehead features erect, smooth, mostly unbranched, cylindrical to four-sided stems. Leaves are found in an opposite pattern and are arranged at 90-degree angles to those below them. Each leaf is lanceolate shaped, toothed, hairless, and attached with little-to-no leaf stalk. Snapdragon-like flowers are found at the top of the stems and bloom from the bottom up. They are tubular with two lips, white with a possible pinkish tinge and about 3 cm in length. Flowers give way to roundish seed pods that contain many small, flattened seeds with broad wings that allow them to float on water.

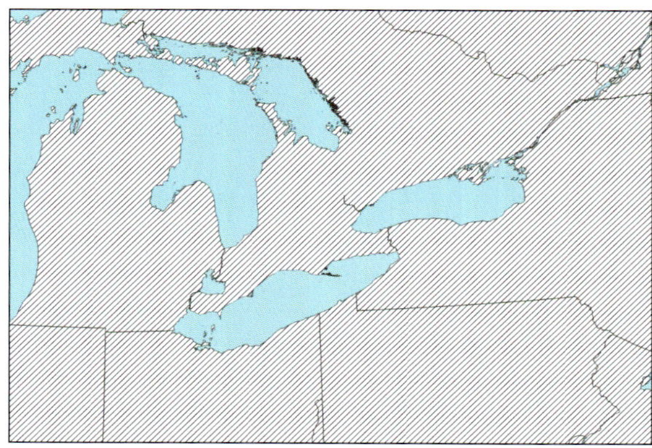

IN THE GARDEN The bold white flowers and upright form of White Turtlehead makes it a valuable addition to native plant gardens. Its stems and seed heads persist and add interest well into the

Leaves

Seed pods

winter months. White Turtlehead has a desirable clumping habit and will tolerate average garden soil, granted it doesn't dry out completely.

SKILL LEVEL Beginner

LIFESPAN Perennial

EXPOSURE Full shade to part shade (if grown in too much shade, plants may need some support)

SOIL TYPE Light, fertile, humus-rich soil

MOISTURE Wet to moist (temporary flooding is tolerated)

HEIGHT 60–90 cm (may reach 180 cm in ideal conditions)

SPREAD 60 cm

BLOOM PERIOD Jul, Aug, Sep

COLOR White, cream

FRAGRANT ❌

SHOWY FRUIT ❌

CUT FLOWER ✅

PESTS No serious insect or disease problems though there is some susceptibility to mildew, particularly if soils are kept on the dry side and/or air circulation is poor

NATURAL HABITAT Margins of sandy or non-sandy wetlands, including swamps, wet open woods, sedge meadows, marshes, and shorelines of ponds or creeks. *Chelone glabra* is said to be an indicator of fens with alkaline groundwater

WILDLIFE VALUE Attracts butterflies, bumblebees, and Ruby-throated Hummingbirds (*Archilochus colubris*)

BUTTERFLY LARVA HOST PLANT FOR Baltimore Checkerspot (*Euphydryas phaeton*)

MOTH LARVA HOST PLANT FOR Verbena Bud Moth (*Endothenia hebesana*), Pink-patched Looper (*Eosphoropteryx thyatyroides*), Turtlehead Borer Moth (*Papaipema nepheleptena*)

USDA HARDINESS ZONES 3–8

PROPAGATION No treatment is necessary if seeds are planted outside in fall. Artificial cold, moist stratification may not be very effective, but you can try chilling for at least six weeks and up to 120 days (four months) and plant in spring. Seeds need light to germinate, so do not cover with soil. Seedlings germinate after one year and flower after two. Roots can be divided in early spring or late fall when the plant is dormant, or new plants can be generated from two-node stem cuttings taken in early summer.

ADDITIONAL INFO Plants can be pruned or pinched back in spring to encourage growth of a shorter sturdier plant. This is usually not necessary unless plants are growing in full shade. Bumblebees are the main pollinators of this plant as they are strong enough to push their way past the tightly closed petals. Once a bumblebee has pushed its way inside, you will see nothing more than two legs sticking out!

SAR STATUS NY – (not ranked)/EV

Claytonia caroliniana
Carolina Spring Beauty

FAMILY *Portulacaceae* (Purslane Family)

ALTERNATE COMMON NAMES Good Morning Spring, Northern Spring Beauty, Rose Elf, Wide-leaved Spring Beauty

PLANT DESCRIPTION Carolina Spring Beauty has an erect, but sometimes sprawling, main stem. Halfway up this stem, you will find a single pair of opposite leaves that each measure up to 7.6 cm long and 2 cm wide. They have smooth edges, a prominent central vein, and a succulent look to them. A cluster of five or more flowers are found at the top of the plant. They each measure just over 1 cm wide and are characterized by five pale-pink-to-white petals with dark pink veins and five stamens with pink tips. Flowers mature into ovoid seed capsules that eject seeds when ripe. Flowers will close at night and on cloudy days.

IN THE GARDEN Carolina Spring Beauty really earns its name by providing some of the earliest and most delightful spring wildflowers. It makes the most of early spring sunshine by emerging early and flowering before the trees above it have even leafed out. It will go dormant by early summer, but, due to its small size, it won't leave a noticeable gap in the garden as other plants grow up around it. Best planted in large clumps, and in ideal conditions it will spread slowly to form colonies. It works well in naturalized lawns if mowing is delayed until after plant goes dormant.

SKILL LEVEL Intermediate

Whole plant | Seed capsules

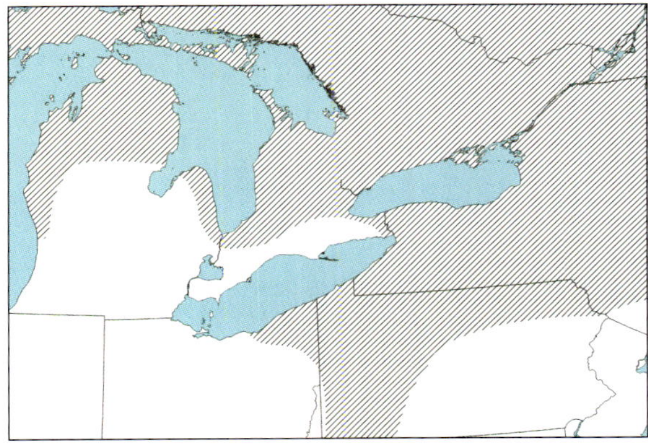

LIFESPAN Perennial

EXPOSURE Part (dappled) shade

SOIL TYPE Tolerates a variety of soils but prefers fertile, humus-rich, well-drained soils

MOISTURE Moist to medium

HEIGHT 20 cm

SPREAD 15 cm

BLOOM PERIOD Apr, May

COLOR Pink and white

FRAGRANT 

SHOWY FRUIT 

CUT FLOWER

PESTS No serious insect or disease problems

NATURAL HABITAT Rich, open woods, alluvial thickets, upland slopes, wetlands, and riparian hardwood forests

WILDLIFE VALUE The bulb-like corms are dug and consumed in spring by chipmunks and mice

BUTTERFLY LARVA HOST PLANT FOR None

MOTH LARVA HOST PLANT FOR None

USDA HARDINESS ZONES 3–8

PROPAGATION As an early spring plant, the seeds need a period of warmth, followed by a period of cold, in order to germinate. They are best directly sown as soon as they are ripe as they should not be allowed to dry out. New plants can also be propagated from the corms, which should be planted 7 to 8 cm deep in the fall.

ADDITIONAL INFO As one of its common names implies (Northern Spring Beauty), this is the more northerly of the two *Claytonia* species, and the two ranges overlap in the southern Great Lakes region. The other one, *C. virginica*, can be most quickly differentiated by its long slender, strap-like leaves.

SAR STATUS N/A

Claytonia virginica
Virginia Spring Beauty

FAMILY *Portulacaceae* (Purslane Family)

ALTERNATE COMMON NAMES Fairy Spuds, Grass Flower, Mayflower, Narrow-leaved Spring Beauty, Spring Beauty

PLANT DESCRIPTION Virginia Spring Beauty features a single smooth, succulent, light-green-to-reddish-green stem. A single pair of opposite leaves is found midway up the stem. These leaves are long and narrow, measuring up to 12 cm long and 6 mm wide. You will notice that each leaf is fleshy with a prominent central vein and a smooth leaf surface and margin. The stem terminates with an open cluster of flowers, each measuring about 8 mm across. Each flower has five white petals with dark-to-light-pink stripes. On cloudy days, or at

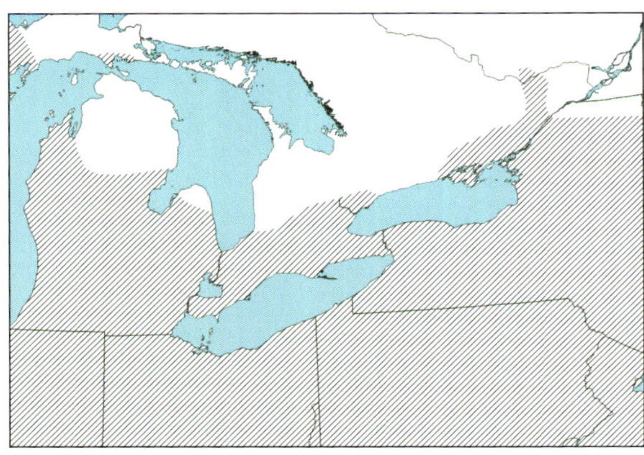

night, the flowers close and nod downwards. Flowers give way to round seed pods that each contain a few round seeds.

IN THE GARDEN The flowers of Virginia Spring Beauty may be small but are sure to delight. It is a true spring ephemeral, meaning it will emerge early in spring and flower before the trees above it have even leafed out. It will go dormant by early summer but, due to its small size, it won't leave a noticeable gap in the garden. Best planted in large clumps, and in ideal conditions, it will spread slowly to form colonies. It works well in naturalized lawns if mowing is delayed until after the plant goes dormant.

SKILL LEVEL Intermediate

LIFESPAN Perennial (spring ephemeral)

EXPOSURE Part (dappled) shade

SOIL TYPE Tolerates a variety of soils but prefers fertile, humus-rich soils

MOISTURE Moist to medium

HEIGHT 10–15 cm

SPREAD 10–15 cm

BLOOM PERIOD Apr, May

COLOR Pink, white

FRAGRANT ✓

SHOWY FRUIT ✗

CUT FLOWER ✗

PESTS No serious insect or disease problems

NATURAL HABITAT Moist, deciduous open forests, woodland edges, flood plains, and sunny streambanks

WILDLIFE VALUE Early source of pollen and nectar for many bees and some butterflies, and the corms are a food source for mice and other small rodents

BUTTERFLY LARVA HOST PLANT FOR None

MOTH LARVA HOST PLANT FOR None

USDA HARDINESS ZONES 3–8

PROPAGATION As an early spring plant, the seeds need a period of warmth, followed by a period of cold, in order to germinate. They are best directly sown as soon as they are ripe and should not be allowed to dry out. New plants can also be propagated from the corms, which should be planted 7 to 8 cm deep in the fall.

ADDITIONAL INFO As one of its common names implies (Narrow-leaved Spring Beauty), this species can be differentiated from its cousin, *Claytonia caroliniana*, by its long slender, strap-like leaves. This is the more southerly of the two *Claytonia* species, and the two ranges overlap in the southern Great Lakes region.

SAR STATUS N/A

Coreopsis lanceolata
Lanceleaf Coreopsis

QUICK GUIDE

PROPAGATION

 **FULL SUN**

 **DRY**

FAMILY *Asteraceae* (Aster Family)

ALTERNATE COMMON NAMES Lanceleaf Tickseed, Long-stalk Coreopsis, Long-stalk Tickseed, Sand Coreopsis, Sand Tickseed, Tickseed Coreopsis

PLANT DESCRIPTION Lanceleaf Coreopsis features both a clump of basal leaves at its base along with opposite leaves along its flower stems. Basal leaves are up to 15 cm long and 2.5 cm across, toothless, oblanceolate (wider towards the tip), and borne on long petioles. They may be hairless or irregularly hairy. Some basal leaves may have one to four narrow lobes towards the base. The leaves found along the stems may be deeply lobed and almost appear to be pinnately compound. A single flower sits atop each flowering stem, measuring about 5 to 7.5 cm wide. Flowers are characterized by

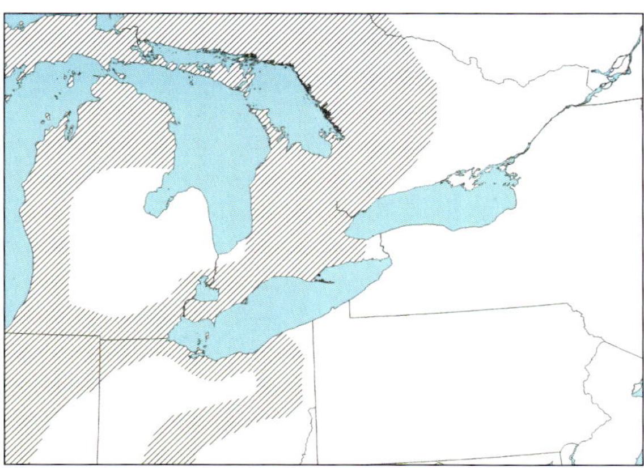

eight yellow petals, each with an unevenly toothed tip. Flowers give way to numerous black, oblong seeds with papery wings.

IN THE GARDEN Lanceleaf Coreopsis is renowned for its profusion of radiant yellow flowers in early summer. These flowers bloom for weeks on end so they are great for extending bloom time in gardens. This is a very dependable species that has the ability to thrive in dry, nutrient-poor soils.

SKILL LEVEL Beginner

LIFESPAN Short-lived perennial, readily self-seeds

EXPOSURE Full sun

SOIL TYPE Any

MOISTURE Dry

HEIGHT 50 cm

SPREAD 45 cm

BLOOM PERIOD May, Jun, Jul, Aug

COLOR Yellow

FRAGRANT ✔

SHOWY FRUIT ✖

CUT FLOWER ✔

PESTS No serious insect or disease problems, though crown rot may occur if grown in moist, poorly drained soils

NATURAL HABITAT Dry open woodlands, prairies, meadows, and pastures, especially in sterile, poor soils

WILDLIFE VALUE Many pollinators, including bees, wasps, butterflies, moths, and beetles, feed on the pollen and nectar, while finches and other songbirds feed on the ripe seeds; rabbits, deer, and groundhogs sometimes browse the foliage

BUTTERFLY LARVA HOST PLANT FOR Southern Dogface Butterfly (*Zerene cesonia*) – though this species rarely visits Ontario

MOTH LARVA HOST PLANT FOR Sunflower Moth (*Homoeosoma electellum*), Common Tan Wave (*Pleuroprucha insulsaria*), Wavy-lined Emerald (*Synchlora aerata*), Dimorphic Gray (*Tornos scolopacinarius*)

USDA HARDINESS ZONES 4–9

PROPAGATION Direct sowing in early spring is easy with good germination. Cover the seeds lightly with soil (or leave uncovered) as light is needed for germination. The seeds will store in refrigerated sealed containers for at least three years. If starting indoors, germination will benefit from 30 to 60 days of cold, moist stratification. Plants are also readily propagated by root division and will benefit from dividing the plant every third year.

ADDITIONAL INFO Frequent deadheading of Lanceleaf Coreopsis will allow it to bloom well into the summer. Although *Coreopsis lanceolata* is a perennial, it will soon die out if soil conditions are overly moist to wet — especially in winter — because crown rot is likely to set in.

SAR STATUS N/A

Coreopsis tripteris
Tall Tickseed

	PART SHADE	FULL SUN
	MOIST	MEDIUM

Flowers

FAMILY *Asteraceae* (Aster Family)

ALTERNATE COMMON NAMES Tall Coreopsis

PLANT DESCRIPTION Tall Tickseed features slender, smooth, cylindrical stems that are unbranched except for along the upper half of the plant. Opposite leaves are found along the full length of the stems. These leaves are odd-pinnate with three to five leaflets, each measuring up to 12.7 cm long and 2 cm wide. Each leaflet has smooth margins, a linear-elliptic shape, a pointed tip, and a wedge-shaped base. Lateral leaflets have no leaf stalk while the end leaflets do have a leaf stalk. Leaflets are further characterized by smooth upper leaf surfaces and finely hairy undersides. The uppermost stems are topped with solitary flowers that collectively form open, loosely flat-topped flower

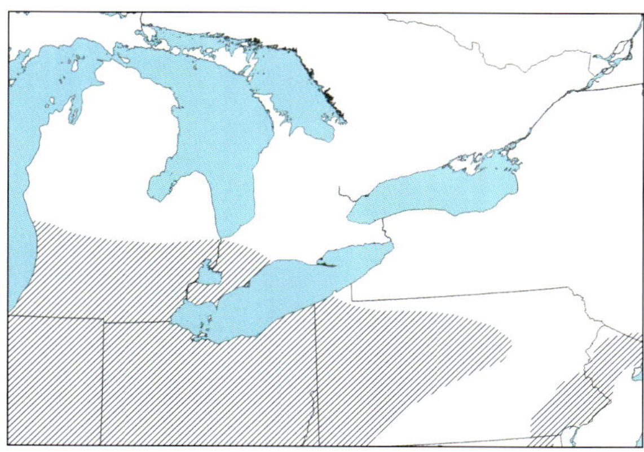

clusters. Each flower is borne on a flower stalk that is up to 25 cm long. These flower stalks may have a couple leafy bracts along them. Individual flowers measure up to 5 cm across and feature eight

"

widely spreading ray florets (petals) surrounding a dense cluster of brown disk florets. Flowers give way to small (4 to 5 mm), oblong, brown seeds with winged sides.

IN THE GARDEN Tall Tickseed is a stately plant with radiant yellow flowers that reach for the sky. It is valued as a dependable structural plant with noteworthy winter interest. It readily self-seeds in open soil and may form colonies if allowed. Plants grown in fertile, moist soil will grow to the higher end of their height range. Consider pairing Tall Tickseed with a companion that can hide the leggy base of the stems.

SKILL LEVEL Beginner

LIFESPAN Perennial

EXPOSURE Part shade to full sun

SOIL TYPE Well drained

MOISTURE Moist to medium (will tolerate some drought)

HEIGHT 60–240 cm

SPREAD 60–240 cm

BLOOM PERIOD Jul, Aug, Sep, Oct

COLOR Yellow

FRAGRANT ❌

SHOWY FRUIT ❌

CUT FLOWER ✅

PESTS No serious insect or disease problems

NATURAL HABITAT Dry to wet prairies and meadows, marshes, oak forests (especially borders and clearings), fields, roadsides, and railroad rights-of-way

WILDLIFE VALUE Flowers are visited by butterflies, skippers, and native bees, and birds are known to feed on the seeds

BUTTERFLY LARVA HOST PLANT FOR Southern Dogface Butterfly (*Zerene cesonia*) — though this species rarely visits Ontario

MOTH LARVA HOST PLANT FOR Sunflower Moth (*Homoeosoma electellum*), Common Tan Wave (*Pleuroprucha insulsaria*), Wavy-lined Emerald (*Synchlora aerata*), Dimorphic Gray (*Tornos scolopacinarius*)

USDA HARDINESS ZONES 3–9

PROPAGATION Easily propagated by surface-sown seeds (requires light for germination) in late fall. The seeds will germinate in a week or two and overwinter as a small cluster of leaves. If planting in the spring, the germination will benefit from 60 days of cold, moist stratification before direct sowing. This coreopsis is also readily propagated by dividing the root clump in early spring or after it has finished flowering in the fall.

ADDITIONAL INFO If grown in light shade, Tall Tickseed tends to be open and leggy with a tendency to lean towards the sun. Grown in full sun, plants tend to be sturdier and have many more blooms. As with other coreopsis, deadheading will extend the bloom period and prevent unwanted seedlings, especially in smaller gardens where this plant can be aggressive, especially if there is adequate moisture.

SAR STATUS ON – S1/(not listed)

Dicentra canadensis
Squirrel Corn

Flowers

Whole plant

FAMILY *Fumariaceae* (Fumitory Family)

ALTERNATE COMMON NAMES Wild Bleeding Heart

PLANT DESCRIPTION Squirrel Corn features compound basal leaves that measure 12 to 22 cm long and 6 to 12 cm wide. These leaves are thrice divided into narrow segments, which give the leaves a lacy look. The leaves are smooth, gray green, and covered in a powdery bloom on the undersides. Reddish-green, erect (but often leaning) flower stalks rise above the leaves and terminate with a cluster of four to eight heart-shaped flowers, each measuring about 2 cm long and just over 1 cm wide. The flowers are characterized by two fused outer petals (forming an elongated heart shape) with upturned lower lips. In addition, two fused interior petals protrude at right angles from the opening of the outer petals. Flowers give way to droopy, green-bean-like seed pods that are divided into two parts lengthways and measure about 1.2 cm

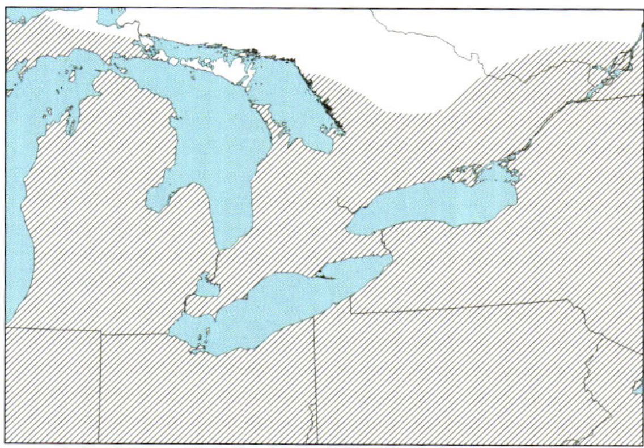

in length. Each seed has an elaiosome (fleshy food structure) attached to it. This provides a desirable food source for ants, who take the seeds back to their nest, thus aiding in dispersal of the seeds. Squirrel Corn is similar to Dutchman's Breeches (*Dicentra cucullaria*), but the latter can be distinguished by its slightly earlier blooms and less-heart-shaped flowers.

IN THE GARDEN Squirrel Corn graces woodland gardens in early spring with its lacy foliage and elegant heart-shaped flowers. This is a spring ephemeral, meaning it will put on an impressive display in early spring but will go dormant by midsummer. It is recommended to pair Squirrel Corn with something that can fill in the gap it leaves behind. This allows gardeners to fit multiple bloom times into one space. Squirrel Corn benefits from a layer of leaflitter around it.

SKILL LEVEL Beginner

LIFESPAN Perennial (spring ephemeral)

EXPOSURE Part shade to full shade (specifically the shade of deciduous trees, rather than from buildings or evergreens)

SOIL TYPE Humus-rich soil, tolerates limestone

MOISTURE Moist

HEIGHT 15–25 cm

SPREAD 25–30 cm

BLOOM PERIOD May

COLOR White

FRAGRANT ✓

SHOWY FRUIT ✗

CUT FLOWER ✓

PESTS No serious insect or disease problems

NATURAL HABITAT Deciduous woods, often among rock outcrops, in rich loam soils, and at the bases of trees

WILDLIFE VALUE The flowers are cross-pollinated by bumblebees, who seek nectar and pollen, and the small tubers attract chipmunks and mice

BUTTERFLY LARVA HOST PLANT FOR None

MOTH LARVA HOST PLANT FOR None

USDA HARDINESS ZONES 4–7

PROPAGATION Seeds should be sown as soon as ripe and should not be allowed to dry out. The root system consists of tiny, yellow pea-like tubers that can be divided when the plant is dormant.

ADDITIONAL INFO Although apparently secure in most of the southern Great Lakes region, Squirrel Corn is at risk in many areas because of loss of habitat through land use change and from invasive species, such as Garlic Mustard (*Alliaria petiolata*), European Buckthorn (*Rhamnus cathartica*), and, in particular, earthworms. The cell sap can cause minor skin irritation when touched by those with sensitive skin.

SAR STATUS N/A

Dicentra cucullaria
Dutchman's Breeches

FAMILY *Fumariaceae* (Fumitory Family)

ALTERNATE COMMON NAMES Bachelor's Breeches, Butterfly Banners, Eardrops, Kitten Breeches, Little Boys Breeches, Monks Head, Soldier's Cap, Staggerweed, White Hearts, Wild Bleeding Heart

PLANT DESCRIPTION Dutchman's Breeches is characterized by compound basal leaves borne on slender, reddish stems. The leaves are smooth, gray green, and thrice divided into narrow segments, which give them a lacy look. Flowers are found in clusters of three to 14 on slender, erect (often leaning) stems. Individual flowers measure 2 cm long and are characterized by two white-to-light-pink outer petals with upwards-pointing spurs and two pale yellow inner petals that flare out like wings. The

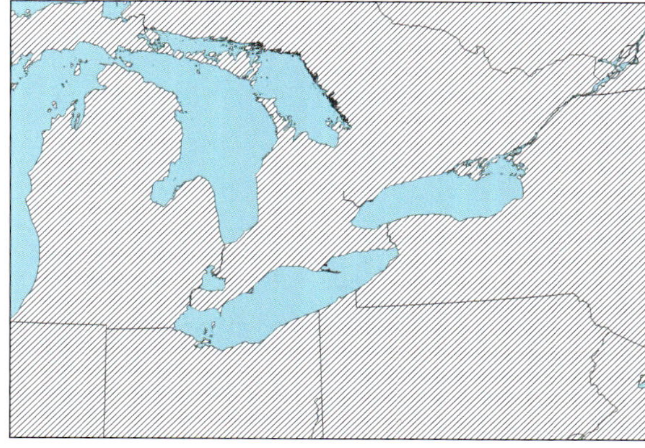

unique structure of the petals makes each flower look like an upside-down pair of pants. Flowers give way to oblong seed pods that are pointed at both ends, inside of which develop small black seeds

Flowers and leaves

about 2 mm long. Each seed has an elaiosome (fleshy food structure) attached to it. This provides a desirable food source for ants, who take the seeds back to their nests, thus aiding in dispersal of the seeds.

Dutchman's Breeches is similar to Squirrel Corn (*Dicentra canadensis*), but the latter has more-heart-shaped flowers.

IN THE GARDEN Dutchman's Breeches show off with delicate, fern-like foliage and charming, intricate flowers. This is a spring ephemeral, meaning it will put on an impressive display in early spring but will go dormant by midsummer. It is recommended to pair Dutchman's Breeches with something that can fill in the gap it leaves behind. Doing so allows gardeners to fit multiple bloom times into one space. Dutchman's Breeches benefits from a layer of leaflitter around it.

SKILL LEVEL Beginner

LIFESPAN Perennial (spring ephemeral)

EXPOSURE Full shade to part shade

SOIL TYPE Humus rich, neutral to alkaline, well drained

MOISTURE Moist (while these plants like an evenly moist soil, they demand very well-drained ground)

HEIGHT 25 cm

SPREAD 30–40 cm

BLOOM PERIOD Apr (late), May

COLOR White

FRAGRANT ✖

Seed pods

SHOWY FRUIT ✖

CUT FLOWER ✔

PESTS No serious pests

NATURAL HABITAT Deciduous woods and clearings, in rich loam soils, often found at the bases of trees

WILDLIFE VALUE Early nectar source and is of special value to bumblebees

BUTTERFLY LARVA HOST PLANT FOR None

MOTH LARVA HOST PLANT FOR None

USDA HARDINESS ZONES 3–7

PROPAGATION As an early-seeding species, the seeds of Dutchman's Breeches require a period of warmth followed by a two-to-three-month period of cold. Seeds should be planted immediately or, if not possible/desirable, then they must be kept moist — store in damp sphagnum moss so they do not dry out. You can also propagate by dividing the crowns and tubers in fall or early spring.

ADDITIONAL INFO Leaflitter is critical for the development and occurrence of Dutchman's Breeches, so much so that any site that has lost its leaflitter will almost certainly not have Dutchman's Breeches. The plant may cause minor skin irritation with repeated contact.

SAR STATUS N/A

Doellingeria umbellata
Flat-topped White Aster

FAMILY *Asteraceae* (Aster Family)

ALTERNATE COMMON NAMES Flat Top Aster, Parasol Aster, Parasol Whitetop, Tall White Aster, Umbellate Aster

PLANT DESCRIPTION Flat-topped White Aster is characterized by upright, green-to-purplish-red stems that emerge from the ground with no basal leaves. Stems are unbranched, except for at the top, where the flowers occur. Leaves are lance shaped, found in an alternate pattern and measure 8 to 15 cm long and 2.5 cm wide. They taper to a point at both ends, are toothless, and are attached to the stem on very short-to-non-existent leaf stalks. Hairiness is highly variable, but leaves are generally sparsely haired. The stem and its branches terminate with open, flat-topped flower clusters measuring up to

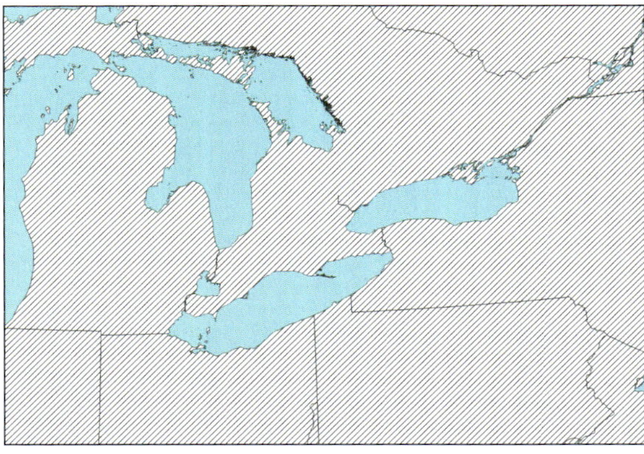

30 cm across and made up of numerous individual flowers. Individual flowers are 4 cm across and have two to 15 unevenly spaced white petals that surround a yellow center that dulls with age. Flowers

are surrounded by three or four layers of short, short-haired, or hairless awl-shaped bracts. Flowers give way to small, dry seeds with tufts of white hairs that carry them off in the wind.

IN THE GARDEN Flat-topped White Aster is one of the earliest asters to bloom and shows off with flat-topped clusters of daisy-like flowers. It has a strong, upright form with stems that persist into the winter months. This aster is usually not a prolific self-seeder, but, in ideal conditions, it may spread rapidly to form colonies. Lower stems may drop leaves in dry conditions, or late in the season, therefore we recommend planting it with a shorter companion that can hide the base.

SKILL LEVEL Beginner

LIFESPAN Perennial

EXPOSURE Full sun to part shade

SOIL TYPE Sandy loam

MOISTURE Wet to medium

HEIGHT 100–210 cm

SPREAD 30–90 cm

BLOOM PERIOD Aug, Sep

COLOR White

FRAGRANT ✖

SHOWY FRUIT ✖

CUT FLOWER ✔

PESTS No serious insect or disease problems

NATURAL HABITAT Moist thickets, woods, boggy low areas, and swamp edges

WILDLIFE VALUE Many native bees, wasps, beetles, and flies as well as butterflies and moths collect nectar and pollen, while deer and rabbits enjoy the foliage and sparrows eat the seeds

BUTTERFLY LARVA HOST PLANT FOR Harris' Checkerspot (*Chlosyne harrisii*), Silvery Checkerspot (*Chlosyne nycteis*), Tawny Crescent (*Phycoides batesii*), Northern Crescent (*Phycoides cocyta*), Pearl Crescent (*Phyciodes tharos*), Painted Lady (*Vanessa cardui*)

MOTH LARVA HOST PLANT FOR At least 58 species of moths, including tiger moths, ribbed cocoon-maker moths, case-bearer moths, twirler moths, geometer moths, leaf-blotch miner moths, slug caterpillar moths, clear-winged moths, flower moths, trumpet leafminer moths, tortrix moths, and 21 species of owlet moths

USDA HARDINESS ZONES 3–9

PROPAGATION Seeds sown in fall will produce blooms the second year. Seeds sown in the spring require 60 days of cold, moist stratification. They also require light to germinate, so do not cover. May also be multiplied by root division.

ADDITIONAL INFO This is one of the tallest native asters in North America. Only New England Aster (*Symphyotrichum novae-angliae*) comes close.

SAR STATUS N/A

Echinacea purpurea
Purple Coneflower

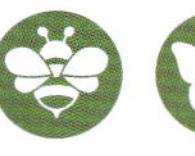

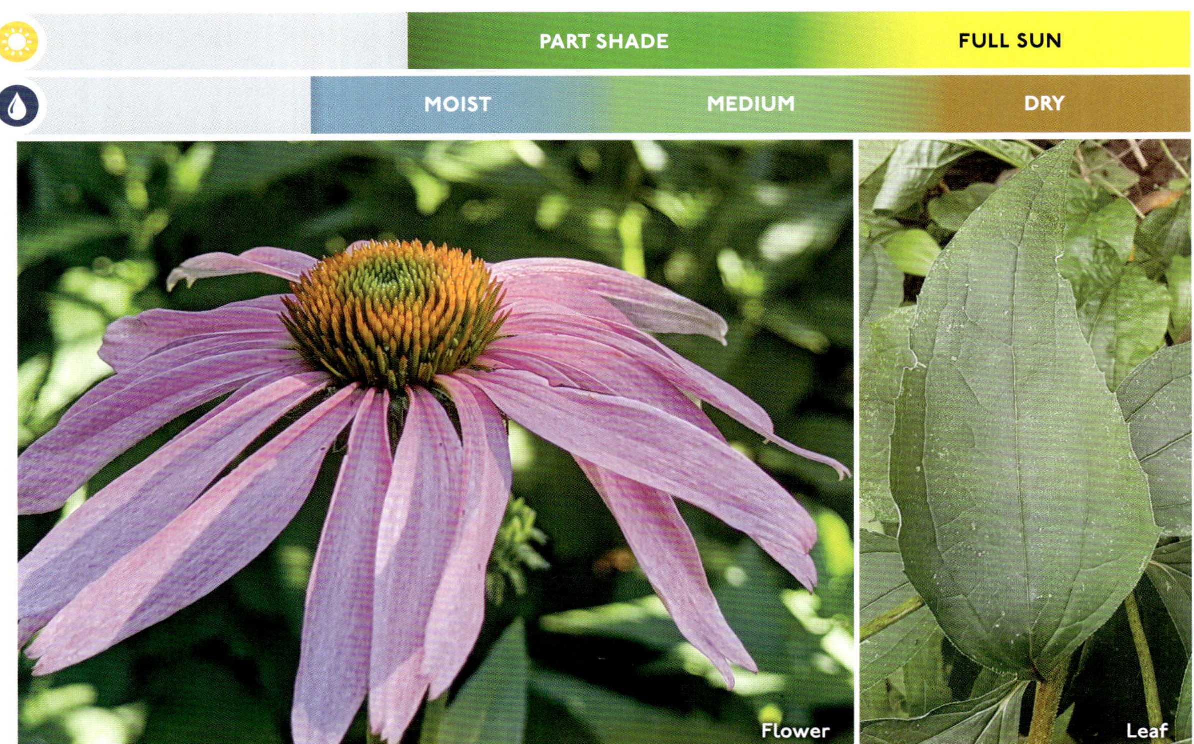

FAMILY *Asteraceae* (Aster Family)

ALTERNATE COMMON NAMES Black Sampson, Comb Flower, Eastern Purple Coneflower, Hedge-flower, Hedgehog, Indian Head, Kansas Snakeroot, Purple Echinacea, Scurvy Root, Snakeroot

PLANT DESCRIPTION Purple Coneflower has mostly unbranched, hairy, light green stems with purple streaks along it. Its leaves are lance shaped with a broad base and a pointed tip. They measure 15 cm long and 7.5 cm wide and have serrated edges, tiny appressed (flattened) hairs, and slightly winged leaf stalks. Stems terminate with solitary flowers, each measuring up to 10 cm across. They are characterized by a reddish-brown, flat-to-conical center disk surrounded by 10 to 20 pink-to-purple petals with notched tips. These petals grow out but curve down with age. Flowers give way to prickly, dark-colored seed heads containing numerous triangular seeds.

IN THE GARDEN Purple Coneflower is a classic wildflower that has earned a place in many gardens, namely for its long-lasting, showy flowers and its resilience. The dark, spiky seed heads stick around through the winter months to provide contrast, texture, and wildlife value to the winter garden.

SKILL LEVEL Beginner

LIFESPAN Perennial

EXPOSURE Full sun to part shade

SOIL TYPE Any well-drained soil

MOISTURE Moist to dry

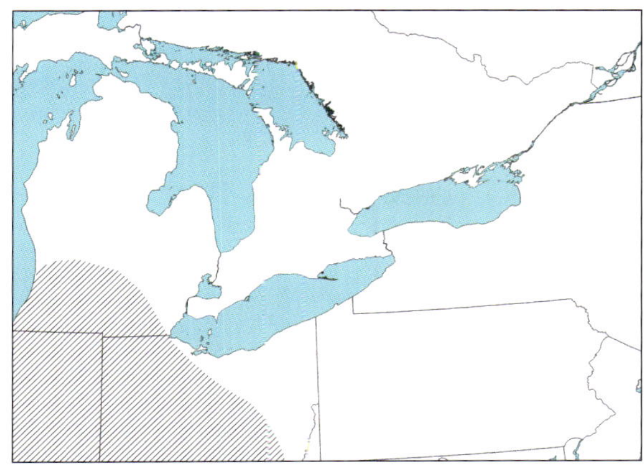

HEIGHT 50–120 cm

SPREAD 45–60 cm

BLOOM PERIOD Jul, Aug, Sep

COLOR Pink (purple)

FRAGRANT ❌

SHOWY FRUIT ❌

CUT FLOWER ✅

PESTS No serious insect or disease problems

NATURAL HABITAT Rocky, open woods, thickets, and prairies, especially near waterways

WILDLIFE VALUE Good source of pollen and nectar for bees and butterflies, and seeds for songbirds

BUTTERFLY LARVA HOST PLANT FOR Silvery Checkerspot (*Chlosyne nycteis*)

MOTH LARVA HOST PLANT FOR Blackberry Looper (*Chlorochlamys chloroleucaria*), Common Eupithecia (*Eupithecia miserulata*), Sunflower Moth (*Homoeosoma electellum*), Wavy-lined Emerald (*Synchlora aerata*)

USDA HARDINESS ZONES 3–9

PROPAGATION Seeds may be direct-sowed in fall or early spring. If starting seeds indoors, 30 days of cold, moist stratification can help germination. The clumps of fibrous roots may be divided in early spring or late fall.

ADDITIONAL INFO Pinch off spent flowers on a regular basis — or use them as cuttings in flower arrangements — to extend the blooming period.

SAR STATUS MI – SX/X

Flower with a bee visitor

Whole plant

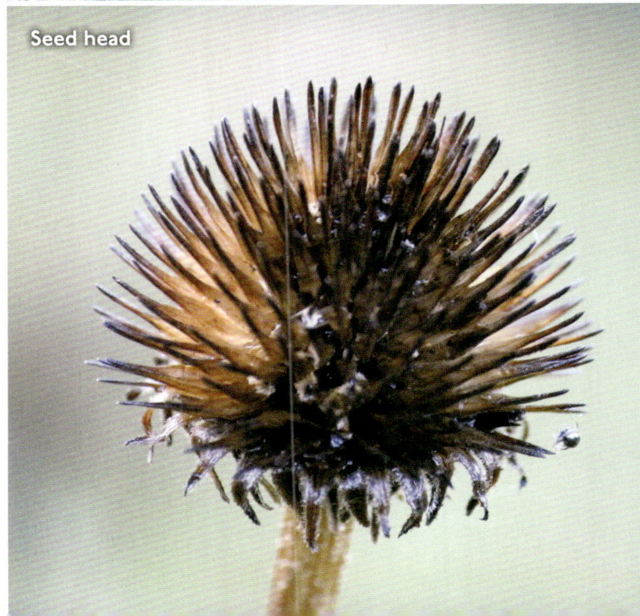
Seed head

Erythronium americanum
Trout Lily

PART SHADE

MOIST — **MEDIUM**

Flower

Leaves

FAMILY *Liliaceae* (Lily Family)

ALTERNATE COMMON NAMES Adder's Tongue, American Trout-lily, Dogtooth Violet, Eastern Trout-lily, Yellow Adder's Tongue, Yellow Dogtooth Violet, Yellow Fawn Lily, Yellow Trout-lily

PLANT DESCRIPTION Trout Lily has one to two waxy basal leaves that measure up to 15 cm long and 5 cm across. These leaves are oval in shape with a taper on both ends, ascending and green with a brownish-purple mottling. From the base of the leaf/leaves rises a smooth, reddish-brown flowering stalk topped with a solitary flower measuring about 2.5 cm long and 4 cm wide. This flower is characterized by six yellow petals (often tinged with purple) and six long, rusty-red stamens protruding from the center. Petals flare backwards with age. Seed pods are oval in shape and measure about 2 cm long. They split open into three parts to release their seeds.

IN THE GARDEN The plants will spread slowly to form colonies but can take many years to flower (I have had Trout Lily under my Sugar Maples in sandy-loam soil for at least 10 years and have yet to have one produce a blossom). A single, very attractive flower will bloom for just a few days on mature plants early in the spring.

SKILL LEVEL Advanced

LIFESPAN Perennial (spring ephemeral)

EXPOSURE Part shade

SOIL TYPE Fertile, humus-rich soil

MOISTURE Moist to medium

HEIGHT 5–20 cm

SPREAD 10–15 cm

BLOOM PERIOD Apr, May

COLOR Yellow

FRAGRANT ✖

SHOWY FRUIT ✖

CUT FLOWER ✖

PESTS No serious insect or disease problems

NATURAL HABITAT Moist woods, on wooded slopes and bluffs, and along streams

WILDLIFE VALUE Early source of nectar and pollen for native bees; source of food for ants, who distribute the seeds

BUTTERFLY LARVA HOST PLANT FOR None

MOTH LARVA HOST PLANT FOR None

USDA HARDINESS ZONES 3–9

PROPAGATION Seeds mature six to eight weeks after flowering and are best sown fresh (they must be kept moist to remain viable). Seeds need a period of warmth followed by a period of cold to break dormancy. Plants started from seed will not flower for at least four years. The easiest way to propagate is

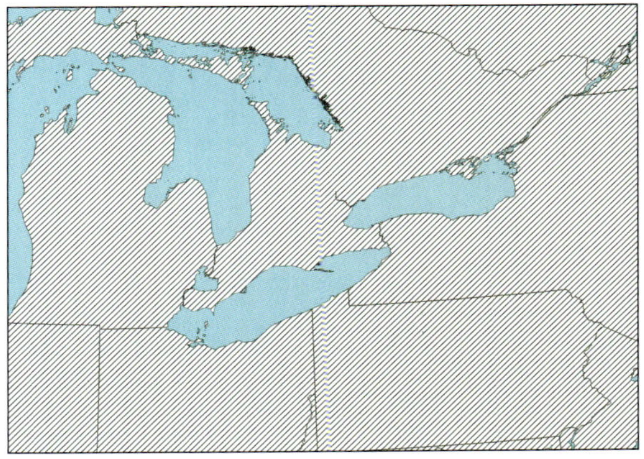

by marking the plants in the spring and digging the corms in late summer. Plant these small bulbs 8 to 10 cm deep in part-to-full shade but also where they will get ample sun in early spring. Mulch well. These native plants do not transplant well and should not be dug from the wild.

ADDITIONAL INFO Trout Lily corms work their way deeper into the soil with the freeze-thaw cycles. Once they reach 10 to 20 cm depth, they will start to convert their energy to leaf and flower production. However, they do not produce flowers every year, so in any given year only a portion of the plants will be in bloom.

SAR STATUS N/A

Eupatorium perfoliatum
Boneset

Flowers with insects

Whole plant

FAMILY *Asteraceae* (Aster Family)

ALTERNATE COMMON NAMES Agueweed, Fever-wort, Indian Sage, Perfoliate Boneset, Perfoliate Thoroughwort, Thoroughwort

PLANT DESCRIPTION Boneset has upright stems that are unbranched, except at the top where the flowers are located. The stems are covered in long, white hairs. Lanceolate leaves are found in an opposite pattern along the stem. You will notice that they are fused around the stem to look like one large leaf. Each leaf measures up to 20 cm long and 5 cm wide and is characterized by pointed tips, serrated edges, and hairs on both upper and lower surfaces. Flat-topped flower clusters are found at the top of the stem and are made up of many smaller clusters of up to 15 flowers. Individual flowers measure about 0.6 cm across and are made up of five petal-like lobes. Flowers give way to small, black seeds with tufts of hairs that allow them to be carried by the wind.

IN THE GARDEN Boneset is a dependable choice for moist sites. It has very distinct, bold leaves with a crinkly texture, and the pure white flowers bloom for a long time. The rigid stems often last through the winter months to extend seasonal interest. Despite its affinity for moist areas, it can tolerate average moisture conditions once established. Boneset typically has a clump-forming habit, but it will freely self-seed in open soil.

SKILL LEVEL Beginner

LIFESPAN Perennial

Seed heads

Leaves

EXPOSURE Full sun to part shade

SOIL TYPE Any moist, well-drained soil

MOISTURE Moist to wet (can even tolerate flooded soils for short times)

HEIGHT 40–150 cm

SPREAD 60–120 cm

BLOOM PERIOD Jul, Aug, Sep

COLOR White

FRAGRANT ✅

SHOWY FRUIT ❌

CUT FLOWER ✅

PESTS Plants are typically pest resistant, though in some years the caterpillars of the Morning Glory Plume Moth (*Emmelina mondactyla*) can defoliate plants, though the plants usually recover without any ill effects; foliage is unpalatable to deer and other herbivores

NATURAL HABITAT Damp prairies, bogs, openings in floodplain forests, and other poorly drained areas, and various kinds of wetlands

WILDLIFE VALUE Nectar and pollen of the flowers attract many kinds of insects

BUTTERFLY LARVA HOST PLANT FOR None

MOTH LARVA HOST PLANT FOR Blackberry Looper Moth (*Chlorochlamys chloroleucaria*), Morning Glory Plume Moth (*Emmelina mondactyla*), Clymene Moth (*Haploa clymene*), Burdock Borer

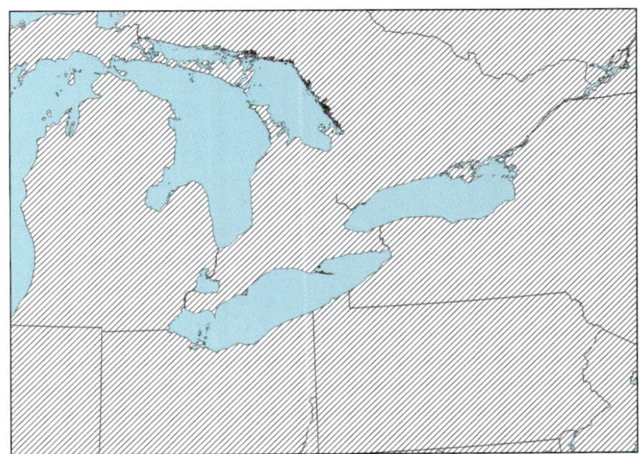

Moth (*Papaipema cataphracta*), Lined Ruby Tiger Moth (*Phragmatobia lineata*)

USDA HARDINESS ZONES 3–8

PROPAGATION If seeds are sown directly, surface sow (seeds need light to germinate) in the fall, and sow thickly as germination rates are typically low. If starting indoors, cold, moist stratification for at least 30 days will increase germination percentages. Seeds will last up to three years if stored in the fridge. Boneset can also be propagated by root division in the fall just as they go dormant or in early spring just as the first shoots appear. Plants may also be started by taking softwood tip cuttings (at least two nodes) in late spring.

ADDITIONAL INFO This plant is highly attractive to beneficial predatory insects in the garden.

SAR STATUS N/A

Eurybia macrophylla
Large Leaf Aster

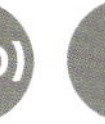

FAMILY *Asteraceae* (Aster Family)

ALTERNATE COMMON NAMES Bigleaf Aster, Large-leaf Wood Aster

PLANT DESCRIPTION Large Leaf Aster features large, heart-shaped basal leaves that measure about 20 cm long, 15 cm wide and are coarsely toothed, sparsely hairy, and borne on long leaf stalks. As the leaves ascend the stem, they decrease in size and become shorter stalked to the point where the uppermost leaves are reduced to stalkless bracts. Leaf stalks are narrowly winged but become broadly winged as they ascend the stem. Stems are upright, rigid, sparsely hairy, and unbranching except for at the top where they branch out to support open, flat-topped flower clusters. Each cluster can have eight to 90 stalked flowers that measure up to 4 cm

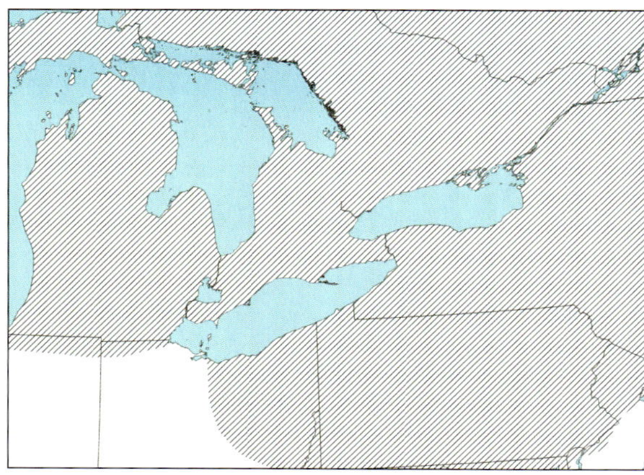

across. Individual flowers are characterized by nine to 20 unevenly spaced, thin ray florets (petals) that surround a yellow center disk, which turns a

brownish red with age. Tightly packed, stout, hairy bracts surround the base of each flower and are found in four to six layers. They are green with a purple tinge and have whitish edges. Flowers give way to dry, brown, narrow seeds with tufts of white hairs that allow them to be carried by the wind.

IN THE GARDEN Large Leaf Aster is valued by gardeners looking for a lush, reliable woodland groundcover with a bold texture. It spreads non-aggressively by rhizomes to form dense colonies over time. It is tolerant of dry shade and will grow in the dense shade under pine or spruce trees but with reduced flowering and vigor.

SKILL LEVEL Beginner

LIFESPAN Perennial

EXPOSURE Full shade to part shade

SOIL TYPE Sandy loam to rich loam

MOISTURE Moist to dry

HEIGHT 15–110 cm

SPREAD 60 cm

BLOOM PERIOD Aug, Sep, Oct

COLOR White, blue

FRAGRANT ❌

SHOWY FRUIT ❌

CUT FLOWER ❌

PESTS No serious insect or disease problems

NATURAL HABITAT Open woods, thickets, and clearings

WILDLIFE VALUE Nectar and pollen of the flowers attract a large number of native bees and other insects, and Ruffed Grouse (*Bonasa umbellus*) and Wild Turkey (*Meleagris gallopavo*) eat the seeds and foliage

BUTTERFLY LARVA HOST PLANT FOR Silvery Checkerspot (*Chlosyne nycteis*), Pearl Crescent (*Phyciodes tharos*)

MOTH LARVA HOST PLANT FOR Aster Borer Moth (*Carmenta corni*), Asteroid Moth (*Cucullia asteroides*), Arcigera Flower Moth (*Schinia arcigera*), others

USDA HARDINESS ZONES 3–7

PROPAGATION Sow seeds outside in fall or, if starting indoors, provide 60 days of cold, moist stratification for any seeds that haven't germinated in three to four weeks. Do not cover the seeds as they require light to germinate. Also can be propagated by stem cuttings taken in late spring or by dividing mature clumps.

ADDITIONAL INFO Individual plants don't flower every year, which means that in a patch of Large Leaf Aster only a few plants will bloom each year, and therefore it is best used as a groundcover for its heart-shaped leaves.

SAR STATUS N/A

Euthamia graminifolia
Grass-leaved Goldenrod

FULL SUN

MOIST

Flowers

Leaves

FAMILY *Asteraceae* (Aster Family)

ALTERNATE COMMON NAMES Common Gold-entop, Flat-top Fragrant Goldenrod, Flat-top Goldenrod, Flat-topped Bushy Goldenrod, Lance-leaved Goldenrod, Narrowleaf Goldenrod, Povertyweed

PLANT DESCRIPTION Grass-leaved Goldenrod has slender stems that are unbranched on the lower stem but typically branch out towards the top. Ascending, lance-shaped leaves are found in an alternate pattern along the stem and measure up to 12.5 cm long and just over 1 cm wide. They gently taper to a sharp tip and are stalkless at the base. Larger leaves have three prominent veins and sometimes two additional, but less prominent, veins. Leaves are either smooth or with fine hairs along major veins. Dense, flat-topped (sometimes

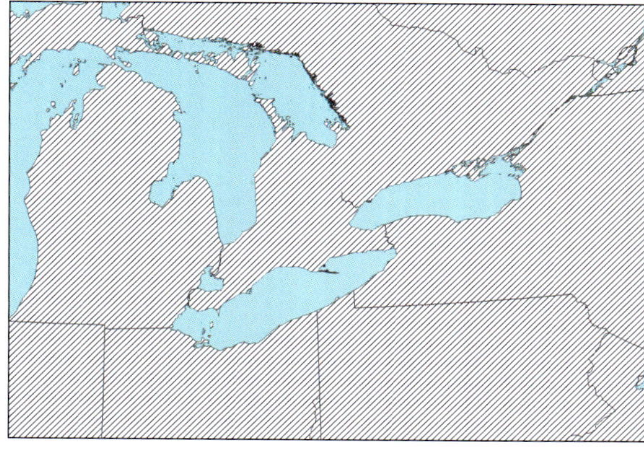

rounded) clusters of yellow flowers are found at the top of the stems and along many side branches. These clusters are made up of many individual flowers that each measure 3 mm across. Flowers

Whole plant

Seed heads

turn into small seeds with hairs that allow them to be carried by the wind.

Grass-leaved Goldenrod is similar to Riddell's Goldenrod (*Solidago riddellii*); however, the latter has larger leaves that sheath the stem and are folded in towards the central vein.

IN THE GARDEN Grass-leaved Goldenrod is a vigorous and adaptable goldenrod that produces bountiful yellow flower clusters. The flowers don't all bloom at once, which helps to extend bloom time in the landscape.

SKILL LEVEL Beginner

LIFESPAN Perennial

EXPOSURE Full sun (will tolerate some light shade)

SOIL TYPE Most well-drained soils

MOISTURE Moist (but will tolerate drier soils once established; fairly drought tolerant)

HEIGHT 150 cm

SPREAD 30–60 cm

BLOOM PERIOD Jul (late), Aug, Sep (early)

COLOR Yellow

FRAGRANT ✓

SHOWY FRUIT ✗

CUT FLOWER ✓

PESTS No serious insect or disease problems, but may be susceptible to rust

NATURAL HABITAT Damp to drier thickets and open areas

WILDLIFE VALUE Attracts many kinds of bees, wasps, butterflies, moths, and other insects, and the seeds are eaten by American Goldfinch (*Spinus tristis*) and Swamp Sparrow (*Melospiza georgiana*)

BUTTERFLY LARVA HOST PLANT FOR None

MOTH LARVA HOST PLANT FOR White-dotted Groundling (*Condica videns*), Gray-hooded Owlet (*Cucullia florea*), Shining Dichomeris Moth (*Dichomeris ochripalpella*), Goldenrod Gall Moth (*Epiblema scudderiana*), Solidago Eucosma (*Eucosma cataclystiana*), and the following moths that do not have common names: *Coleophora intermediella, Dichomeris leuconotella, Dichomeris levisella, Epiblema desertana*

USDA HARDINESS ZONES 2–9

PROPAGATION Seeds need light to germinate and germination may be increased with 60 days of cold, moist stratification. However, stratification is not absolutely necessary as Grass-leaved Goldenrod seeds have a high germination success even without it. This plant spreads by long rhizomes, which can be divided while the plant is dormant.

ADDITIONAL INFO Grass-leaved Goldenrod can be aggressive, spreading by rhizomes, and therefore may not be suitable for small gardens. This species is not a true goldenrod (*Solidago* spp.), and for that reason, it is now classified in a different genus, as *Euthamia graminifolia*.

SAR STATUS N/A

Eutrochium maculatum
Spotted Joe Pye Weed

QUICK GUIDE

PROPAGATION

	PART SHADE	FULL SUN

WET	MOIST	

Flowers

Whole plant

FAMILY *Asteraceae* (Aster Family)

ALTERNATE COMMON NAMES Joe Pye Weed, Purple Boneset, Purple Joe Pye Weed, Spotted Boneset, Spotted Trumpetweed

PLANT DESCRIPTION Spotted Joe Pye Weed has rigid, upright stems that are spotted with purple (sometimes they are solid purple) and covered in fine hairs. The stems are unbranched except for the flower stalks at the top of the plant. Leaves are found in whorls of four to five along the stem, with each leaf measuring up to 22 cm long and 5 cm wide. They are each attached by short leaf stalks and further characterized by coarsely toothed margins and prominent veins. At the top of the plant, you will find flat-topped flower clusters reaching up to 15 cm across. Individual flowers are 8 mm long, lack

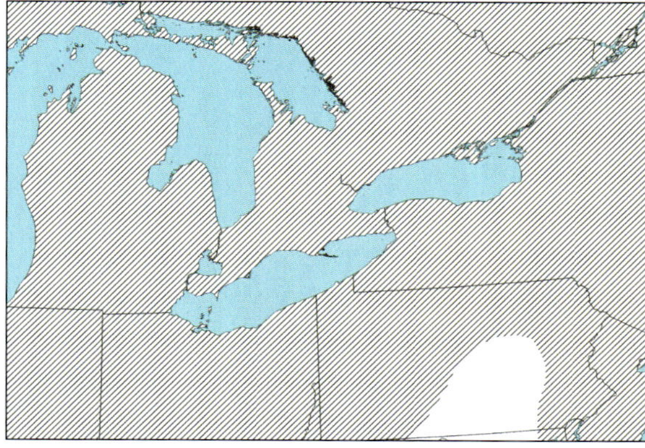

petals and have two long styles protruding from them. Flowers turn into small, brown seeds, each with a tuft of hair that allows them to be carried by the wind.

IN THE GARDEN Despite being a common and widespread plant, Spotted Joe Pye Weed still deserves a special spot in your garden. Its tall, sturdy stems support impressive flower clusters that act as a beacon to bring in pollinators from far and wide. The stems and seed heads persist into winter months to extend seasonal interest.

SKILL LEVEL Beginner

LIFESPAN Perennial

EXPOSURE Full sun to part shade

SOIL TYPE Sand, loam, clay

MOISTURE Moist to wet

HEIGHT 120–180 cm

SPREAD 60–120 cm

BLOOM PERIOD Jul, Aug, Sep

COLOR Pink, purple

FRAGRANT  (the flowers have a light vanilla fragrance that becomes more intense when crushed)

SHOWY FRUIT 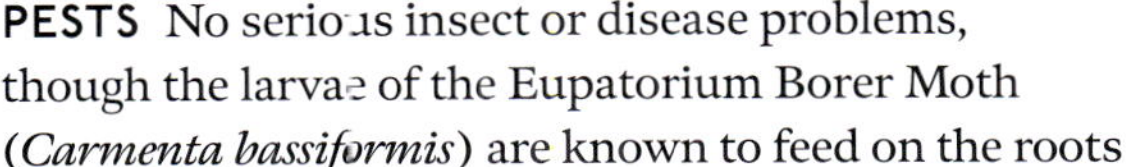

CUT FLOWER ✔

PESTS No serious insect or disease problems, though the larvae of the Eupatorium Borer Moth (*Carmenta bassiformis*) are known to feed on the roots

NATURAL HABITAT Grows in wet meadows, on the edges of lakes and streams, in roadside ditches, and in other wet places throughout its range

WILDLIFE VALUE An important food source for butterflies, bumblebees, green metallic sweat bees, and skippers

BUTTERFLY LARVA HOST PLANT FOR None

MOTH LARVA HOST PLANT FOR Eupatorium Borer Moth (*Carmenta bassiformis*), Common Pug (*Eupithecia miserulata*), Ruby Tiger Moth (*Phragmatobia fuliginosa*), Three-lined Flower Moth (*Schinia trifascia*)

USDA HARDINESS ZONES 2–9

PROPAGATION Plants may be started from seed, which benefit from 30 days of cold, moist stratification (sources vary from 14 to 60 days) and need light to germinate; they can take two to three weeks to sprout. More commonly, Spotted Joe Pye Weed is propagated by root division in spring as new growth starts, or in the fall. It can also be propagated by root cuttings taken in the spring. Stem cuttings are also easily rooted.

ADDITIONAL INFO Spotted Joe Pye Weed is similar to Sweet Joe Pye Weed (*Eutrochium purpureum*), but the latter doesn't have purple-spotted stems. Instead, it has green stems with purple coloring at the leaf nodes. *E. purpureum* also has more rounded flower clusters than *E. maculatum* and has a preference for part shade.

SAR STATUS N/A

Eutrochium purpureum
Sweet Joe Pye Weed

QUICK GUIDE

PROPAGATION

| FULL SHADE | PART SHADE | FULL SUN |

| MOIST | MEDIUM |

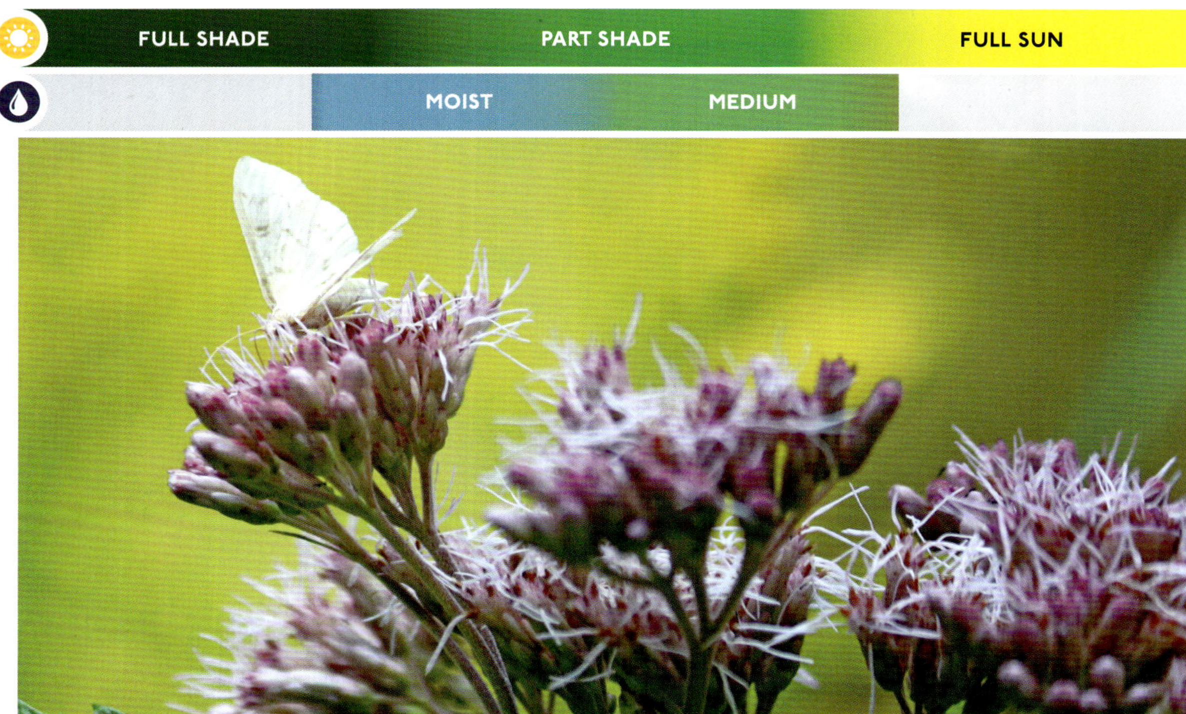

FAMILY *Asteraceae* (Aster Family)

ALTERNATE COMMON NAMES Gravel Root, Green Stemmed Joe Pye Weed, Kidney Root, Purple Boneset, Purple Joe Pye Weed, Sweetscented Joe Pye Weed, Trumpet Weed

PLANT DESCRIPTION Sweet Joe Pye Weed has tall, unbranched stems. They are smooth and light green, with purple coloration where the leaves meet the stem. Leaves are whorled in groups of three to five but usually four. Each leaf is up to 25 cm long and 10 cm wide with coarsely toothed margins and a taper on both ends. Leaves are smooth on top and smooth to finely hairy underneath. Stems terminate with dome-shaped flower clusters made up of many small, petal-less flowers, each with two long styles protruding from them. Flowers turn into small,

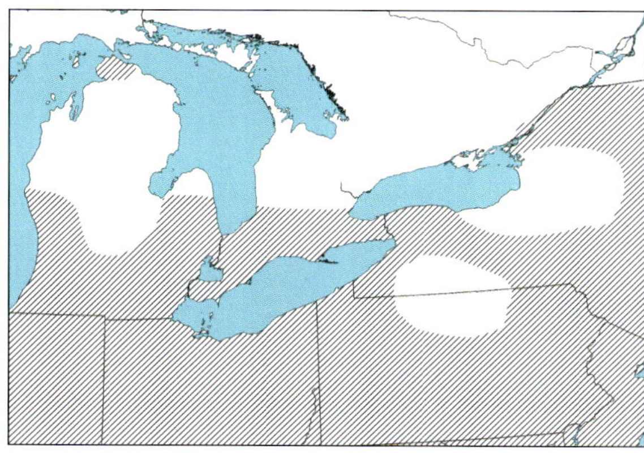

slender seeds, each outfitted with tufts of hairs that allow them to be carried by the wind.

IN THE GARDEN The towering pink flower clusters of Sweet Joe Pye Weed have a delightful vanilla

Leaves

Whole plant

Seed heads

scent and are irresistible to pollinators. Its height, clumping habit, and shade tolerance make it a great structural plant for shady gardens. The rigid stems and ornamental seed heads persist into the winter months to extend garden interest.

SKILL LEVEL Beginner

LIFESPAN Perennial

EXPOSURE Full sun to full shade (grows best in part shade; if grown in moist conditions, it can handle full sun, but it does not do well in hot, dry conditions)

SOIL TYPE Well drained — prefers moist, fertile, humus-rich soils that do not dry out

MOISTURE Moist to medium

HEIGHT 30–200 cm

SPREAD 60–120 cm

BLOOM PERIOD (Jul), Aug, Sep

COLOR Pink, purple

FRAGRANT ✓

SHOWY FRUIT ✗

CUT FLOWER ✓

PESTS No serious insect or disease problems, although somewhat susceptible to powdery mildew

NATURAL HABITAT Moist prairies, the edges of forested areas, and wooded slopes

WILDLIFE VALUE An important source of nectar for bees and butterflies

BUTTERFLY LARVA HOST PLANT FOR None

MOTH LARVA HOST PLANT FOR Eupatorium Borer Moth (*Carmenta bassiformis*), Common Plume Moth (*Emmelina monodactyla*), Common Pug (*Eupithecia miserulata*), Red Groundling (*Perigea xanthioides*), Ruby Tiger Moth (*Phragmatobia fuliginosa*), Three-Lined Flower Moth (*Schinia trifasciata*)

USDA HARDINESS ZONES 3–8

PROPAGATION Sweet Joe Pye Weed seeds require light to germinate, so do not cover. No treatment is necessary if sowing outdoors in the fall, but germination can be enhanced with 30 to 60 days of cold stratification (wet or dry). Sow seeds thickly as germination is usually low. Plants may also be started from softwood cuttings taken in late spring, or by dividing the plants in the fall as they go dormant or in the spring just as shoots first appear.

ADDITIONAL INFO Leaves may scorch if soils are allowed to dry out. Because Sweet Joe Pye Weed blooms on the new season's growth, you can have shorter, bushier plants if you cut the plants back to about 10 to 20 cm in the spring and again by half in June. Cut back to just above a whorl of leaves.

Sweet Joe Pye Weed is similar to Spotted Joe Pye Weed (*Eutrochium maculatum*), but the latter has purple-spotted stems and sometime pure purple stems. *E. maculatum* also has more flattened flower clusters and a preference for full sun.

SAR STATUS N/A

Fragaria vesca
Woodland Strawberry

QUICK GUIDE

PROPAGATION

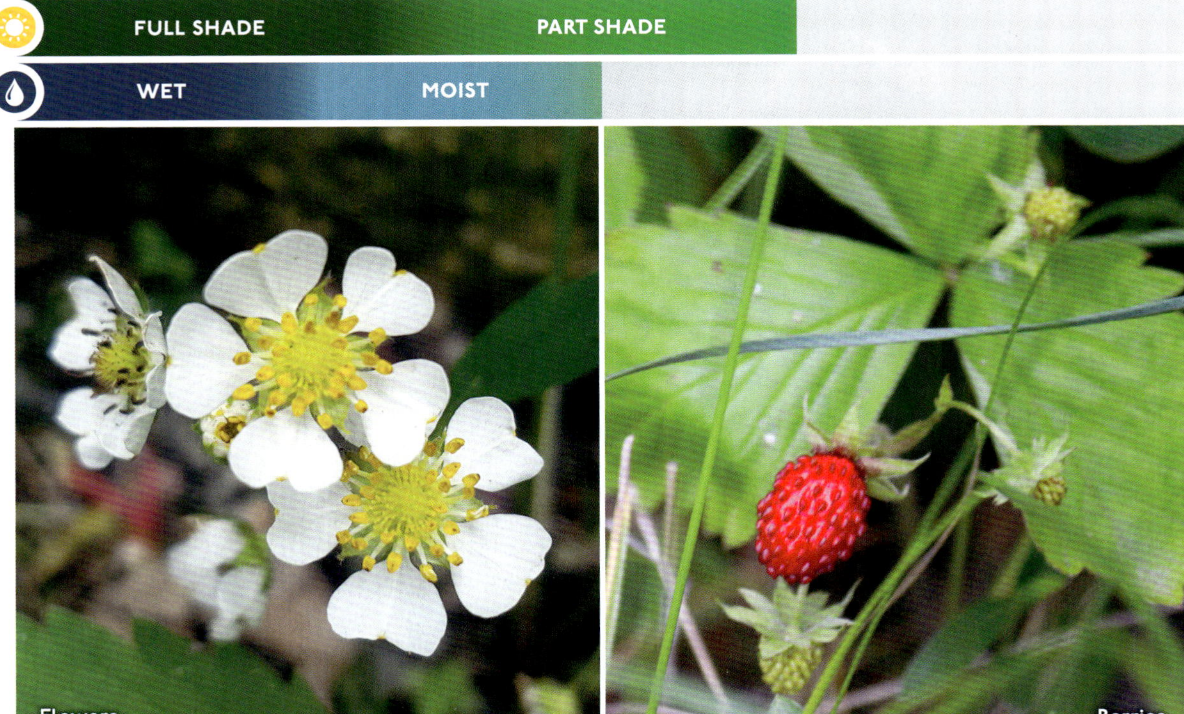

FAMILY *Rosaceae* (Rose Family)

ALTERNATE COMMON NAMES California Strawberry, Starvling Strawberry, Wild Strawberry, Wood Strawberry

PLANT DESCRIPTION Woodland Strawberry is characterized by above-ground runners (stolons) and basal leaves. These leaves are compound into groups of three, with each leaflet measuring up to 3.8 cm long and up to 2.5 cm wide. Leaflets are oval to egg shaped with a tapered base, coarsely toothed, and mostly hairless. The tooth at the end of each leaflet is equal to or slightly longer than teeth on either side of it. Flowers rise above the leaves on slender stems and are found in clusters of two to five. Five rounded petals surround a yellow center. Sepals are sharply pointed and are the same length or longer

than the petals. Small red strawberries are in the shape of elongated ovals and have tiny, raised seeds on the surface.

IN THE GARDEN Woodland Strawberry is a reliable

ground cover for shady areas and will spread indefinitely by runners. Its charming white flowers and vibrant red berries brighten up shady corners of the garden. Great for erosion control or as a no-mow lawn alternative. It's very vigorous, so give it some room to spread.

SKILL LEVEL Beginner

LIFESPAN Perennial

EXPOSURE Full shade to part shade

SOIL TYPE Loam, sand, clay

MOISTURE Moist to wet

HEIGHT 10–20 cm

SPREAD 100 cm (with runners)

BLOOM PERIOD May, Jun

COLOR White

FRAGRANT ❌

SHOWY FRUIT ✅ (edible red berry)

CUT FLOWER ❌

PESTS Relatively resistant to the many diseases and insect pests that often plague commercial strawberries

NATURAL HABITAT Rich, shady woodlands

WILDLIFE VALUE Many bees rely on the nectar and pollen, and many birds and small mammals and some turtles feed on the strawberries

BUTTERFLY LARVA HOST PLANT FOR Grizzled Skipper (*Pyrgus centaureae*), Gray Hairstreak (*Strymon melinus*)

MOTH LARVA HOST PLANT FOR Garden Webworm (*Achyra rantalis*), Smeared Dagger Moth (*Acronicta oblinita*), Strawberry Leafroller Moth (*Ancylis comptana fragariae*), Red-headed Ancylis (*Ancylis muricana*), Omnivorous Leafroller Moth (*Archips purpurana*), Celypha Moth (*Celypha cespitana*), Blackberry Looper Moth (*Chlorochlamys chloroleucaria*), Common Marbled Carpet (*Dysstroma truncata*), Wild Strawberry Seed Borer (*Grapholita angleseana*), Spotted Straw (*Heliothis turbatus*), Green Cloverworm (*Hypena scabra*), Drab Brown Wave

(*Lobocleta ossularia*), Olivaceous Olethreutes (*Olethreutes olivaceana*), Variegated Leafroller Moth (*Platynota flavedana*), Garden Tortrix (*Ptycholoma peritana*), Purple-lined Sallow (*Pyrrhia exprimens*), Clandestine Dart (*Spaelotis clandestina*), Sparganothis Leafroller Moth (*Sparganothis sulfureana*), Strawberry Crown Borer (*Synanthedon bibionipennis*), *Tinagma obscurofasciella* (no common name)

USDA HARDINESS ZONES 3–10

PROPAGATION Germination of wild strawberry seeds is very poor and may be enhanced with 60 days of cold, moist stratification. Vegetative propagation — by separating rooted plantlets in spring or early summer, or by taking cuttings of stolon internodes — is the most effective method of multiplying plants.

ADDITIONAL INFO Easily confused with the closely related *Fragaria virginiana* (Wild Strawberry), the two can be distinguished by carefully observing the leaf and the fruit. On *F. vesca* the terminal tooth on the leaf is more or less equal in size to the side teeth but are almost always smaller on *F. virginiana*. As well on *F. vesca* the leaflets are sparsely hairy, more prominently veined with large teeth, whereas *F. virginiana* is more softly veined and toothed and hairier overall. Woodland Strawberry fruit tends to be oblong with the seeds raised on the surface, while Wild Strawberry fruit is more round, with seeds in shallow pits. Finally, *F. vesca* usually has fewer flowers than *F. virginiana* and these are usually positioned above the leaves, while *F. virginiana* flower stems are typically shorter than the leaf stems.

SAR STATUS N/A

Fragaria virginiana
Wild Strawberry

QUICK GUIDE

PROPAGATION

PART SHADE   FULL SUN

MOIST   MEDIUM

Flowers

Berry

FAMILY *Rosaceae* (Rose Family)

ALTERNATE COMMON NAMES Blue-leaf Strawberry, Common Wild Strawberry, Scarlet Strawberry, Virginia Strawberry

PLANT DESCRIPTION Wild Strawberry features above-ground runners and basal leaves. Leaves are compound into three coarsely toothed leaflets, each measuring about 6 cm long and 3.8 cm wide. They are oval with rounded tips and wedge-shaped bases. You will notice the tooth at the tip of the leaf is much smaller than the teeth around it and doesn't extend past them. Flowers are found in clusters and are usually shorter than the leaves around them. Individual flowers are 1.2 cm to 2 cm wide and are characterized by five white, oval petals surrounding a yellow center. There are also five sharply pointed

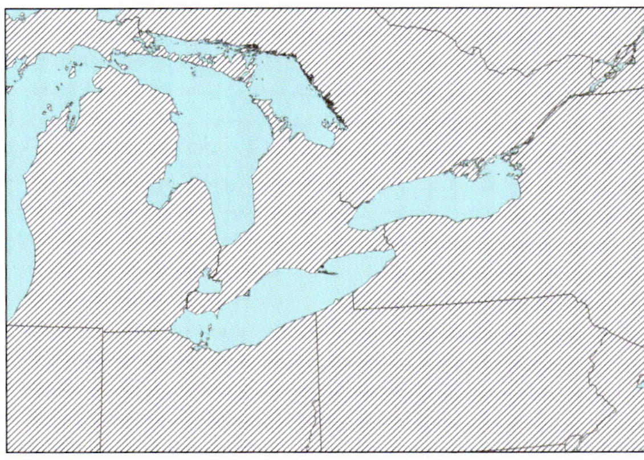

sepals that are the same length or shorter than the petals. Flowers give way to small, round berries with seeds found in shallow pits on the surface.

Wild Strawberry (*Fragaria virginiana*) is very

similar to Woodland Strawberry (*F. vesca*). See "Additional Info" under *F. vesca* for a description of the differences.

IN THE GARDEN Wild Strawberry is a dependable ground cover for sunny sites. The bright white flowers are found in abundance and the semi-evergreen leaves provide excellent fall color. Suitable as a no-mow lawn alternative and for erosion control. Very vigorous, so give it some space to spread.

SKILL LEVEL Beginner

LIFESPAN Perennial

EXPOSURE Full sun to part shade

SOIL TYPE Sand, loam, clay

MOISTURE Moist to medium

HEIGHT 5–15 cm

SPREAD 60+ cm

BLOOM PERIOD Apr, May

COLOR White

FRAGRANT ✗

SHOWY FRUIT ✓ (edible red berry)

CUT FLOWER ✗

PESTS Seldom bothered by the pests and diseases that affect commercial strawberries

NATURAL HABITAT Fields, prairies, and woodland edges

WILDLIFE VALUE Many bees rely on the nectar and pollen, and many birds and small mammals and some turtles feed on the strawberries

BUTTERFLY LARVA HOST PLANT FOR Gray Hairstreak (*Strymon melinus*), Grizzled Skipper (*Pyrgus centaureae*)

MOTH LARVA HOST PLANT FOR Garden Webworm (*Achyra rantalis*), Smeared Dagger Moth (*Acronicta oblinita*), Strawberry Leafroller Moth (*Ancylis comptana fragariae*), Red-headed Ancylis (*Ancylis muricana*), Omnivorous Leafroller Moth (*Archips purpurana*), Celypha Moth (*Celypha cespitana*), Blackberry Looper Moth (*Chlorochlamys chloroleucaria*), Common Marbled Carpet (*Dysstroma truncata*), Wild Strawberry Seed Borer (*Grapholita angleseana*), Spotted Straw (*Heliothis*

turbatus), Green Cloverworm (*Hypena scabra*), Drab Brown Wave (*Lobocleta ossularia*), Olivaceous Olethreutes (*Olethreutes olivaceana*), Variegated Leafroller Moth (*Platynota flavedana*), Garden Tortrix (*Ptycholoma peritana*), Purple-lined Sallow (*Pyrrhia exprimens*), Clandestine Dart (*Spaelotis clandestina*), Sparganothis Leafroller Moth (*Sparganothis sulfureana*), Strawberry Crown Borer (*Synanthedon bibionipennis*), *Tinagma obscurofasciella* (no common name)

USDA HARDINESS ZONES 3–9

PROPAGATION Germination of wild strawberry seeds is very poor and may be enhanced with 60 days of cold, moist stratification. Seeds need light to germinate and should not be covered. For new plants, separate rooted plantlets in spring or early summer, or take cuttings of stolon internodes.

ADDITIONAL INFO Plants may slow down or go dormant in hot summer months.

SAR STATUS N/A

Gentiana andrewsii
Bottle Gentian

Flowers

Leaf

FAMILY *Gentianaceae* (Gentian Family)

ALTERNATE COMMON NAMES Andrew's Gentian, Blind Gentian, Cloistered Heart, Closed Bottle Gentian, Closed Gentian, Fringe-top Bottle Gentian, Gall Flower, Prairie Closed Gentian, Sampson's Snakeroot

PLANT DESCRIPTION Bottle Gentian has smooth, unbranched stems that are round and light green to purple. Opposite, stalkless leaves are found along the stem with the uppermost set of leaves being whorled. Leaves are broadly lanceolate, glossy on top and become larger as they ascend the stem, reaching up to 10 cm long and 5 cm wide. The stem terminates with a cluster of tubular flowers, but there may also be secondary flower clusters emerging from leaf axils. Each

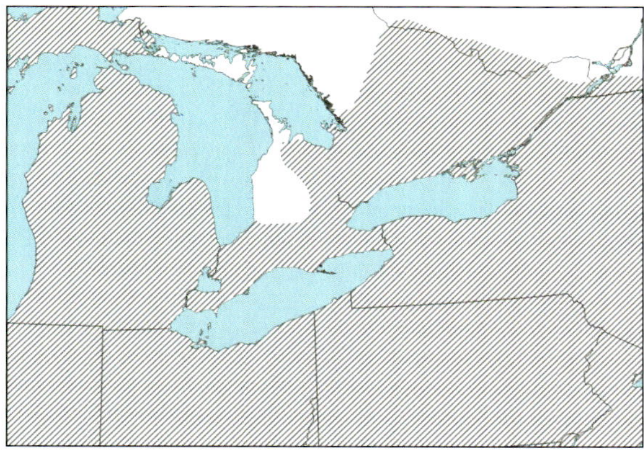

flower measures 2.5 cm to 4 cm long and has five fused petals with tiny teeth around their tips. The flowers never open and resemble closed buds even when in full bloom. Each flower turns into

a papery capsule that splits open to release many tiny seeds. Each seed has papery wings that allow them to be carried by water or wind.

IN THE GARDEN Bottle Gentian is valued by gardeners for its intriguing flowers and bold leaves. The flowers add a welcome touch of blue to the late-summer garden, while the leaves take on shades of purple and burgundy in the fall. Bottle Gentian is not competitive, so choose its companions wisely. Slow growing but well worth the wait!

SKILL LEVEL Beginner

LIFESPAN Perennial

EXPOSURE Full sun to part shade

SOIL TYPE Humus-rich, slightly acidic, sandy loam

MOISTURE Moist to wet

HEIGHT 30–60 cm

SPREAD 30–50 cm

BLOOM PERIOD Aug, Sept, (Oct)

COLOR Blue, purple

FRAGRANT ❌

SHOWY FRUIT ❌

CUT FLOWER ✅

PESTS Mature plants are rarely bothered by foliar disease or leaf-chewing insects

NATURAL HABITAT Moist and shaded sites, meadows, damp prairies, and along shores

WILDLIFE VALUE Especially valuable to bumblebees, just about the only insect with enough strength to force its way into the closed flower

BUTTERFLY LARVA HOST PLANT FOR None

MOTH LARVA HOST PLANT FOR Verbena Bud Moth (*Endothenia hebesana*)

USDA HARDINESS ZONES 3–6

PROPAGATION Germination of seed requires cold, moist stratification for at least 60 days, and exposure to light (surface sow). Bottle Gentian are said to be difficult to start from seed, though William Cullina states that sowing outdoors in the fall produces excellent results. Some sources suggest plants may be propagated by dividing the root crowns in fall or early spring, but this is apparently tricky to do without killing the plant.

ADDITIONAL INFO This plant tends to lean at maturity, so plant among sturdier plants for support. If left undisturbed, plants in optimum growing conditions will naturalize over time into large clumps. Hummingbirds have been observed nectaring on this plant. You can tell if a hummingbird has visited because the beak forces the flower open and it doesn't fully close up afterward.

SAR STATUS NY – (not ranked)/EV

Geranium maculatum
Wild Geranium

Flower

Whole plant

FAMILY *Geraniaceae* (Geranium Family)

ALTERNATE COMMON NAMES American Cranesbill, Cranesbill, Spotted Cranesbill, Spotted Geranium

PLANT DESCRIPTION Wild Geranium features a mounding clump of basal leaves. Each leaf measures up to 12 cm across and is deeply divided into three to seven wedge-shaped lobes that may be further divided at the tips with coarse teeth. Flowering stems rise directly from the roots and feature two small leaves at the base of each flower cluster. Both leaves and stems are hairy. Stems terminate with clusters of two to five flowers. Each flower measures up to 3.8 cm wide and has five petals with dark veins along them. Flowers turn into beaked seed pods that measure up to 3.8 cm long. When dried, the seed pods release stored elastic energy and catapult seeds away from the mother plant.

IN THE GARDEN Wild Geranium is cherished for its lush mound of leaves and elegant purple flowers. It works well in formal or natural gardens alike, and the leaves turn a vibrant red/maroon color in the fall. Check the flowers at dusk or dawn and you may notice solitary bees spending the night in the flowers.

SKILL LEVEL Beginner

LIFESPAN Perennial

EXPOSURE Full sun to full shade (prefers part shade but produces most flowers in full sun)

SOIL TYPE Almost any soil that isn't too wet (prefers moist, humus-rich soils, but also tolerates poor soils)

MOISTURE Moist to dry

HEIGHT 30 cm

SPREAD 30–60 cm

BLOOM PERIOD May, Jun, Jul

COLOR Violet (pink, purple)

FRAGRANT ✓

SHOWY FRUIT ✗

CUT FLOWER ✓

PESTS No serious insect or disease problems

NATURAL HABITAT Mesic deciduous woodlands, savannas, and meadows in wooded areas

WILDLIFE VALUE The nectar and pollen of the flowers attract a variety of native bees, and chipmunks feed on the seeds

BUTTERFLY LARVA HOST PLANT FOR None

MOTH LARVA HOST PLANT FOR Omnivorous Leafroller Moth (*Archips purpurana*), Tobacco Budworm Moth (*Heliothis virescens*)

USDA HARDINESS ZONES 3–8

PROPAGATION Seeds should be planted fresh

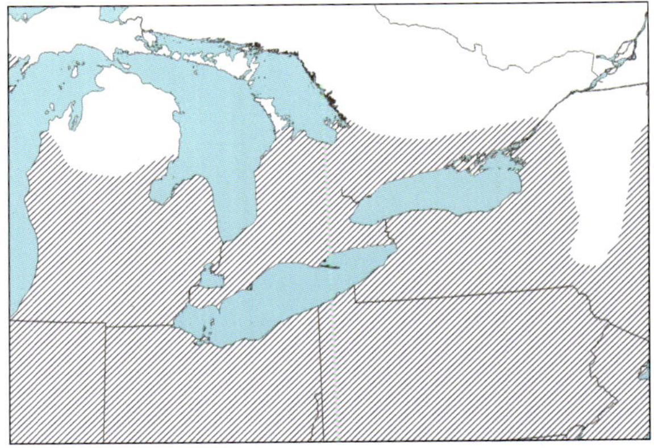

or kept moist and must be stratified in order to germinate, with higher germination rates the longer the cold period. Sowing outdoors in the fall is the easiest, with no artificial stratification needed. Wild geranium can also be propagated by dividing the roots in early spring or fall, cutting them where they form right angles. In the wild, plants don't usually bloom until their second or third year, but in a garden setting, they often bloom in the first year.

ADDITIONAL INFO For best performance, plant in full sun or light shade, in rich soil with plenty of organic matter and lots of moisture. Once established, *Geranium maculatum* requires little maintenance. Since the flowers don't usually repeat bloom, even deadheading is not necessary.

SAR STATUS N/A

Geum fragarioides
Barren Strawberry

FAMILY *Rosaceae* (Rose Family)

ALTERNATE COMMON NAMES Appalachian Barren Strawberry, Barren-ground Strawberry, Dry Strawberry, Northern Barren Strawberry

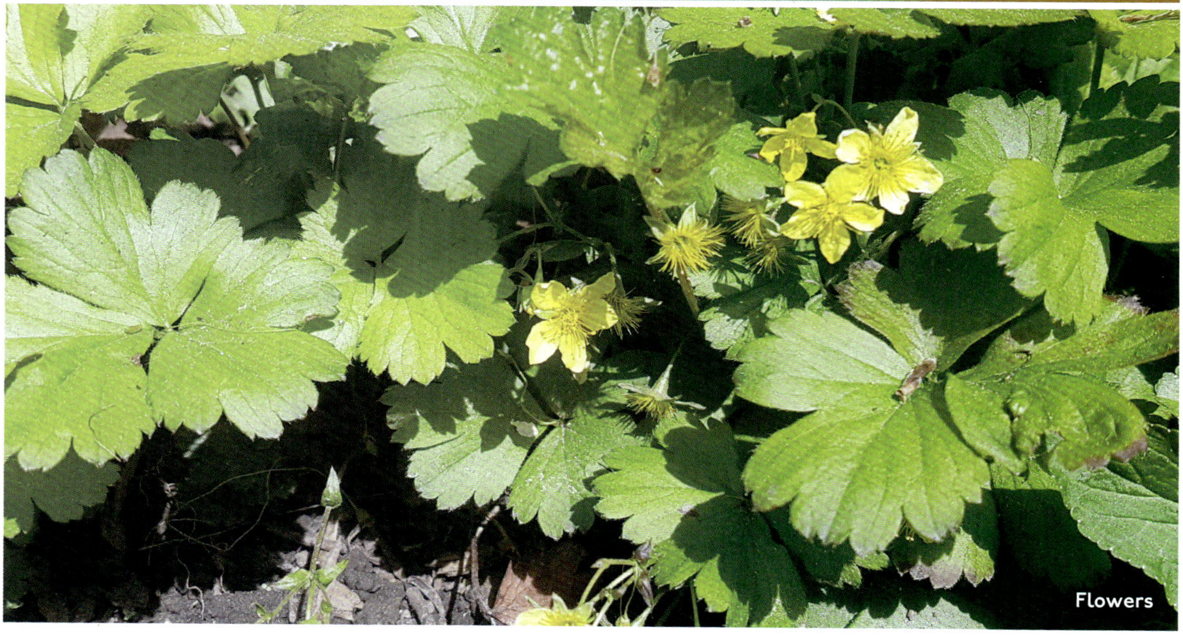

PLANT DESCRIPTION Barren Strawberry features glossy basal leaves borne on hairy leaf stalks. Each leaf is about 4 cm long by 4 cm wide and palmately compound into groups of three wedge-shaped leaflets. Each leaflet is three lobed and has teeth on the upper parts. The leaf stalks are hairy, while the leaves themselves are finely hairy to smooth. Yellow flowers are found in open clusters atop hairy, branching stems. They each measure just over 1 cm wide and feature five rounded, yellow petals surrounding many yellow stamens. Note how the petals alternate with sharply pointed, green sepals that are shorter than the petals themselves. Flowers mature into seed heads made of up to six dry seeds.

IN THE GARDEN Barren Strawberry spreads non-aggressively via rhizomes to form a dense, weed-choking groundcover. Its glossy leaves stay green throughout the year, and in the spring, it dots the landscape with cheery yellow flowers.

SKILL LEVEL Beginner

LIFESPAN Perennial

EXPOSURE Full sun to part shade

SOIL TYPE Average, well-drained soil, but prefers humus-rich, slightly acidic soils; tolerates rocky/shallow soil

MOISTURE Dry to moist

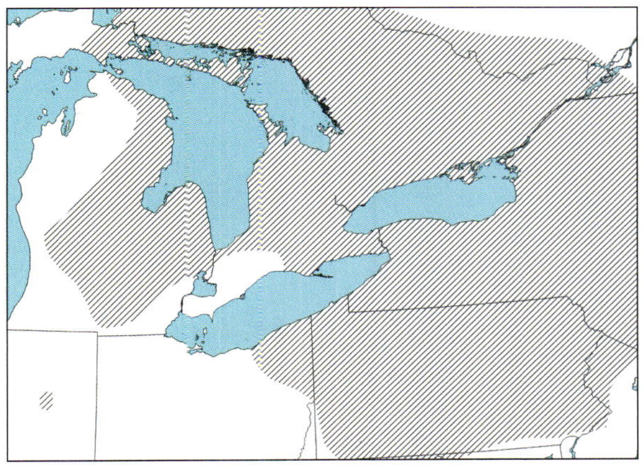

HEIGHT 20 cm

SPREAD 15–30 cm

BLOOM PERIOD May

COLOR Yellow

FRAGRANT ✗

SHOWY FRUIT ✗

CUT FLOWER ✗

PESTS No serious insect or disease problems, though slugs are occasional visitors

NATURAL HABITAT Woods, thickets, and clearings

WILDLIFE VALUE An early-season source of nectar and pollen for native bees

BUTTERFLY LARVA HOST PLANT FOR None

MOTH LARVA HOST PLANT FOR *Tinagma obscurofasciella* (no common name)

USDA HARDINESS ZONES 4–7

PROPAGATION No pretreatment required, but surface sow as the seeds need light to break dormancy. Clumps of mature plants may be divided in early spring or fall, or you can dig up a few runners and transplant these.

ADDITIONAL INFO This plant dislikes hot summer weather, and it requires some protection from the afternoon sun. This plant should not be confused with the similar Siberian Barren Strawberry (*Waldsteinia ternate*), which is often sold in garden centers as a native species.

SAR STATUS IN – S3/T

Leaf

Whole plant

Seed head

Geum rivale
Water Avens

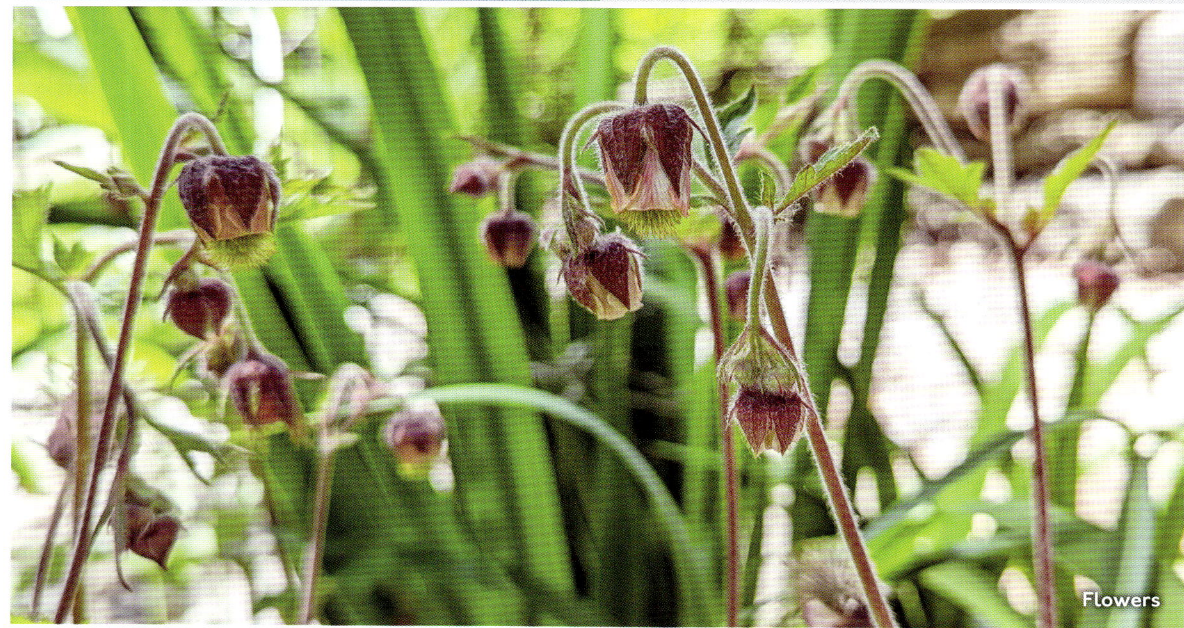

FAMILY *Rosaceae* (Rose Family)

ALTERNATE COMMON NAMES Bennet, Chocolate Root, Chocolate-plant, Cure All, Drooping Avens, Indian Chocolate, Maiden Hair, Nodding Avens, Purple Avens, River Avens, Throatroot, Throatwort, Water Flower

PLANT DESCRIPTION Water Avens have one or several erect, hairy stems and have both basal leaves and stem leaves. Basal leaves are up to 20 cm long and 9 cm wide and are compound into five to seven leaflets, with the smallest of them being found at the base and the largest of them being found at the tip. Leaflets are coarsely and unevenly toothed, hairy and rough textured. Alternate compound leaves occur along the lower part of each stem. These leaves have three leaflets and are smaller than the

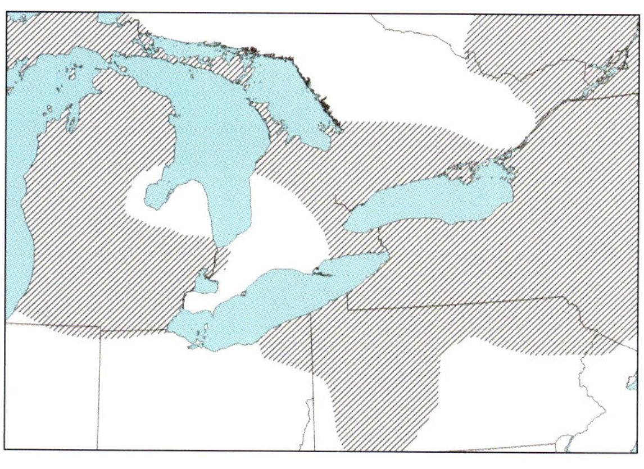

basal leaves. Stems rise above the basal leaves and are topped with open, droopy flower clusters. Each flower measures up to 2.5 cm across and is characterized by five conspicuously veined, purple-to-pink

petals encased by five larger, triangular, dark purple sepals. In the center of each flower, you will notice many yellow stamens. Once pollinated, the flowers go from droopy to upright and give way to seeds with feathery plumes that allow them to be blown by the wind.

IN THE GARDEN The intricate, droopy flowers of Water Avens will add a touch of elegance to your garden and are best enjoyed up close. They provide a long bloom time and turn into ornamental, fluffy seed heads. Water Avens maintains a clumping form and looks its best when planted en masse.

SKILL LEVEL Beginner

LIFESPAN Perennial

EXPOSURE Full sun to part shade

SOIL TYPE Various kinds of soil are tolerated, providing it is well drained

MOISTURE Moist to wet

HEIGHT 20–45 cm

SPREAD 15–30 cm

BLOOM PERIOD May (late), Jun, Jul, Aug (early)

COLOR Purple (brown-purple calyx and cream to purple-pink petals)

FRAGRANT 

SHOWY FRUIT ✓ (seed heads make an interesting garden feature)

CUT FLOWER ✗

PESTS No serious insect or disease problems

NATURAL HABITAT Bogs and damp meadows, stream sides, pond edges, damp deciduous woodlands, and swamps

WILDLIFE VALUE They are pollinated primarily by bumblebees

BUTTERFLY LARVA HOST PLANT FOR None

MOTH LARVA HOST PLANT FOR *Tinagma obscurofasciella* (no common name)

USDA HARDINESS ZONES 3–7

PROPAGATION Seeds need no pretreatment. Direct sow in fall or spring. Plants may be propagated by dividing in spring or fall every three to four years, which will also help to maintain the vigor of the plant.

ADDITIONAL INFO The fragrant flowers were once used to flavor ales, and the roots can be boiled to make a chocolate-like drink. Not always reliably winter hardy in areas north of Zone 5 and may be short-lived in heavy clay soils and/or hot summer climates. Deadheading spent flowers will encourage additional bloom. In ideal conditions, Water Avens can reproduce somewhat aggressively by its rhizomes, which makes it a good candidate for erosion control for moist areas.

SAR STATUS IN – S1/E; OH – S3/T(P)

Geum triflorum
Prairie Smoke

Flower

Whole plant

FAMILY *Rosaceae* (Rose Family)

ALTERNATE COMMON NAMES Grandpa's Whiskers, Old Man's Whiskers, Purple Avens, Three Flowered Avens, Three-sisters, Torchflower

PLANT DESCRIPTION Prairie Smoke features a rosette of basal leaves that each measure about 13 cm long and 4.5 cm wide. They are compound into seven or more wedge-shaped leaflets that may have much smaller secondary leaflets in between them. Each leaflet is coarsely toothed and covered in fine hairs on the upper surface, while the lower surface has hairs primarily along major veins. Naked flowering stems rise above the leaves and are reddish brown and finely hairy. Each stem is topped with a cluster of three droopy, bell-shaped flowers that each measure up to 2.5 cm in length and width.

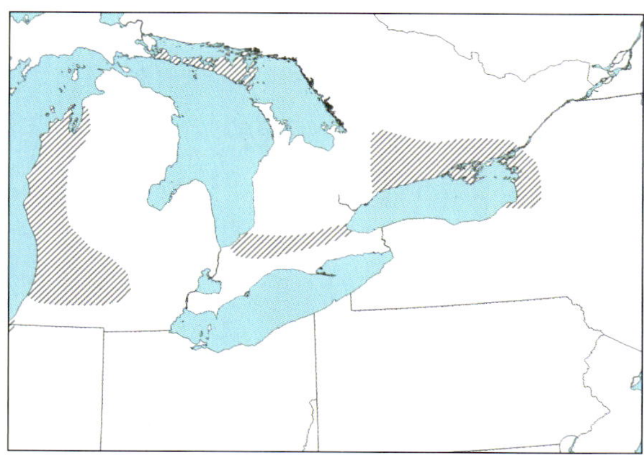

They are characterized by five white-to-pink petals that are mostly hidden under five reddish-pink sepals. Five narrow floral bracts also flare outwards from the sepals. The flowers barely open, even when

in full bloom. Flowers give way to dry seeds, each with a feathery tail. The seed heads look like smoke being blown in the wind.

IN THE GARDEN Prairie Smoke is an absolute delight to have in the garden. The nodding, pink flowers rise above fern-like leaves and provide an early-season buffet for bumblebees. When the flowers fade away, wispy seed heads steal the show as they fill the landscape with pastel hues of pink. An excellent choice for a sunny border or along a pathway. It will maintain a nice clumping form and is not a big self-seeder.

SKILL LEVEL Beginner

LIFESPAN Perennial

EXPOSURE Full sun to part shade

SOIL TYPE Well-drained barren soil that is rocky, gravelly, or sandy

MOISTURE Dry to moist (will not tolerate wet soil)

HEIGHT 10–45 cm

SPREAD 15–25 cm

BLOOM PERIOD May, Jun

COLOR Pink (red, purple)

FRAGRANT ✓ (seeds have been used to make perfume)

SHOWY FRUIT ✓ (seed heads are very showy)

CUT FLOWER ✓

PESTS No serious insect or disease problems, but root rot can be a problem in poorly drained soils, particularly in winter

NATURAL HABITAT Dry, open woods, prairies, open slopes, and meadows

WILDLIFE VALUE Bumblebees seek nectar from the flowers; these insects are strong enough to force their way into the flowers

BUTTERFLY LARVA HOST PLANT FOR None

MOTH LARVA HOST PLANT FOR *Tinagma obscurofasciella* (no common name)

USDA HARDINESS ZONES 3–7

PROPAGATION Prairie Smoke seeds do not require any pretreatment, though up to 60 days of cold, moist stratification may increase the germination rate if starting plants indoors. Plants can also be grown from rhizome cuttings or division of mature plants in late summer or early spring.

ADDITIONAL INFO Prairie Smoke does not like to be crowded by taller perennials.

SAR STATUS MI – S2/T; NY – S2/T

Helenium autumnale

Sneezeweed

  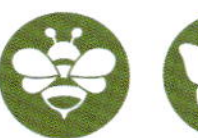

Flowers

FAMILY *Asteraceae* (Aster Family)

ALTERNATE COMMON NAMES Autumn Sneezeweed, Common Sneezeweed, Fall Sneezeweed, Helen's Flower, Swamp Sunflower

PLANT DESCRIPTION Sneezeweed is an upright plant with winged, green stems. Alternate leaves are attached to this stem without leaf stalks, and each leaf measures up to 15 cm long and 4 cm wide. They are lance-shaped, being widest just past the middle, with a pointed tip. You will notice these leaves have irregularly toothed margins. Stems terminate with clusters of numerous stalked flowers, each measuring 2.5 to 5 cm wide. They are characterized by a slightly flattened, globular center disk surrounded by 10 to 15 yellow ray florets (petals). These rays are wedge-shaped and have three teeth at their tips. Flowers fade into globular seed heads that contain finely hairy black seeds.

IN THE GARDEN Sneezeweed makes up for its unfortunate common name by being a showstopper in the garden. It is unlikely to make you sneeze, but it will provide you with abundant, cheerful yellow blooms and a very long bloom time. As the flowers fade and drop their petals, they leave behind globular seed heads that persist through winter to provide extended seasonal interest. Herbivore resistant.

SKILL LEVEL Beginner

LIFESPAN Perennial

EXPOSURE Full sun to part shade

SOIL TYPE Clay soils with loam or silt that is relatively high in organic material

MOISTURE Moist

HEIGHT 50–130 cm

SPREAD 60–90 cm

BLOOM PERIOD Aug, Sep, Oct

COLOR Yellow

FRAGRANT ✖

SHOWY FRUIT ✖

CUT FLOWER ✔

PESTS Larvae of the Aster Borer Moth (*Papaipema impecuniosa*) may bore through its stems and feed on the pith

NATURAL HABITAT Roadsides, fields, along streams and ditches, in seepage areas, and around ponds and lakes

WILDLIFE VALUE Attracts butterflies and native bees

BUTTERFLY LARVA HOST PLANT FOR None

MOTH LARVA HOST PLANT FOR Aster Borer Moth (*Papaipema impecuniosa*)

USDA HARDINESS ZONES 3–9

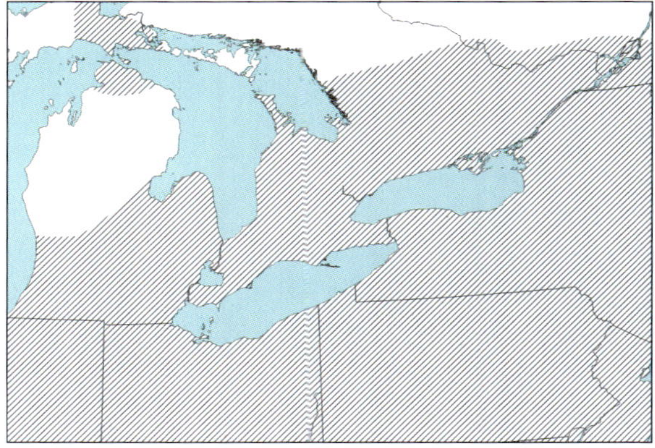

PROPAGATION Surface sow seeds directly; stratification is not required. Seeds need light to germinate. Mature plants may be divided in spring. New plants may also be started from cuttings.

ADDITIONAL INFO *Helenium autumnale* was called Sneezeweed because the flowers were pulverized to make a snuff that induced sneezing. In the garden, avoid fertilizing this plant to prevent lanky, weak stems. Pruning in the spring will induce branching and denser growth.

SAR STATUS N/A

Helianthus divaricatus
Woodland Sunflower

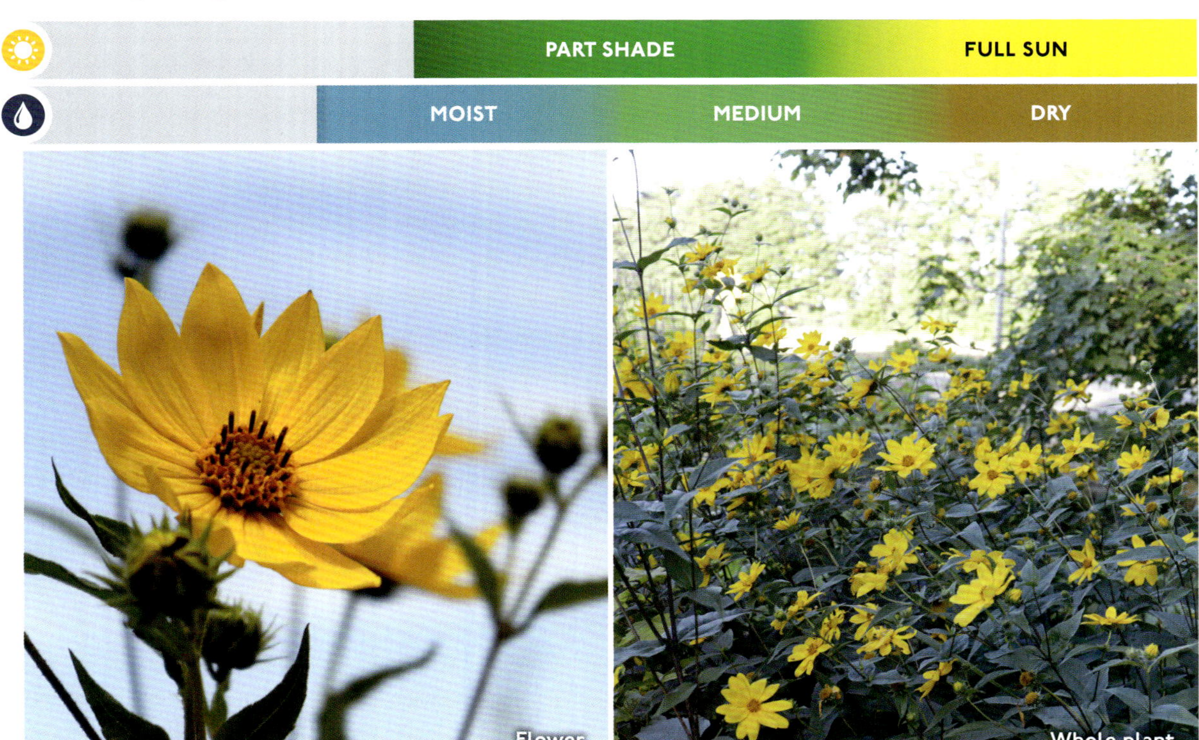

Flower

Whole plant

FAMILY *Asteraceae* (Aster Family)

ALTERNATE COMMON NAMES Rough Sunflower

PLANT DESCRIPTION Woodland Sunflower has rigid, upright stems that are unbranched, except for where flowers occur. These stems are light green to dark purple and smooth or sparsely covered in short, stiff hairs. Alternate leaves are attached directly to the stem in an opposite pattern and are rotated 90 degrees horizontally from the leaves below them. Leaves are lance shaped to ovate with a rounded base and pointed tip and measure up to 15 cm long and 5 cm across. Leaf margins may be toothless or have widely spaced teeth. Stems terminate with 2.5 to 7.5 cm wide flowers borne on slender stalks. Each flower is characterized by eight to 15 bright yellow, widely spreading ray florets

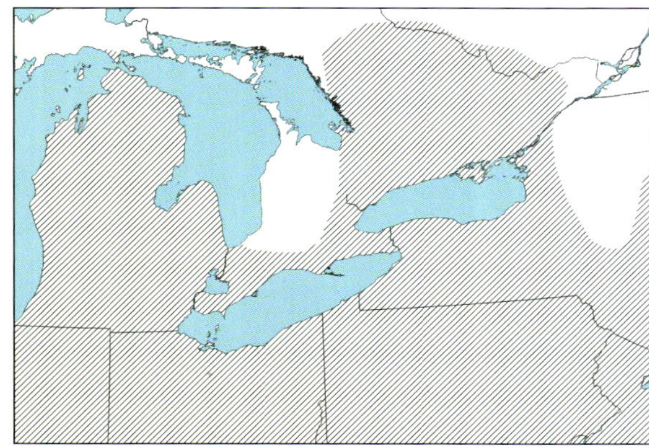

(petals) surrounding a slightly darker center disk. Flowers are replaced by globular seed heads containing numerous black seeds.

IN THE GARDEN The radiant yellow flowers of

Woodland Sunflower bloom in profusion and lighten up partly shaded areas of the garden. Its adaptability and long bloom time mean it will quickly become a favorite in your garden. The flowers fade to globular seed heads that persist well into the winter months to extend seasonal interest and wildlife value.

SKILL LEVEL Intermediate

LIFESPAN Perennial

EXPOSURE Full sun to part shade

SOIL TYPE Sandy, loamy or rocky

MOISTURE Moist to dry

HEIGHT 30–180 cm

SPREAD 30–90 cm

BLOOM PERIOD Jun, Jul, Aug, Sep

COLOR Yellow

FRAGRANT ❌

SHOWY FRUIT ❌

CUT FLOWER ✅

PESTS No serious insect or disease problems

NATURAL HABITAT Dry, open woodland or savanna

WILDLIFE VALUE The nectar and pollen of the flowers attract a wide variety of insects and native bees, and the seeds are eaten by many small birds, squirrels, and mice

BUTTERFLY LARVA HOST PLANT FOR Silvery Checkerspot (*Chlosyne nycteis*), Gorgone Checkerspot (*Chlosyne gorgone*), Painted Lady (*Vanessa cardui*)

MOTH LARVA HOST PLANT FOR Common Looper Moth (*Autographa precationis*), Banded Sunflower Moth (*Cochylis hospes*), Leaf-blotch Miner Moth (*Cremastobombycia ignota*), Cream-edged Dichomeris Moth (*Dichomeris flavocostella*), *Dichomeris leuconotella* (no common name), Common Pug (*Euphithecia miserulata*), American Angle Shades (*Euplexia benesimilis*), Dark-sided Cutworm (*Euxoa messoria*), Red-backed Cutworm (*Euxoa ochrogaster*), Dingy Cutworm (*Feltia jaculifera*), Sunflower Leafminer (*Frechinia helianthiales*), Arge Tiger Moth (*Grammia arge*), Sunflower Moth (*Homoeosoma electellum*), Giant Leopard Moth (*Hypercompe scribonia*), Common Pinkband (*Ogdoconta cinereola*), Burdock Borer Moth (*Papaipema cataphracta*), Sunflower Borer Moth (*Papaipema necopina*), Ruby Tiger Moth (*Phragmatobia fuliginosa*), Frothy Moth (*Plagiomimicus spumosum*), Isabella Tiger Moth (*Pyrrharctia isabella*), Sparganothis Leafroller Moth (*Sparganothis sulfureana*), Yellow Woolly Bear Moth (*Spilosoma virginica*), Sunflower Bud Moth (*Suleima helianthana*), a number of tortrix moths

USDA HARDINESS ZONES 3–8

PROPAGATION Seeds direct sown in the fall will germinate in the spring. If starting the seeds indoors, they require 30 days of cold, moist stratification. Divide every three to four years to control spread and maintain vigor.

ADDITIONAL INFO Spreads over time by creeping rhizomes to form colonies. Woodland Sunflower is a vigorous spreader and therefore may not be suitable for small planting areas.

SAR STATUS N/A

Helianthus giganteus
Tall Sunflower

QUICK GUIDE

PROPAGATION

FULL SUN

WET　　　MOIST

Flower

Whole plant

FAMILY *Asteraceae* (Aster Family)

ALTERNATE COMMON NAMES Giant Sunflower, Swamp Sunflower

PLANT DESCRIPTION Tall Sunflower features reddish-purple stems covered with spreading white hairs. The stems are unbranched, except for the upper third of the plant. Alternate leaves are attached to the stem with little-to-no leaf stalk and measure up to 18 cm long and 5 cm wide. The lowest leaves occur in an opposite pattern. The upper surface of each leaf is rough from tiny hairs, but the lower surface is more softly hairy. Stems terminate with yellow flowers that each measure 5 to 7.5 cm across and are characterized by 10 to 20 petals (ray florets) surrounding a slightly darker

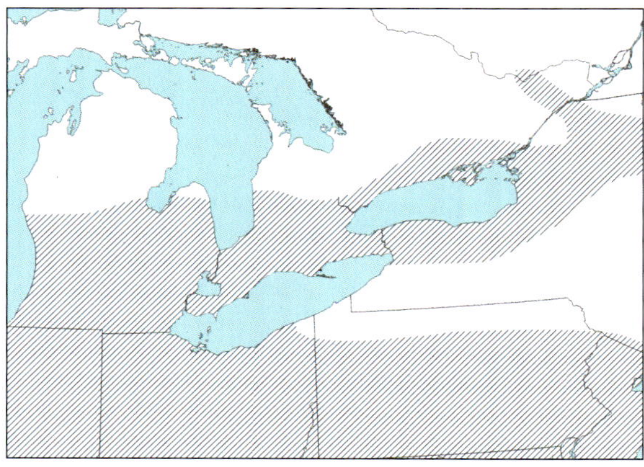

center disk. Each flower is backed by long, narrow, hairy bracts. The center disks become seed heads containing many dry, oblong, flattened seeds.

IN THE GARDEN Tall Sunflower produces an abundance of cheery yellow flowers that reach for the sky. It makes an excellent structural plant given its tall, rigid stems. These stems persist through the winter months to feed birds and add garden interest.

SKILL LEVEL Beginner

LIFESPAN Perennial

EXPOSURE Full sun (to part shade in ideal conditions)

SOIL TYPE Sandy

MOISTURE Wet to moist (to medium in ideal conditions)

HEIGHT 90–270 cm

SPREAD 120–180 cm

BLOOM PERIOD Jul, Aug, Sep

COLOR Yellow

FRAGRANT 

SHOWY FRUIT 

CUT FLOWER ✓

PESTS No serious insect or disease problems

NATURAL HABITAT Moist woods, marshes, and swamps

WILDLIFE VALUE The nectar and pollen attract many native bees, and the seeds are eaten by several birds during the fall and winter

BUTTERFLY LARVA HOST PLANT FOR Gorgone Checkerspot (*Chlosyne gorgone*), Silvery Checkerspot (*Chlosyne nycteis*), Painted Lady (*Vanessa cardui*)

MOTH LARVA HOST PLANT FOR Common Looper Moth (*Autographa precationis*), Banded Sunflower Moth (*Cochylis hospes*), Leaf-blotch Miner Moth (*Cremastobombycia ignota*), Cream-edged Dichomeris Moth (*Dichomeris flavocostella*), *Dichomeris leuconotella* (no common name), Common Pug (*Euphithecia miserulata*), American Angle Shades (*Euplexia benesimilis*), Dark-sided Cutworm (*Euxoa messoria*), Red-backed Cutworm (*Euxoa ochrogaster*), Dingy Cutworm (*Feltia jaculifera*), Sunflower Leafminer (*Frechinia helianthiales*), Arge Tiger Moth (*Grammia arge*), Sunflower Moth (*Homoeosoma electellum*), Giant Leopard Moth (*Hypercompe scribonia*), Common Pinkbank (*Ogdoconta cinereola*), Burdock Borer Moth (*Papaipema cataphracta*), Sunflower Borer Moth (*Papaipema necopina*), Rigid Sunflower Borer Moth (*Papaipema rigida*), Ruby Tiger Moth (*Phragmatobia fuliginosa*), Frothy Moth (*Plagiomimicus spumosum*), Isabella Tiger Moth (*Pyrrharctia isabella*), Sparganothis Leafroller Moth (*Sparganothis sulfureana*), Yellow Woolly Bear Moth (*Spilosoma virginica*), Sunflower Bud Moth (*Suleima helianthana*), a number of tortrix moths

USDA HARDINESS ZONES 4–8

PROPAGATION Seeds directly sown in the fall and allowed to overwinter need no other treatment. If starting indoors, seeds benefit from at least 30 days of cold, moist stratification. Tall Sunflower can also be propagated by stem cuttings taken before flowering. However, the easiest way to generate new plants is to divide the clump in early spring.

ADDITIONAL INFO The size of individual plants is highly variable, depending on the level of moisture, nutrients, and competition from other plants. Tall Sunflower can be aggressive in the garden, spreading by rhizomes, and therefore it may not be suitable for smaller gardens. The large stalks of Tall Sunflower growing in wetlands are sometimes used by beavers in the construction of their dams and lodges.

SAR STATUS N/A

Helianthus tuberosus
Jerusalem Artichoke

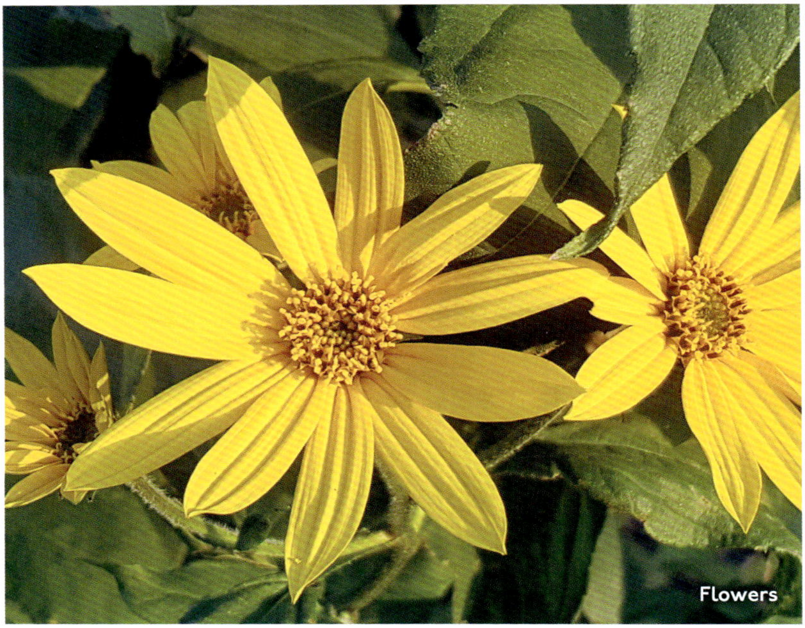

Flowers

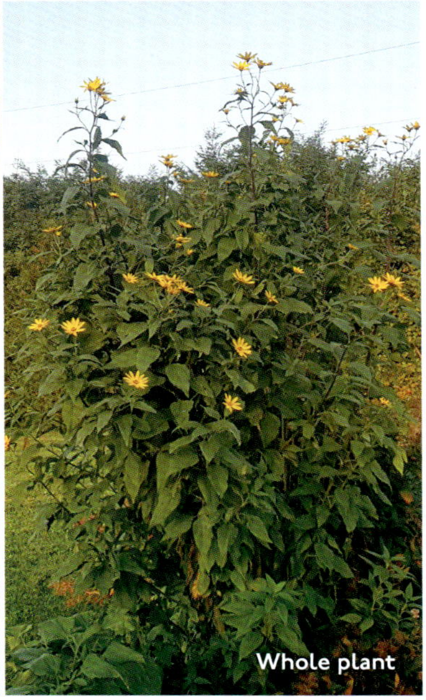

Whole plant

FAMILY *Asteraceae* (Aster Family)

ALTERNATE COMMON NAMES Canada Potato, Earth-apple, Girasole, Sunchoke, Sunflower Artichoke, Sunroot, Tuberous Sunflower

PLANT DESCRIPTION Jerusalem Artichoke has upright, rigid stems that are light green to reddish brown and covered in stiff hairs. The stems are unbranched except for towards the top where the flowers are found. Shallowly toothed leaves occur in an opposite arrangement on the lower part of the plant and become alternate as they ascend the stem. They are up to 25 cm long and 12 cm wide, lance-shaped to ovate with a pointed tip, rounded base, and rough texture on top. Leaves are borne on winged leaf stalks ranging from 2 cm to 7.5 cm long, becoming shorter as they ascend the stem. Stems terminate with flowers that measure up to 9 cm wide and are characterized by 10 to 20 yellow ray florets (petals) surrounding a slightly darker center disk. At the base of each flower are two to three sets of overlapping bracts, each being 1.2 cm long, hairy, and pointed. Flowers become dry seed heads, each containing flattened and slightly downy seeds.

IN THE GARDEN Jerusalem Artichoke is a very robust sunflower, putting on a dramatic display of yellow blooms in early fall. The tall, rigid stems persist throughout the winter months to extend wildlife value and garden interest. Best suited to large gardens or naturalized areas, where it can be allowed to spread.

SKILL LEVEL Beginner

LIFESPAN Perennial

EXPOSURE Full sun to part shade

SOIL TYPE Prefers loose, well-drained soil, but will tolerate poor soils.

MOISTURE Moist to dry (but do not plant in areas that are consistently wet as wet soil will rot the tubers)

HEIGHT 300 cm

SPREAD 100 cm

BLOOM PERIOD Aug, Sep, Oct

COLOR Yellow

FRAGRANT ✓

SHOWY FRUIT ✗

CUT FLOWER ✓

PESTS Powdery mildew

NATURAL HABITAT Open areas and moist thickets, prairie remnants along railroads, moist meadows along rivers, woodland borders, and disturbed areas (where it is mostly found)

WILDLIFE VALUE Several native bees are attracted to the flowers, and the seeds are an important source of food for many birds and small mammals; when growing near streams or ponds, the stems and leaves are used by beavers and muskrats for dam and den building

BUTTERFLY LARVA HOST PLANT FOR Gorgone Checkerspot (*Chlosyne gorgone*), Silvery Checkerspot (*Chlosyne nycteis*), Painted Lady (*Vanessa cardui*)

MOTH LARVA HOST PLANT FOR Common Looper Moth (*Autographa precationis*), Banded Sunflower Moth (*Cochylis hospes*), Leaf-blotch Miner Moth (*Cremastobombycia ignota*), Cream-edged Dichomeris Moth (*Dichomeris flavocostella*), *Dichomeris leuconotella* (no common name), Common Pug (*Eupithecia miserulata*), American Angle Shades (*Euplexia benesimilis*), Dark-sided Cutworm (*Euxoa messoria*), Red-backed Cutworm (*Euxoa ochrogaster*), Dingy Cutworm (*Feltia jaculifera*), Sunflower Leafminer (*Frechinia helianthiales*), Arge Tiger Moth (*Grammia arge*), Sunflower Moth (*Homoeosoma electellum*), Giant Leopard Moth (*Hypercompe scribonia*), Common Pinkbank (*Ogdoconta cinereola*), Burdock Borer Moth (*Papaipema cataphracta*), Sunflower Borer Moth (*Papaipema necopina*), Ruby

Seed head

Leaves

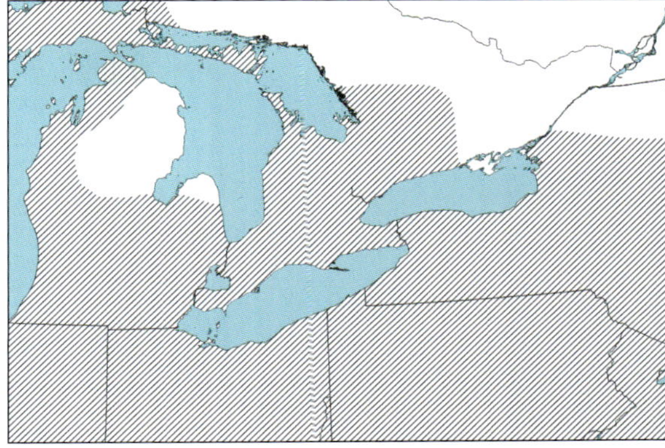

Tiger Moth (*Phragmatobia fuliginosa*), Frothy Moth (*Plagiomimicus spumosum*), Isabella Tiger Moth (*Pyrrharctia isabella*), Sparganothis Leafroller Moth (*Sparganothis sulfureana*), Yellow Woolly Bear Moth (*Spilosoma virginica*), Sunflower Bud Moth (*Suleima helianthana*), a number of tortrix moths

USDA HARDINESS ZONES 3–9

PROPAGATION Although possible to grow from seed, it is rarely done because this plant is so easy to grow from tubers or pieces of tubers (also, seeds tend to have low viability). Place your tubers in the ground, root-down and stalk-up, around 12.5 cm deep, and cover with soil. The best time to plant is in the spring after the danger of deep freezing has passed, but these hardy plants can be planted just about any time the ground isn't frozen.

ADDITIONAL INFO Jerusalem Artichoke is a very aggressive spreader in the garden. In my own garden, I planted them in half of a plastic 45-gallon drum, buried in the ground, to prevent spreading into my lawn. Jerusalem Artichoke has been grown commercially for use as a human food source, for livestock feed, and for ethanol production. Cultivated varieties yield white tubers that are clustered near the main stem as opposed to wild types, which produce reddish elongated tubers at the end of long rhizomes.

SAR STATUS N/A

Heliopsis helianthoides
False Sunflower

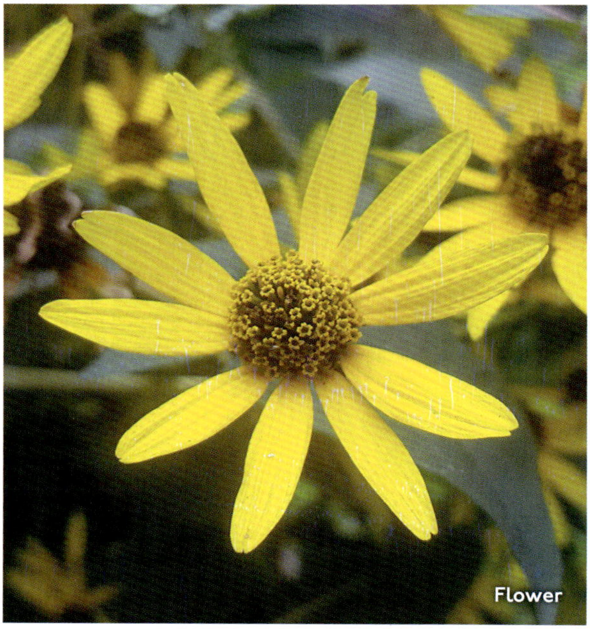

Flower

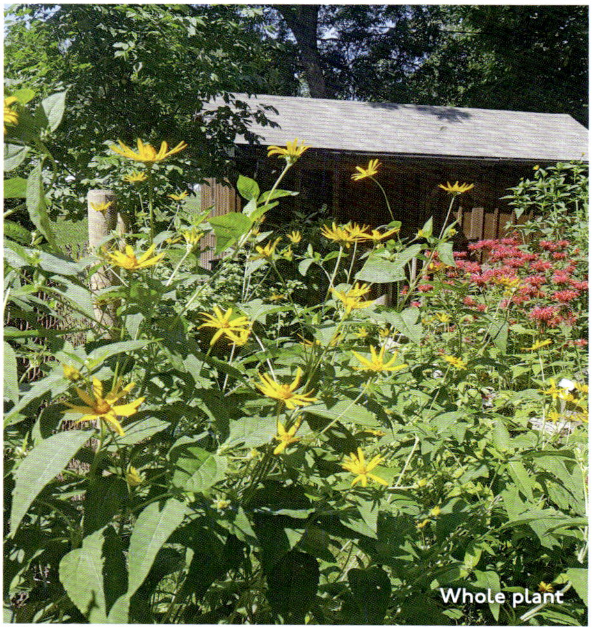

Whole plant

FAMILY *Asteraceae* (Aster Family)

ALTERNATE COMMON NAMES Early Sunflower, Oxeye Daisy, Oxeye Sunflower, Smooth Oxeye, Smooth Sunflower, Sweet Oxeye, Sweet Smooth Oxeye

PLANT DESCRIPTION False Sunflower features light-green-to-reddish, branching stems that are rough textured from short hairs. Leaves are borne on 2.5 cm long leaf stalks and occur in an opposite pattern, measuring up to 12.7 cm long and 8.8 cm wide. They are egg shaped to lance shaped with a pointed tip, serrated edges, and a rough upper surface with a less-hairy lower surface. Upper stems terminate with one to 15 stalked flowers that each measure 8.8 cm across. They are characterized by eight to 20 yellow ray florets (petals) surrounding a

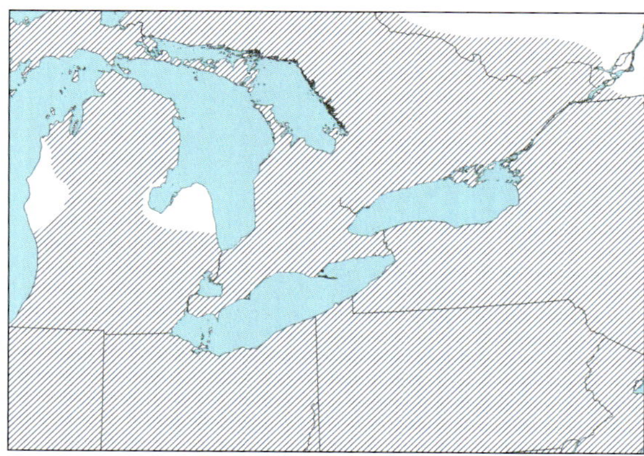

darker yellow center disk. The best way to identify this plant is to look at the bracts behind the flower. Look for two layers of hairy, appressed bracts that alternate between long and short.

IN THE GARDEN While not a true sunflower, False Sunflower still provides the same energy and charisma that we would expect from a sunflower. It blooms earlier than sunflowers and keeps going for longer, so it's great for extending bloom time in the landscape. In addition, the rigid stems persist into the early winter months.

SKILL LEVEL Beginner

LIFESPAN Short-lived perennial

EXPOSURE Full sun to part shade

SOIL TYPE Poor-to-average well-drained soils

MOISTURE Moist to dry

HEIGHT 120–180 cm

SPREAD 90–150 cm

BLOOM PERIOD Jun, Jul, Aug, Sep

COLOR Yellow

FRAGRANT ✖

SHOWY FRUIT ✖

CUT FLOWER ✔

PESTS A number of aster-feeding aphids of the genus *Uroleucon*

NATURAL HABITAT Floodplains and fields and at the edge of forests

WILDLIFE VALUE False Sunflower is a favorite of many different bees, but the Ground Nesting Bee (*Holcopasites heliopsis*) is a specialist pollinator that visits *Heliopsis helianthoides* almost exclusively; seeds attract finches and other small birds, and stems provide winter cover for beneficial insects

BUTTERFLY LARVA HOST PLANT FOR Gorgone Checkerspot (*Chlosyne gorgone*), Silvery Checkerspot (*Chlosyne nycteis*), Painted Lady (*Vanessa cardui*)

MOTH LARVA HOST PLANT FOR Tischeriid Moth (*Astrotischeria heliopsisella*), Sunflower Moth (*Homoeosoma electellum*), Rigid Sunflower Borer Moth (*Papaipema rigida*)

USDA HARDINESS ZONES 4–9

PROPAGATION False Sunflower seeds germinate readily if sown outside in fall or winter. If starting seeds indoors, germination will be improved after 30 to 60 days of cold, moist stratification. Seeds are slow to germinate and can take a month to sprout. Plants may be divided in the fall, though if the clumps are large, this can be a challenge. Plants may also be started from cuttings — five to seven node cuttings taken from the top 30 cm of growth in late spring root easily.

ADDITIONAL INFO Staking may be necessary, especially in shade. Deadhead to extend the blooming season. This is a very prolific self-seeder so it may not be suitable for small gardens.

SAR STATUS N/A

Hepatica acutiloba
Sharp-lobed Hepatica

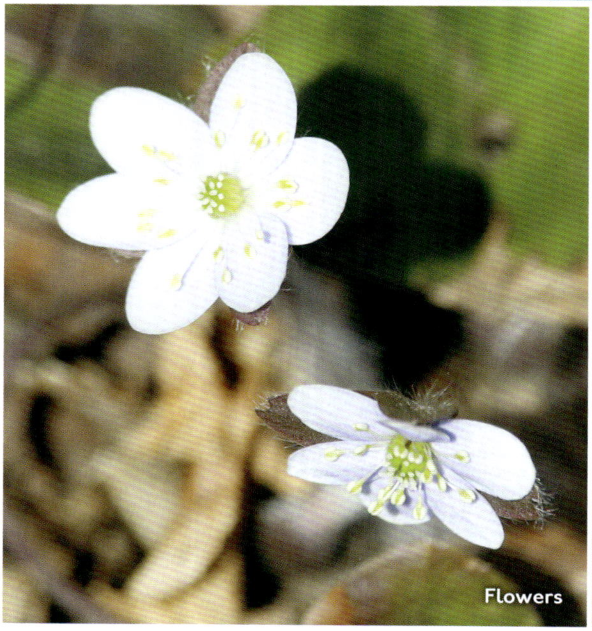

Flowers

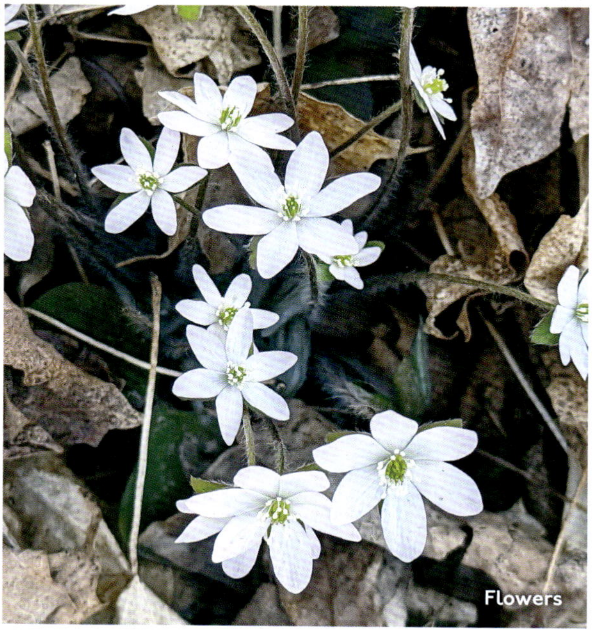

Flowers

FAMILY *Ranunculaceae* (Buttercup Family)

ALTERNATE COMMON NAMES Liverleaf, Mountain Hepatica

PLANT DESCRIPTION Sharp-lobed Hepatica is a low, stemless plant with three-lobed basal leaves. Leaves reach 7 cm long and wide on hairy stalks that reach up to 15 cm long. Each lobe is egg shaped with a pointy tip that distinguishes it from Round-lobed Hepatica (*Hepatica americana*). A solitary flower is borne at the end of each hairy, leafless flower stalk. Flowers contain five to 12 petals, measure up to 3 cm across, and are backed by three hairy bracts. The flower stalks emerge before new leaf growth. Flowers mature into clusters of green, beaked achenes (dry fruit) that droop when ripe.

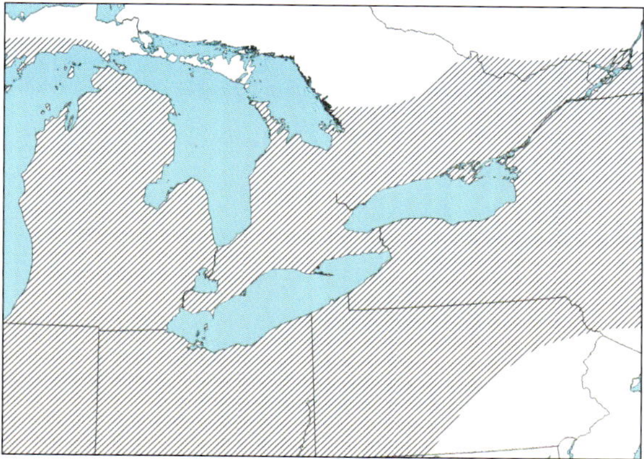

IN THE GARDEN Sharp-lobed Hepatica is among the first flowers to bloom in the spring, often flowering before the trees above have leafed out. Best

planted in big clumps to add a delicate, cheerful statement to a shade garden. It's slow to establish, but it will quickly become one of the plants you most look forward to in the spring.

SKILL LEVEL Beginner

LIFESPAN Perennial

EXPOSURE Full shade to part shade

SOIL TYPE Well-drained, semi-rich calcareous soil with a neutral pH

MOISTURE Moist to medium

HEIGHT 15 cm

SPREAD 10–15 cm

BLOOM PERIOD Apr, May

COLOR White (pink, purple)

FRAGRANT ✖

SHOWY FRUIT ✖

CUT FLOWER ✖

PESTS No serious insect or disease problems

NATURAL HABITAT Rich deciduous or mixed woods, often in calcareous soils

WILDLIFE VALUE Early pollen source for native bees

BUTTERFLY LARVA HOST PLANT FOR None

MOTH LARVA HOST PLANT FOR None

USDA HARDINESS ZONES 3–9

PROPAGATION Seeds should be sown immediately or stored moist (damp sphagnum moss works well), as they will not tolerate drying out. If starting indoors, they will benefit from 30 days of cold, moist stratification. Plants will not bloom until three years old or more. Plants may be divided in the fall, but it is important to make sure you do not break the leaves off as they are needed to keep this evergreen plant alive through the winter. Plant so the leaf buds are just at the soil surface, then mulch lightly. Divisions, however, are slow to increase. When dividing a clump, it is best to leave two to three buds in each division.

ADDITIONAL INFO This plant is deer and rabbit resistant. It will tolerate somewhat dry conditions, but too much sun will damage the leaf edges.

SAR STATUS N/A

Hepatica americana
Round-lobed Hepatica

Flowers

FAMILY *Ranunculaceae* (Buttercup Family)

ALTERNATE COMMON NAMES Liverleaf, Liver-wort Leaf

PLANT DESCRIPTION Round-lobed Hepatica is a low, stemless plant with leathery, three-lobed basal leaves. The leaves reach about 7 cm wide and equally as long and usually stay green throughout winter. Each anemone-like flower measures up to 3 cm across with five to 12 petals that are backed by three hairy bracts. A solitary flower sits atop each hairy, leafless stalk. The flower stalks emerge before new leaf growth. Flowers mature into clusters of green, beaked achenes (dry fruit) that droop when ripe.

IN THE GARDEN Round-lobed Hepatica is one of the earliest wildflowers to bloom in spring. The dainty flowers are in full bloom well before the trees above them leaf out. Plant it in big clumps to add a delicate, cheerful statement to a shade garden. It's slow to establish, but it will quickly become one of the plants you most look forward to in the spring.

SKILL LEVEL Beginner

LIFESPAN Perennial

EXPOSURE Full shade to part shade

SOIL TYPE Well-drained, organic-rich, slightly acid (pH <6.8)

MOISTURE Dry

HEIGHT 15 cm

SPREAD 10–15 cm

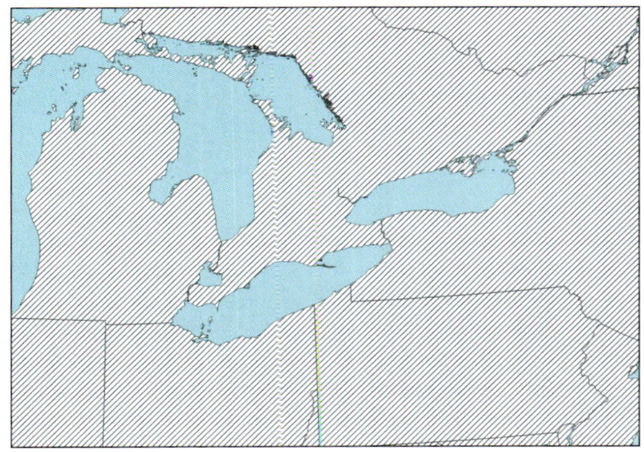

BLOOM PERIOD Apr. May

COLOR Blue (pink, white)

FRAGRANT ✓ (fresh, delicate scent)

SHOWY FRUIT ✗

CUT FLOWER ✗

PESTS No serious insect or disease problems

NATURAL HABITAT Rich, mesic to dry, deciduous, pine, and sometimes spruce woods

WILDLIFE VALUE Round-lobed Hepatica is an early pollen source for native bees, and the early summer fruits are eaten by chipmunks and other rodents; the seeds have a nutrient-rich elaiosome, which ants take back to their nests to consume, then discard the seeds

BUTTERFLY LARVA HOST PLANT FOR None

MOTH LARVA HOST PLANT FOR None

USDA HARDINESS ZONES 3–7

PROPAGATION Seeds should not be allowed to dry out. Planting fresh is best, but if starting indoors, keep moist and provide three to four weeks of cold, moist stratification. Plants may be divided in the fall.

ADDITIONAL INFO Until recently, both Round-lobed and Sharp-lobed Hepatica were placed in the genus *Anemone*, and many sources still identify them as such.

SAR STATUS N/A

Houstonia longifolia
Long-leaved Bluet

PART SHADE | FULL SUN

MEDIUM | DRY

Flowers

FAMILY *Rubiaceae* (Madder Family)

ALTERNATE COMMON NAMES Longleaf Summer Bluet, Pale Bluet

PLANT DESCRIPTION Long-leaved Bluets feature four-sided, light green stems that are hairless or slightly hairy along their margins. They may be branched or unbranched. Basal leaves may be present, but they usually die back by flowering time. Opposite, oblanceolate leaves occur along the stems, measuring up to 1 cm long and under 0.6 cm across. They are smooth with hairy margins. You may note that secondary leaves sometimes emerge from the leaf axils (where the leaf meets the stem) of the main leaves. Upper stems terminate with clusters of two to three tubular flowers. Each one measures 6 mm across and features four pointed lobes and four stamens with creamy-white or yellow tips. The inside of the flower has noticeable hairs. Flowers are replaced by globular, two-sectioned seed capsules each containing many small seeds.

IN THE GARDEN Long-leaved Bluets are excellent performers in tough, gravelly sites and make for an elegant addition to the garden. They also have a nice long bloom time. The plant spreads non-aggressively by self-seeding.

SKILL LEVEL Beginner

LIFESPAN Short-lived perennial

EXPOSURE Full sun to part shade

SOIL TYPE Sand, gravel (seems to prefer rather sterile soil)

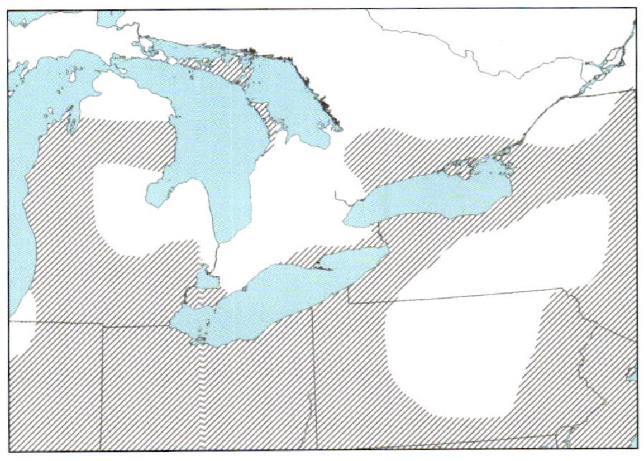

MOISTURE Dry to medium (will tolerate moist soil with adequate drainage)

HEIGHT 6–25 cm

SPREAD 10 cm

BLOOM PERIOD Jun, Jul, Aug

COLOR White (pale blue to pale pink)

FRAGRANT ❌

SHOWY FRUIT ❌

CUT FLOWER ❌

PESTS No serious insect or disease problems

NATURAL HABITAT Rocky woods, fields and prairies in dry soil, rocky ledges, and gravelly riverbanks

WILDLIFE VALUE Pollinated primarily by small bees that seek nectar and collect pollen from the flowers

BUTTERFLY LARVA HOST PLANT FOR None

MOTH LARVA HOST PLANT FOR Spotted Thyris (*Thyris maculata*)

USDA HARDINESS ZONES 3–8

PROPAGATION Seeds require light to germinate — surface sow outdoors in the fall/winter — no other pretreatment is necessary. Long-leaved Bluets can be divided in the spring or fall and are easily transplanted.

ADDITIONAL INFO Easy to cultivate, but it does not tolerate competition from taller vegetation.

SAR STATUS N/A

Hydrophyllum virginianum
Virginia Waterleaf

FAMILY *Hydrophyllaceae* (Waterleaf Family)

ALTERNATE COMMON NAMES Eastern Waterleaf, John's Cabbage, Shawnee Salad

PLANT DESCRIPTION Virginia Waterleaf features reddish-green-to-reddish-brown stems that are either smooth or with fine, flattened hairs. Leaves are attached in an alternate pattern and are pinnately divided into three to seven lobes that are each coarsely toothed with pointed tips. Each leaf measures 15 cm long and 10 cm wide and is usually finely hairy. Stalks terminate with loose, rounded flower clusters that each measure 5 cm wide and contain eight to 20 bell-shaped flowers. Each flower is about 1.2 cm long and characterized by five lobes and five protruding hairy stamens. Flower color is

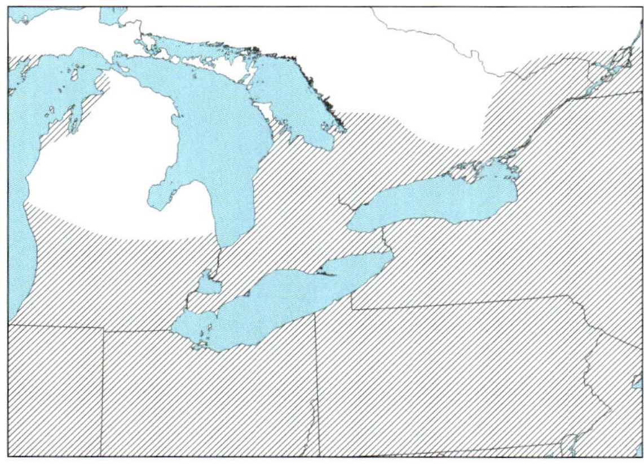

highly variable from white to purple. Flowers turn into seed capsules that split open to release their seeds.

IN THE GARDEN Virginia Waterleaf is a reliable ground cover that puts on a verdant display of leaves early in the spring. Delicate-looking flower clusters rise above the leaves in late spring to dot the landscape with pastel-purple hues.

SKILL LEVEL Beginner

LIFESPAN Perennial

EXPOSURE Part shade to full (light) shade (will tolerate full sun with sufficient moisture)

SOIL TYPE Rich, loamy soil with decaying leaves

MOISTURE Moist to medium (not drought tolerant — it will go dormant in hot, dry weather)

HEIGHT 30–60 cm

SPREAD 30–60 cm

BLOOM PERIOD May (late), Jun (early)

COLOR White, purple

FRAGRANT ✗

SHOWY FRUIT ✗

CUT FLOWER ✓

PESTS No serious insect or disease problems

NATURAL HABITAT Rich upland woods, shady floodplains, and moist clearings

WILDLIFE VALUE The nectar and pollen of the flowers attract bumblebees and other native bees; an andrenid bee called *Andrena geranii* is a specialist pollinator of *Hydrophyllum* species

BUTTERFLY LARVA HOST PLANT FOR None

MOTH LARVA HOST PLANT FOR None

USDA HARDINESS ZONES 4–9

PROPAGATION Virginia Waterleaf is easily grown from seed, but it typically will not flower until the second or third year. Seeds are best sown when fresh, though they will store if kept slightly moist (e.g., in damp peat moss in a resealable plastic bag). They require 60 days of cold, moist stratification to germinate. Division of the plant is the most practical method of starting new plants. Divide the rhizome in fall or when the plants go dormant, making sure each section has a leaf bud and some roots.

ADDITIONAL INFO This plant is a great groundcover for shade, but it can be a bit aggressive for small areas. The leaves may be solid green, but the earliest leaves to emerge will develop white spots that resemble water stains on the upper surface as they age.

SAR STATUS N/A

Impatiens capensis
Spotted Jewelweed

The flower of *Impatiens capensis*

The leaves of *Impatiens capensis*

FAMILY *Balsaminaceae* (Touch-me-not Family)

ALTERNATE COMMON NAMES Orange Balsam, Orange Jewelweed, Spotted Touch-me-not, Wild Balsam

PLANT DESCRIPTION Spotted Jewelweed is a heavily branched plant with smooth, succulent stems that are reddish green and nearly translucent. Oval-to-egg-shaped leaves are borne in an alternate pattern and measure up to 7.5 cm long and almost 4 cm wide. They are smooth to the touch with widely spaced, broad teeth. Flowers measure 2.5 cm long by 2 cm wide and emerge from upper leaf axils (where the leaf meets the stem) in small clusters of one to three flowers. Each flower is tubular in shape with two broad lower lobes and one smaller upper lobe. Sticking out from the back of each flower is a long, narrow nectar spur that curls back underneath the flower. Color can vary, but the flowers are usually orange with red spots on the front petals. Note that these red spots may be very dense or even completely absent, depending on the specimen. Flowers give way to thin, green seed pods that pop open from the slightest touch to spread their seeds away from the mother plant.

This species is very similar to Pale Touch-me-not (*Impatiens pallida*); however, the latter has fewer, but larger, yellow flowers with a shorter spur that bends down rather than sitting parallel with the flower. It also has more finely toothed leaves and is a much larger plant overall. *I. pallida* also seems to prefer soils on the sandier end of the spectrum, while *I. capensis* seems to favor heavier soils.

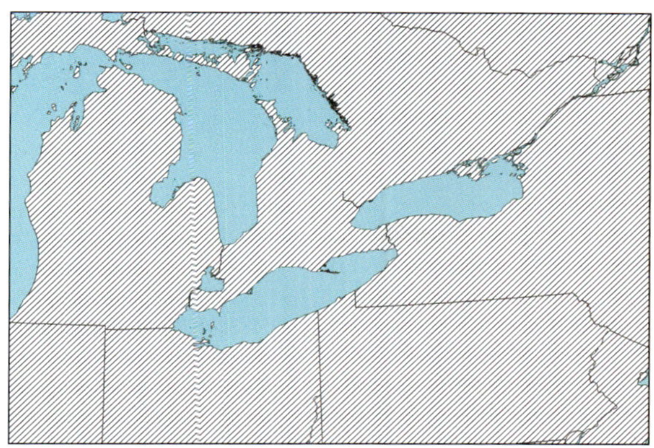

Impatiens capensis range

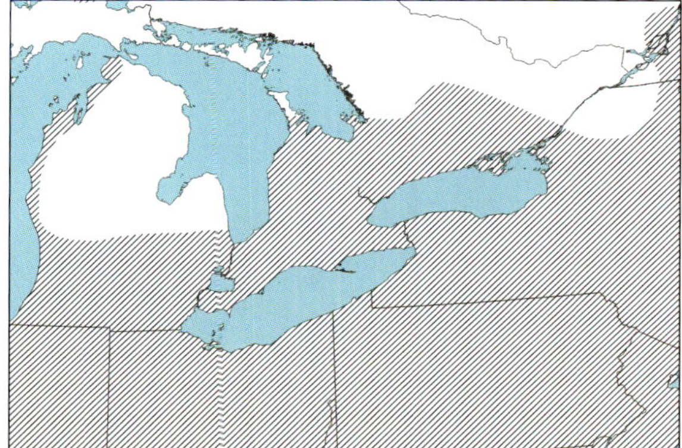

Impatiens pallida range

The flower of *Impatiens pallida*

IN THE GARDEN The vibrant orange flowers of Spotted Jewelweed dangle gracefully between its lush foliage, blooming for months on end. It will eagerly self-seed and quickly cover shady, moist areas with beauty and wildlife value.

SKILL LEVEL Beginner

LIFESPAN Annual

EXPOSURE Full shade to part shade

SOIL TYPE Fertile clay, loam, sand with an abundance of organic material

MOISTURE Moist to wet (submergence of the roots by floodwater is tolerated for up to two weeks without apparent ill effects)

HEIGHT 90–150 cm

SPREAD 45–75 cm

BLOOM PERIOD Jun, Jul, Aug, Sep, Oct (till frost)

COLOR Orange

FRAGRANT ❌

SHOWY FRUIT ❌

CUT FLOWER ✅

PESTS No serious insect or disease problems

NATURAL HABITAT Shady wetlands

WILDLIFE VALUE Hummingbirds and butterflies seek nectar, and several native bees collect pollen (the plant is listed by the Xerces Society as of special value to bumblebees); deer will browse the foliage, while mice and many birds eat the seeds

BUTTERFLY LARVA HOST PLANT FOR None

MOTH LARVA HOST PLANT FOR Obtuse Euchlaena (*Euchlaena obtusaria*), Pink-Legged Tiger Moth (*Spilosoma latipennis*), White-Striped Black (*Trichodezia albovittata*)

USDA HARDINESS ZONES 2–11

PROPAGATION Seeds are best sown when fresh as they do not tolerate drying out. Spotted Jewelweed seeds need light to germinate and a period of cold, moist stratification, followed by a warm, moist period, then another period of moist and cold. They typically require two years to germinate in the wild, though depending on the winter conditions, they may germinate after the first winter.

ADDITIONAL INFO The juice from Jewelweed stems contains a compound called lawsone, which has shown to have antihistamine and anti-inflammatory properties. It is said to relieve itching from Poison Ivy, mosquito bites, and Stinging Nettle and has also been used to treat athlete's foot.

The plant gets one of its common names, "Touch-me-not," because when the ripe seed pods are touched even lightly, the pods' explosive spring-action projects the seeds for a distance of a meter or more.

SAR STATUS N/A

Iris versicolor
Blue Flag Iris

FAMILY *Iridaceae* (Iris Family)

ALTERNATE COMMON NAMES American Blue Flag, Dagger Flower, Flag Lily, Harlequin Blueflag, Large Blue Iris, Larger Blue Flag, Multi-colored Blue Flag, Northern Blue Flag, Northern Iris, Poison Flag, Snake Lily, Water Flag

PLANT DESCRIPTION Blue Flag Iris features sword-like basal leaves that are usually erect, but larger leaves may be slightly spreading. They are about 2.5 cm wide at the base, taper gradually to a pointed tip and are often purple at the base. Smooth flower stalks emerge from the base of the plant and are topped by one to a few flowers that each measure up to 10 cm across. These flowers are a very familiar iris shape with three sepals, three petals, and three stamens. The sepals spread outwards from

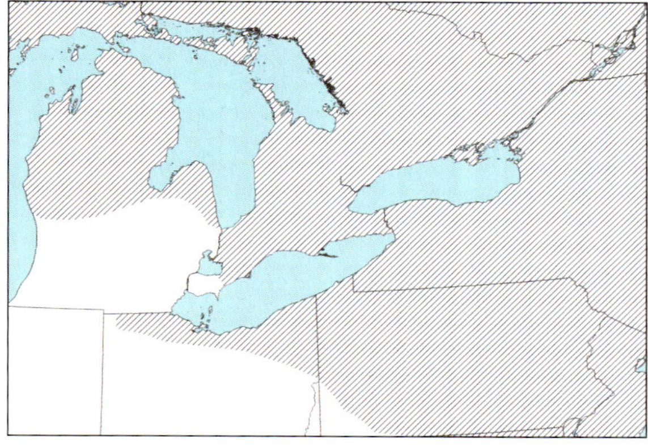

the center of the flower and each one has a patch of yellow and white at the base with purple veins fanning out from it. The upper lip of this sepal curves up like a shoehorn and forms an open tubular shape

Leaves

Seed pods

Split seed pods with seeds

with the bottom lip. The petals are located between the sepals, measure two-thirds the length of the sepals, and are violet-blue with dark purple veins. Flowers are replaced by angular, oblong seed capsules that split open to release its seeds.

IN THE GARDEN Blue Flag Iris steals the show in early summer with its intricate, jewel-toned flowers and its bold, sword-like leaves. The flowers are relished by hummingbirds, but deer and other herbivores rarely touch this plant. In addition, the angular seed pods add excellent winter interest.

SKILL LEVEL Beginner

LIFESPAN Perennial

EXPOSURE Full sun to part shade

SOIL TYPE Prefers clay and mucky soils but will grow in most soils (I have seen them growing in a gravelly road shoulder)

MOISTURE Wet to moist (but will tolerate short periods of drought)

HEIGHT 60–90 cm

SPREAD 60–75 cm

BLOOM PERIOD May, Jun, Jul

COLOR Blue/purple

FRAGRANT 

SHOWY FRUIT 

CUT FLOWER ✓

PESTS Susceptible to a number of insect pests, including the larvae of the Iris Borer Moth (*Macronoctua onusta*), thrips, and aphids; potential disease problems include various rots (rhizome rot, crown rot, bacterial soft rot), leaf spot, and leaf/blossom blight, while aphids can spread mosaic virus

NATURAL HABITAT Marshes, swamps, shorelines, wet meadows, margins of ponds and creeks, sedge meadows, and borders of wetland forests

WILDLIFE VALUE Attracts butterflies and native bees; hummingbirds seek nectar from the flowers

BUTTERFLY LARVA HOST PLANT FOR None

MOTH LARVA HOST PLANT FOR Virginia Ctenucha (*Ctenucha virginica*), Iris Borer Moth (*Macronoctua onusta*), Burdock Borer Moth (*Papaipema cataphracta*), Agreeable Tiger Moth (*Spilosoma congrua*)

USDA HARDINESS ZONES 2–7

PROPAGATION Seeds should be sown when fresh or, if sowing later, should be stored in a cool, moist setting. Most sources say they do not tolerate drying out, though William Cullina claims they do just fine stored dry. They require at least 120 days of cold, moist stratification to germinate and will take two years till they produce flowers. To propagate vegetatively, the roots can be divided in early summer and potted or planted along the water's edge.

ADDITIONAL INFO In smaller water features, consider growing this in large pots submerged to the rim.

SAR STATUS N/A

Liatris cylindracea
Ontario Blazing Star

QUICK GUIDE

PROPAGATION

FULL SUN

MEDIUM · **DRY**

Flowers with a Monarch Butterfly

Whole plant

FAMILY *Asteraceae* (Aster Family)

ALTERNATE COMMON NAMES Barrelhead Gayfeather, Cylindrical Blazing Star, Dwarf Blazing Star, Dwarf Gayfeather, Dwarf Liatris, Few Headed Blazing Star, Ontario Gayfeather, Ontario Liatris, Slender Blazing Star

PLANT DESCRIPTION Ontario Blazing Star has an unbranched, smooth central stem that is usually green but sometimes purple tinged near flower clusters. Stalked basal leaves measure up to 22 cm long and 8 mm wide and are lance shaped, sometimes being widest just past the middle. Stem leaves are stalk-less and found in an alternate pattern, but they may appear to be whorled due to their density. Note that they become smaller as they ascend the stem and are hairless with smooth margins. One to 28 composite,

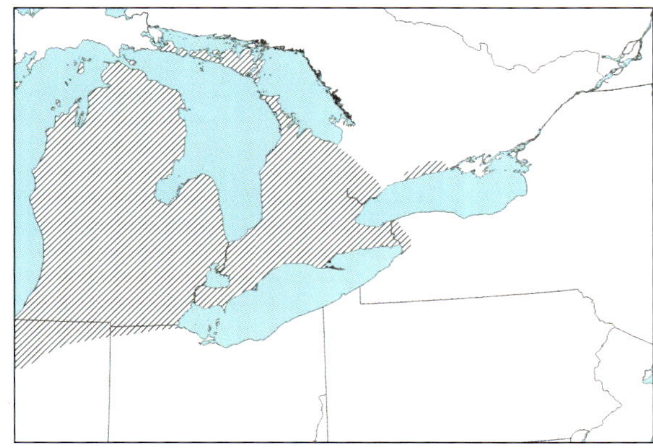

pinkish-purple flower heads are found in a spike-like raceme along the top of the stem. Each flower head is up to 2.5 cm across and made up of 15 to 25 individual flowers. Each flower has five lobes and a divided style

protruding from it. Behind each flower, you will notice five to seven layers of flattened, scale-like bracts. They are oblong to oval in shape with pointed tips. Flowers become dry seeds with tufts of hairs that allow them to be carried by the wind.

Ontario Blazing Star is easily distinguished from other *Liatris* species by its scale-like bracts and narrow flower heads.

IN THE GARDEN Ontario Blazing Star features the gorgeous purple flowers of a *Liatris* but in a more compact form! Being the shortest of the *Liatris* species, it makes a really showy addition to garden borders. Despite its delicate look, this is a very drought-tolerant plant that can also tolerate the poorest of soils.

SKILL LEVEL Beginner

LIFESPAN Perennial

EXPOSURE Full sun

SOIL TYPE Thrives in poor, well-drained soils (intolerant of wet soils in winter)

MOISTURE Dry to medium

HEIGHT 20–30 cm (rarely to 60 cm)

SPREAD 15–30 cm

BLOOM PERIOD Jul (mid), Aug, Sep (early)

COLOR Pink, mauve

FRAGRANT ✖

SHOWY FRUIT ✖

CUT FLOWER ✔

PESTS No serious insect or disease problems

NATURAL HABITAT Dry prairies and savannas, and dry open woods

WILDLIFE VALUE A pollinator magnet, the nectar attracts many butterflies and bees (the Xerces Society reports that it is of special value to bumblebees) and even hummingbirds. Many herbivores eat the leaves and stems, and squirrels and other small rodents will dig up the corms and eat them. In my southwestern Ontario garden, this flower consistently attracts more nectaring Monarch Butterflies (*Danaus plexippus*) than any other plant

BUTTERFLY LARVA HOST PLANT FOR None

MOTH LARVA HOST PLANT FOR Liatris Borer Moth (*Carmenta anthracipennis*), Liatris Flower Moth (*Schinia sanguinea*)

USDA HARDINESS ZONES 4–7

PROPAGATION Direct sow seeds in late fall by pressing them into the surface of the soil. If starting indoors or saving seeds for spring planting, they will require 60 days of cold, moist stratification for effective germination. Mature corms may be divided with care before growth starts in the spring.

ADDITIONAL INFO Ontario Blazing Star won't compete well so keep tall, aggressive plants away. Readily self-seeds in suitable conditions, but first year seedlings have only one linear, grass- or sedge-like leaf and may be inadvertently weeded out.

SAR STATUS ON – S3/(not listed); NY – S1/E; OH – S3/T

Liatris spicata
Dense Blazing Star

PART SHADE | FULL SUN

MOIST

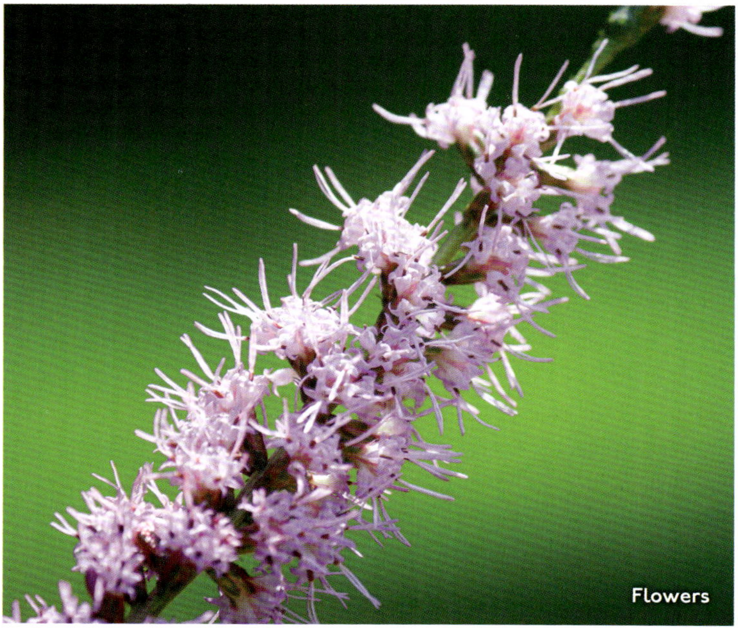

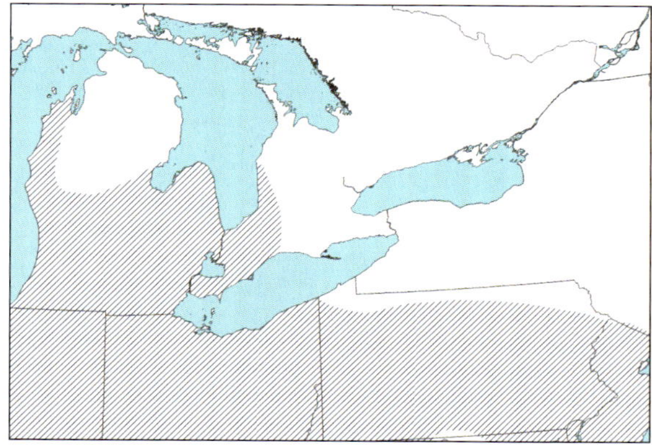

Flowers

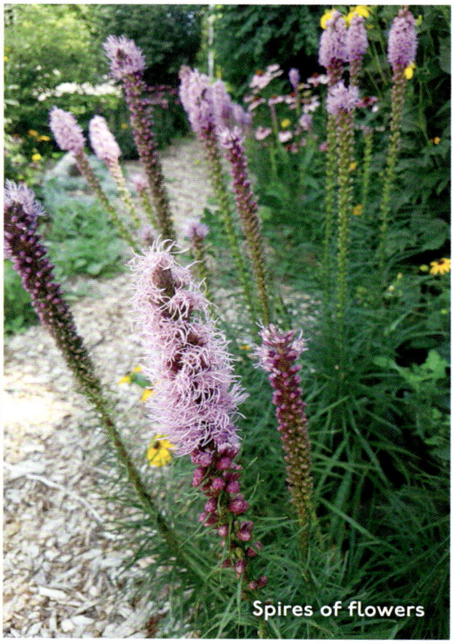

Spires of flowers

FAMILY *Asteraceae* (Aster Family)

ALTERNATE COMMON NAMES Button Snake-root, Button Snakewort, Dense Gayfeather, Dense Liatris, Florist's Gayfeather, Marsh Blazing Star, Marsh Gayfeather, Marsh Liatris, Sessile-headed Blazing Star, Spiked Blazing Star

PLANT DESCRIPTION Dense Blazing Star has an unbranched central stem that is slightly hairy and light green to purplish. Leaves are lance shaped with smooth surfaces and have a prominent central vein. The basal leaves measure up to 25 cm long and 1.2 cm wide, while the alternate leaves are attached directly to the stem and become increasingly smaller as they ascend. Note that they may appear whorled due to their dense sprrangement. Stems terminate with a spire of densely packed purple (rarely white) flower heads that measure about 1.2 cm wide. Each flower head is made up of four to 10 individual flowers, each with five lobes and a long, divided style protruding from it. At the base of each flower

170

Leaves

Whole plant

head, you will notice bracts that are smooth, oval, flattened, and green to purple. Flowers are replaced by seeds with fluffy tufts of hairs that allow them to be carried by the wind.

IN THE GARDEN Dense Blazing Star is the most floriferous of the blazing stars, sending up dramatic spires of purple flowers. It is reliable and tough and has very attractive seed heads.

SKILL LEVEL Beginner

LIFESPAN Perennial

EXPOSURE Full sun to part shade

SOIL TYPE Any average, well-drained soil

MOISTURE Moist

HEIGHT 60–90 cm (occasionally to 150 cm)

SPREAD 30–45 cm

BLOOM PERIOD Jul, Aug, Sep

COLOR Purple

FRAGRANT ❌

SHOWY FRUIT ❌

CUT FLOWER ✅

PESTS No serious insect or disease problems

NATURAL HABITAT Wet meadows and swampy areas

WILDLIFE VALUE Nectar source for hummingbirds and butterflies and seed source for small birds and mammals

BUTTERFLY LARVA HOST PLANT FOR None

MOTH LARVA HOST PLANT FOR Liatris Borer

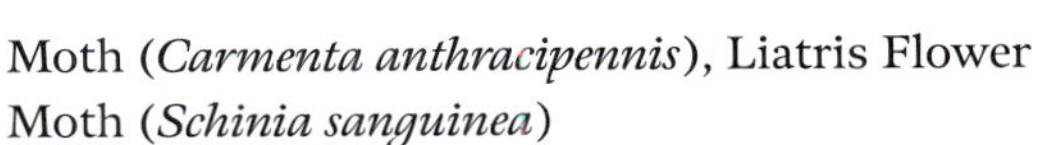
Seed heads

Moth (*Carmenta anthracipennis*), Liatris Flower Moth (*Schinia sanguinea*)

USDA HARDINESS ZONES 3–9

PROPAGATION Fall sow or, if starting in the spring or indoors, seeds should be nicked with a knife or gently rubbed with sandpaper to scarify the hard seed coat, then cold, moist stratified for 60 days. Seeds usually take three to four weeks to germinate. This *Liatris* species can also be propagated by division in the spring.

ADDITIONAL INFO The tall stalks of this plant often need staking without the support of nearby plants. *Liatris spicata* may be short-lived in the average garden, but it readily self-seeds in suitable conditions. First year seedlings look a lot like grass or sedge leaves to the inexperienced eye and may be inadvertently weeded out.

SAR STATUS ON – S2/T

Lilium canadense
Canada Lily

PART SHADE · FULL SUN

MOIST · MEDIUM

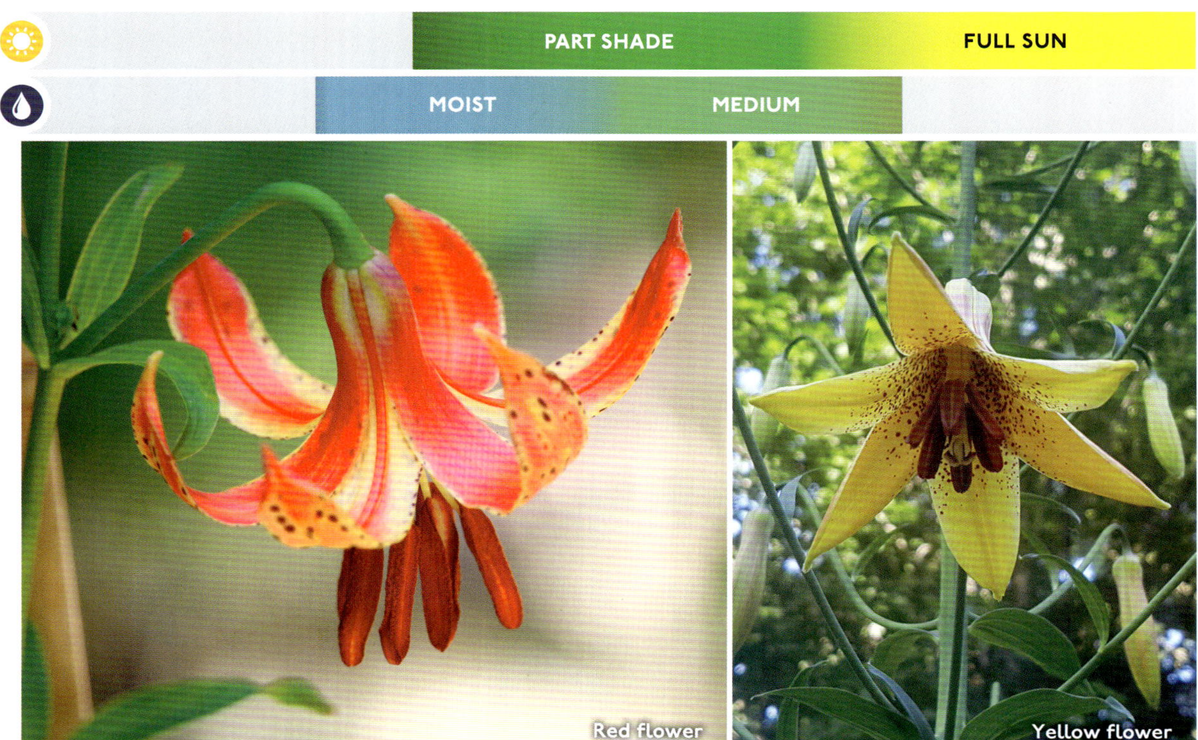

Red flower

Yellow flower

FAMILY *Liliaceae* (Lily Family)

ALTERNATE COMMON NAMES Meadow Lily, Wild Yellow Lily, Yellow Wood Lily

PLANT DESCRIPTION Canada Lily features smooth, light green stems that are unbranched, except at the top where the flowers are found. Leaves are distributed along the stem in whorls of three to eight with some smaller alternate leaves occurring along the upper portion of the stem. Each leaf is up to 15 cm long, 2.5 cm wide, smooth, toothless, and narrowly ovate. Stems terminate with up to 20 nodding, trumpet-shaped flowers borne on long stalks and can range in color from reddish orange to yellow. These flowers are up to 10 cm across and feature six tepals that flare backwards (but not past the base of the flower), six stamens,

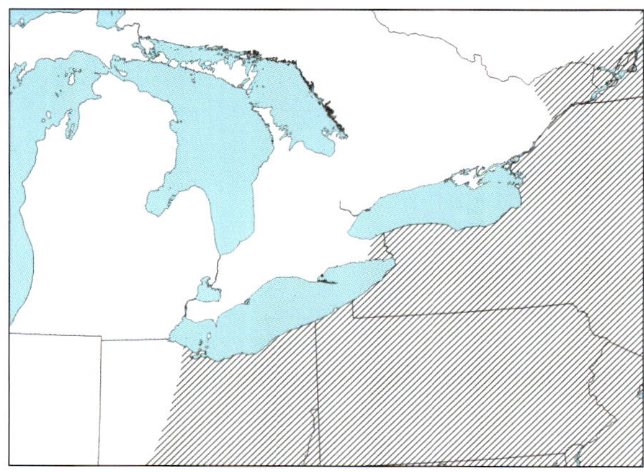

and dark dots on the inside of the tepals. Flowers become oblong, 5 cm long seed capsules that are divided into three cells containing flat seeds.

Leaves

Whole plant

Seed capsule

IN THE GARDEN Canada Lily adorns the summer garden with trumpet-like flowers that hang gracefully from the plant. Besides its blooms, it is valued for its clumping habit and interesting whorled foliage.

SKILL LEVEL Beginner to intermediate

LIFESPAN Perennial

EXPOSURE Full sun to part shade (prefers dappled shade)

SOIL TYPE Rich loamy or slightly sandy soil

MOISTURE Moist to medium

HEIGHT 125–210 cm

SPREAD 15–20 cm

BLOOM PERIOD Jun, Jul, Aug

COLOR Red, orange, or yellow

FRAGRANT ✓

SHOWY FRUIT ✗

CUT FLOWER ✓

PESTS The caterpillars of several moth species feed on the leaves, stems, and corms of Canada Lily, and the introduced Lily Leaf Beetle or Red Lily Beetle (*Lilioceris lilii*) feeds on its leaves

NATURAL HABITAT Wet meadows, moist rich woods, streamsides, and wetlands, and along wet roadsides and railroads

WILDLIFE VALUE The nectar attracts large butterflies, particularly the Great Spangled Fritillary (*Speyeria cybele*) and various swallowtail butterflies. Some bees collect pollen from the flowers, but they are ineffective at cross-pollination because of their small size. A number of mammalian herbivores browse on the foliage, and voles and chipmunks are known to eat the corms

BUTTERFLY LARVA HOST PLANT FOR None

MOTH LARVA HOST PLANT FOR Carrion Flower Moth (*Acrolepiopsis incertella*), Burdock Borer Moth (*Papaipema cataphracta*), Golden Borer Moth (*Papaipema cerina*), Common Borer Moth (*Papaipema nebris*), Sparganothis Leafroller Moth (*Sparganothis sulfureana*)

USDA HARDINESS ZONES 4–8

PROPAGATION Canada Lily seeds must undergo a period of one to two months of warmth, at which time they will swell and become a small bulb. These then need another 60 to 90 days of cold before they begin to sprout. Seedlings typically go dormant by midsummer. Plants grown from seed will take five to six years before they flower. Propagation is easiest from division of the scaly bulb, which can be dug as soon as the plant goes dormant in late summer.

ADDITIONAL INFO Canada Lily is primarily pollinated by the Ruby-throated Hummingbird (*Archilochus colubris*).

SAR STATUS ON – S1/(not listed); IN – S3/T; NY – (not ranked)/EV

Lilium michiganense
Michigan Lily

Flower

Whole plant

FAMILY *Liliaceae* (Lily Family)

ALTERNATE COMMON NAMES Turk's Cap Lily

PLANT DESCRIPTION Michigan Lily features erect, unbranched stems that are light green and smooth. Whorled leaves are attached directly to the stem in groups of three to nine, with each leaf measuring 12.5 cm long and 2.5 cm wide. You may notice some alternate leaves towards the top of the plant. Leaves are elliptical with pointed tips, are smooth to the touch, and have prominent veins. Stems terminate with one to eight nodding flowers borne on long stalks. Each flower is up to 7.5 cm across. Flowers are characterized by six orange tepals with dark purple dots. These tepals are strongly recurved to the point that their tips are often near the base of the flower. Furthermore, six stamens and one style strongly protrude from the center of the flower. Flowers turn into oblong seeds pods that measure just under 4 cm long and have three compartments containing flattened seeds.

IN THE GARDEN Michigan Lily is truly a delight to have in the garden! It maintains a neat clumping habit and puts on a captivating display of nodding, orange flowers.

SKILL LEVEL Beginner

LIFESPAN Perennial

EXPOSURE Full sun to part shade

SOIL TYPE Deep, fertile, cool soil, sandy loam (mulch helps keep the root zone cool)

MOISTURE Moist (can withstand some drought once established)

HEIGHT 60–180 cm

SPREAD 30–60 cm

BLOOM PERIOD Jun (mid), Jul

COLOR Orange

FRAGRANT ✗

SHOWY FRUIT ✗

CUT FLOWER ✓

PESTS No serious insect or disease problems

NATURAL HABITAT Wet meadows, fens, swamps, and moist savannas or open woods

WILDLIFE VALUE The large showy flowers attract hummingbirds and the larger day-flying insects, such as sphinx moths, hummingbird moths (*Hemaris* spp.), and the larger butterflies, which all act as pollinators; chipmunks and voles sometimes eat the bulbs

BUTTERFLY LARVA HOST PLANT FOR None

MOTH LARVA HOST PLANT FOR Carrion Flower Moth (*Acrolepiopsis incertella*), Burdock Borer Moth (*Papaipema cataphracta*), Golden Borer Moth (*Papaipema cerina*), Sparganothis Leafroller Moth (*Sparganothis sulfureana*)

USDA HARDINESS ZONES 4–8

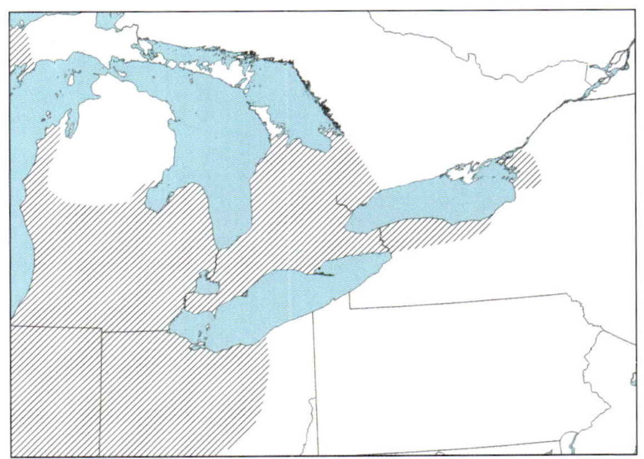

PROPAGATION Keep seeds moist as drying decreases viability. Michigan Lily seeds must undergo a period of one to two months of moist warmth, at which time they will swell and become a small bulb. These then need 60 to 90 days of moist cold before they begin to sprout. Otherwise, plants sown in the spring will take a year to sprout. Plants grown from seed will take five to six years before they flower. Propagation is easiest from division of the bulb, which can be dug as soon as the plant goes dormant in late summer.

ADDITIONAL INFO If given too much water and fertilized, Michigan Lily tends to flop over. It can produce dense stands under cultivation, though these tend not to persist for very long for reasons that are unclear.

SAR STATUS NY – S1/E

Lilium philadelphicum
Wood Lily

Flower

Whole plant

FAMILY *Liliaceae* (Lily Family)

ALTERNATE COMMON NAMES Prairie Lily, Red Lily, Western Orange-cup Lily, Western Red Lily

PLANT DESCRIPTION Wood Lily features smooth, erect stems that range in color from light green to reddish to light yellow. Leaves are attached directly to the stem in an alternate pattern with a single whorl of three to six leaves at the top of the stem. Each leaf is up to 10 cm long, 2.5 cm wide, elliptical with a pointed tip, smooth, toothless, and strongly ascending (pointing upwards). Stems terminate with one to three flowers (rarely four to five) that measure about 6 cm wide and feature six ascending sepals, six stamens, and a red style. You will notice that the base of each sepal abruptly narrows and is yellow with dark spots. Flowers are replaced by oblong seed capsules up to 7 cm long. These capsules have three compartments that contain flat seeds.

IN THE GARDEN The stunning flowers of Wood Lily add fiery red/orange accents to the summer garden, while its whorled leaves add an interesting texture. It has a very elegant look overall and maintains a clumping habit. Pair it with plants of similar height as it does not like competition. It's slow to establish but well worth the wait.

SKILL LEVEL Beginner to intermediate

LIFESPAN Perennial

EXPOSURE Full sun to part shade

SOIL TYPE Well-drained, humus-rich soils

MOISTURE Dry

Leaves   Seed capsule

HEIGHT 30–90 cm (occasionally to 120 cm)

SPREAD 18–25 cm

BLOOM PERIOD Jun, Jul, Aug

COLOR Reddish orange

FRAGRANT ❌

SHOWY FRUIT ❌

CUT FLOWER ✅

PESTS No serious insect or disease problems

NATURAL HABITAT Prairies and deciduous forest openings

WILDLIFE VALUE The flowers are cross-pollinated primarily by large butterflies, including swallow-tail butterflies (*Papilionidae*), Monarch Butterflies (*Danaus plexippus*), and Great Spangled Fritillaries (*Speyeria cybele*), all of which seek nectar; other visitors include the Ruby-throated Hummingbird (*Archilochus colubris*), hummingbird moths (*Hemaris* spp.), and sweat bees, which seek pollen. It is the host plant for the caterpillars of several moth species. Squirrels and other rodents will eat the bulbs

BUTTERFLY LARVA HOST PLANT FOR None

MOTH LARVA HOST PLANT FOR Burdock Borer Moth (*Papaipema cataphracta*), Golden Borer Moth (*Papaipema cerina*), Stalk Borer Moth (*Papaipema nebris*), Sparganothis Leafroller Moth (*Sparganothis sulfureana*)

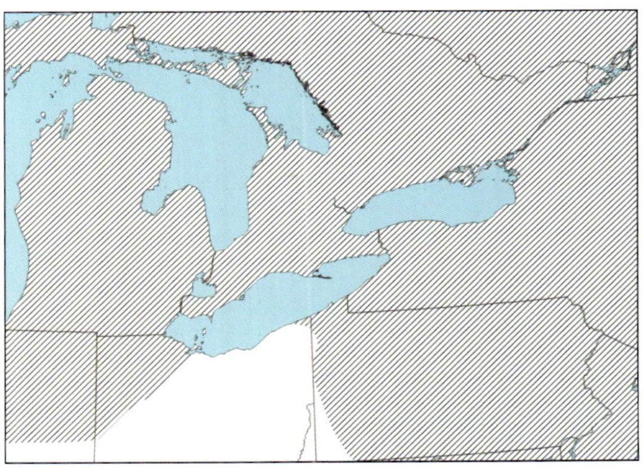

USDA HARDINESS ZONES 2–7

PROPAGATION Seeds need 60 days of cold, moist stratification, though some sources suggest they also need a warm, moist period first (as with *Lilium canadense* or *L. michiganense*). Wood Lilies grown from seed are very slow growing and will take five to seven years to bloom. Propagation is easiest from division of the scaly bulb, which can be dug as soon as the plant goes dormant in late summer.

ADDITIONAL INFO Regional varieties of this species exist, therefore it is important to use a reputable supplier as close to home as possible. Cross-pollination is necessary for Wood Lilies to produce seeds, so plant several bulbs if seeds are desired.

SAR STATUS IN – SU/WL; NY – (not ranked)/EV; OH – S2/E

Lobelia cardinalis
Cardinal Flower

Flowers

Flowers with a hummingbird visitor

FAMILY *Campanulaceae* (Bellflower Family)

ALTERNATE COMMON NAMES Indian Pink

PLANT DESCRIPTION Cardinal Flower has an unbranched central stem that is light green and variably hairy. Leaves are attached to the stem in an alternate pattern and are up to 15 cm long and 4 cm wide but usually only get to half this size. The lower leaves have short stalks while the upper leaves are stalkless. Each leaf is coarsely toothed, sharply pointed, and usually hairless. Stems terminate with spike-like clusters of tubular, ascending red flowers, each measuring up to 4 cm long and 2.5 cm wide. The upper lip of each flower has two lobes that spread out sideways while the lower lip is divided into three lobes. A red style with a hooked tip rises above the upper lobes. Flowers turn into small capsules containing many tiny seeds.

IN THE GARDEN Cardinal Flower is nothing short of a showstopper! In midsummer, it sends up magnificent spikes of scarlet red flowers that add a strong vertical presence to the landscape. Fortunately, herbivores tend to avoid this plant.

SKILL LEVEL Beginner to intermediate

LIFESPAN Short-lived perennial

EXPOSURE Full sun to full shade (but does best in part shade)

SOIL TYPE Clay to sandy, limestone-based soil and humus-rich soil

MOISTURE Moist to wet (needs constant moisture to thrive)

HEIGHT 30–120 cm

SPREAD 30–60 cm

BLOOM PERIOD Jul, Aug, Sep

COLOR Red

FRAGRANT ❌

SHOWY FRUIT ❌

CUT FLOWER ✅

PESTS No serious insect or disease problems, though snails and slugs may munch on the leaves

NATURAL HABITAT Wet ditches and other low damp areas, seasonally inundated depressions in open woodlands, and along streambanks

WILDLIFE VALUE An important nectar source for hummingbirds and swallowtail butterflies (*Papilionidae*)

BUTTERFLY LARVA HOST PLANT FOR None

MOTH LARVA HOST PLANT FOR Pink-washed Looper Moth (*Enigmogramma basigera*)

USDA HARDINESS ZONES 3–9

PROPAGATION Surface sow as the very tiny seeds need light to break dormancy. Seeds require 60 days of cold, moist stratification to germinate if starting indoors or spring sowing. Surface sow in the fall. Basal offsets may be separated to start new plants. Plants may also be started by taking stem cuttings (be sure to include one or two nodes), but to ensure the plants have developed a good basal

rosette, the earlier these cuttings are taken in the season, the better. New plants may also be propagated by layering; in midsummer, carefully bend the plant over and pin it to the ground, lightly covering the plant with soil. New roots will grow along the stem and new shoots will emerge, which can be transplanted in the fall.

ADDITIONAL INFO Though this short-lived perennial typically only lives for two to three years, you can extend its life in your garden by dividing it or moving it every year or two. This plant is at its finest when growing with minimal competition in forested wetlands. Note that commercial garden centers often sell cultivars of this plant that may or may not be as valuable to wildlife as the true species.

SAR STATUS NY – (not ranked)/EV

Lobelia siphilitica
Blue Lobelia

Flowers

Flower stalk

FAMILY *Campanulaceae* (Bellflower Family)

ALTERNATE COMMON NAMES Blue Cardinal Flower, Giant Blue Lobelia, Great Blue Lobelia

PLANT DESCRIPTION Blue Lobelia has erect, unbranched stems that are light green and smooth to sparsely hairy. Alternate leaves measure up to 15 cm long and 5 cm wide and are attached to the stem with no leaf stalk. Leaves are elliptical to lance shaped, irregularly toothed, and smooth to sparsely hairy and have pointed tips. Stems terminate with spike-like clusters of tubular flowers that are densely packed together. Each flower measures up to 4 cm long and is angled upwards. Flowers are characterized by three pointed, descending lower lobes and two small upper lobes that can be upright but are usually curved back. A curved style

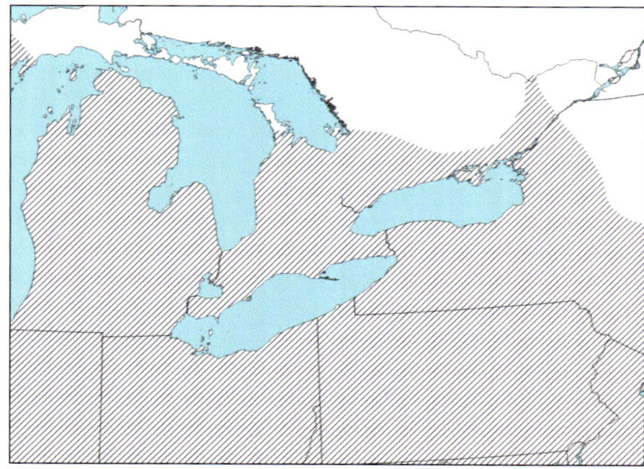

protrudes from between the two upper lobes. In addition, you will notice a small white patch at the base of the lower lobes. Flowers turn into small,

Leaves

Whole plant

Seed capsules

two-chambered seed capsules that contain numerous tiny seeds.

IN THE GARDEN Blue Lobelia is valued by gardeners for its rich, jewel-toned blue flowers — a color that can be hard to come by in the garden. The spiked flower clusters add a great vertical presence to the garden, and it is adaptable to a variety of light conditions. Herbivores tend to ignore it.

SKILL LEVEL Beginner to intermediate

LIFESPAN Short-lived perennial (though somewhat longer lived than Cardinal Flower)

EXPOSURE Part shade (full sun is tolerated if the soil is kept constantly moist)

SOIL TYPE Fertile and loamy

MOISTURE Wet to moist (not at all drought tolerant; keep soil moist, watering if necessary in average soil conditions; withstands occasional flooding)

HEIGHT 60–120 cm

SPREAD 30–60 cm

BLOOM PERIOD Jul, Aug, Sep

COLOR Blue

FRAGRANT ❌

SHOWY FRUIT ❌

CUT FLOWER ✅

PESTS No serious insect or disease problems, though snails and slugs may munch on the foliage

NATURAL HABITAT Wetlands and edges of streams

WILDLIFE VALUE The nectar attracts butterflies and native bees and occasionally hummingbirds

BUTTERFLY LARVA HOST PLANT FOR None

MOTH LARVA HOST PLANT FOR Pink-washed Looper Moth (*Enigmogramma basigera*)

USDA HARDINESS ZONES 4–8

PROPAGATION Surface sow in the fall (seeds need light to break dormancy), or cold, moist stratify for 60 days if starting indoors or in the spring. Seeds typically take about two weeks to germinate. Blue Lobelia will produce offsets around the base that will generate their own roots. These can be removed with a sharp knife in the spring or fall, being careful to retain their roots, and transplanted. These small offsets are delicate, so care should be taken not to bury them under thick mulch. This plant can also be propagated by making 10 to 15 cm long cuttings from actively growing stems, removing the bottom leaves, and dipping the stems in rooting hormone before placing in pots of soil. Once you see new leaves forming, in two to three weeks, you will know that it has developed new roots.

ADDITIONAL INFO Pinch back the plants to make them bushier.

SAR STATUS N/A

Lupinus perennis
Wild Lupine

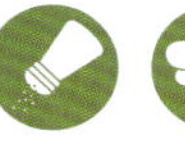

FAMILY *Fabaceae* (Pea Family)

ALTERNATE COMMON NAMES Indian Beet, Old Man's Bonnet, Perennial Lupine, Sundial Lupine, Wild Blue Lupine

PLANT DESCRIPTION Wild Lupine has light-green-to-reddish stems that are angular and smooth to hairy. Leaves alternate along the stem on long leaf stalks and are compound into seven to 11 hairy leaflets that all fan out from a central point. Each leaflet measures 6 cm long and is oblanceolate (lance shaped with the pointed end at the base) with toothless margins, a rounded tip, and a sharp point at the end. Stems terminate with a spike-like flower cluster that measures up to 25 cm long. Each flower is pea shaped, about 2.5 cm long, and borne on short stalks. Flowers are replaced by pea-like seed pods that each contain two or more hard, dry seeds. As the pods dry, they catapult the seeds away from the mother plant. Spent seed pods are twisted.

Wild Lupine is similar to the non-native Large-leaved Lupine (*Lupinus polyphyllus*) that is commonly found in gardens across our region. They can be differentiated by the fact that *L. polyphyllus* is a much larger plant overall, having nine to 17 leaflets and much longer flower spikes. Both *L. perennis* and *L. polyphyllus* will readily hybridize with each other.

IN THE GARDEN The floriferous blue/purple flower spikes of Wild Lupine rise above palm-shaped leaves and make for a classy addition to the late-spring garden. Its ability to fix atmospheric nitrogen allows it to thrive in poor soils while benefiting plants around it. Herbivores usually ignore this plant.

SKILL LEVEL Intermediate

LIFESPAN Short-lived perennial (spring ephemeral)

EXPOSURE Full sun to part shade

SOIL TYPE Sandy — requires good drainage, but is very adaptable

MOISTURE Dry to moist

HEIGHT 30–60 cm

SPREAD 45 cm

BLOOM PERIOD May, Jun, Jul

COLOR Blue, purple

FRAGRANT ✓

SHOWY FRUIT ✗

CUT FLOWER ✓

PESTS Slugs and snails can be problematic, as can powdery mildew and aphids

NATURAL HABITAT Sand prairies, openings in sandy woodlands, sandy savannas, edges of sandy woodlands, stabilized sand dunes, and powerline clearances in sandy areas (in other words, if your soil isn't sandy, you're not likely to grow lupines successfully)

WILDLIFE VALUE Of special value to bumblebees; birds and small mammals will feed on the seeds, and in my southwestern Ontario garden, it is often nibbled on by rabbits

BUTTERFLY LARVA HOST PLANT FOR Frosted Elfin Butterfly (*Callophrys irus*) — listed as extirpated in ON and imperiled throughout the rest of the southern Great Lakes region — Wild Indigo Duskywing (*Erynnis baptisiae*), Persius Duskywing (*Erynnis persius*), Silvery Blue (*Glaucophsyche lygdamus*), Karner Blue Butterfly (*Lycaeides melissa* sub. *samuelis*) — classified as extirpated in ON and PA, critically imperiled in NY, OH, and IN, and imperiled in MI

MOTH LARVA HOST PLANT FOR Clover Looper Moth (*Caenurgina crassiuscula*), Phyllira Tiger Moth (*Grammia phyllira*), Placentia Tiger Moth (*Grammia placentia*), Bella Moth (*Utetheisa bella*), Sweet Clover Root Borer (*Walshia miscecolorella*)

USDA HARDINESS ZONES 3–8

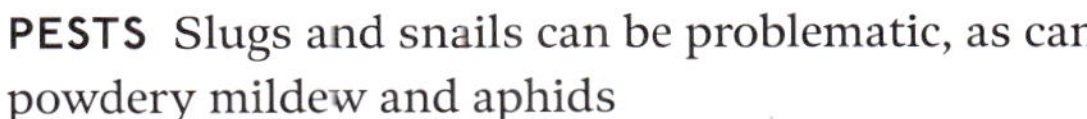

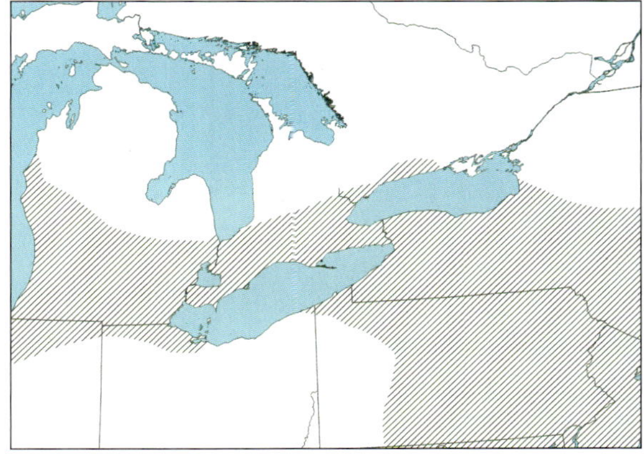

PROPAGATION Scarify seeds by rubbing between two sheets of sandpaper to scrape the seed coat, then plant 3 mm deep in late fall, or scarify and refrigerate for 10 days if starting seeds indoors or in the spring. Another pretreatment strategy involves pouring hot water (80°C/180°F) over the seeds and letting them soak for a couple of days until you notice them starting to swell. The seeds are best treated with an appropriate rhizobium inoculant before sowing, especially in poorer soils. Plants have a long taproot and do not transplant well; for this reason, they should be moved from pots to the garden as soon as possible in the spring, or sown directly into the garden.

ADDITIONAL INFO This plant is sometimes considered a spring ephemeral; it will grow and bloom in the spring and then go dormant by midsummer, so plan for summer and fall-blooming plants nearby in the garden to fill in where the Lupine is asleep.

SAR STATUS ON – S2/(not listed); NY – S3/R; OH – S3/T(P); PA – S3/R

Maianthemum canadense
Canada Mayflower

	FULL SHADE	PART SHADE	

	WET	MOIST	

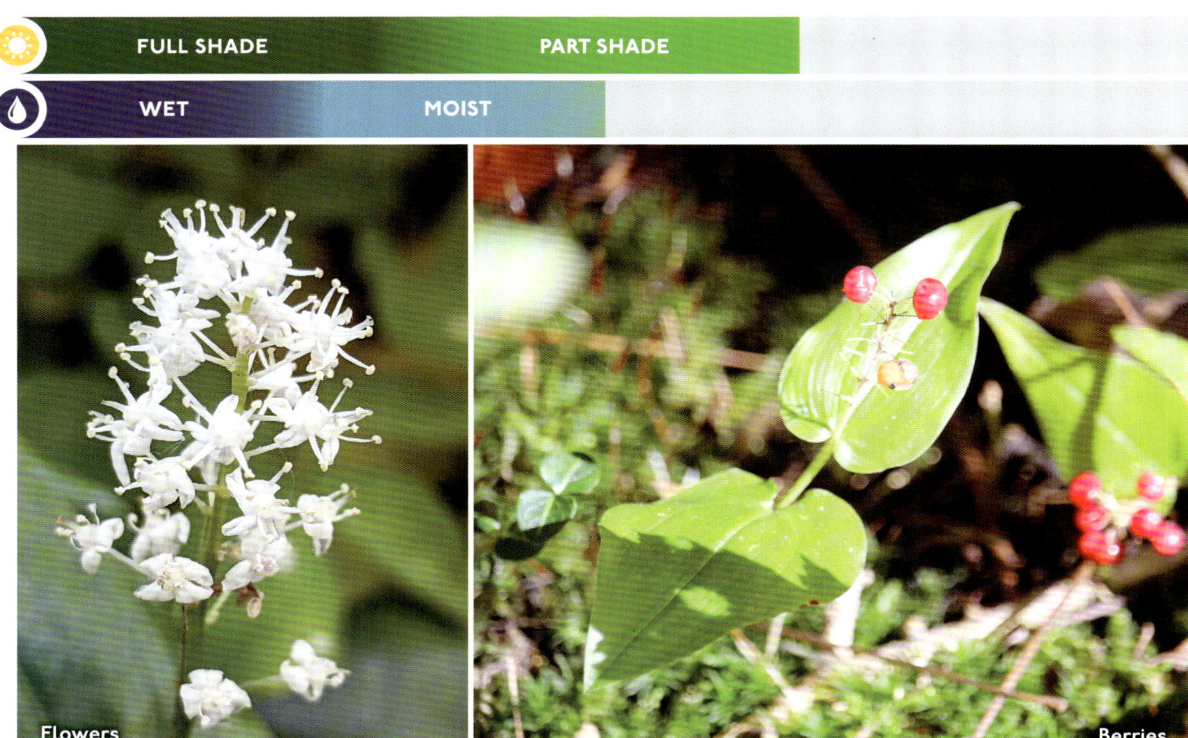

Flowers

Berries

FAMILY *Liliaceae* (Lily Family)

ALTERNATE COMMON NAMES Canadian May-lily, False Lily-of-the-valley, Scurvy Berries, Two-leaved Solomon's Seal, Western False Lily-of-the-valley, Wild Lily-of-the-valley

PLANT DESCRIPTION Canada Mayflower appears in two forms. The first is a non-flowering form that consists of a single basal leaf on a 5 cm long stalk. The second is a flowering form where plants have a zigzagged, central stem with two to three (usually two) alternate leaves. The stem is green and smooth to sparsely hairy. All leaves are heart shaped with pointed tips, toothless, and hairless. Stem leaves clasp the stem with no stalk. The largest leaves measure up to 7.5 cm long and 5 cm wide. Star-like flowers are found in a densely packed, 6 cm long cluster at the tip of the stem. Each flower is just under 1 cm across and features four white petals and four protruding stamens with creamy tips. Flowers are replaced by globular, red berries that are speckled with pale red. Each berry usually contains two seeds.

IN THE GARDEN Canada Mayflower lightens up shade gardens in the spring with its fragrant, star-like flower clusters. It makes a great ground-cover with its bold, glossy leaves, and in the right conditions it will slowly spread by rhizomes to form colonies. Herbivores tend to ignore this plant.

SKILL LEVEL Beginner to intermediate

LIFESPAN Perennial

EXPOSURE Full shade to part (dappled) shade

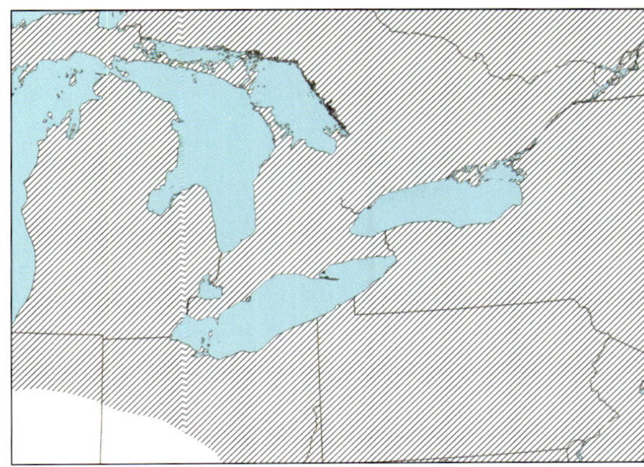

SOIL TYPE Acidic soil containing peat, sand, or rocky material (e.g., sandstone)

MOISTURE Moist to wet

HEIGHT 10–25 cm

SPREAD 15 cm

BLOOM PERIOD May

COLOR White

FRAGRANT ✓

SHOWY FRUIT ✓

CUT FLOWER ✗

PESTS The introduced Lily Leaf Beetle or Red Lily Beetle (*Lilioceris lilii*) is known to feed on the foliage

NATURAL HABITAT Deciduous and mixed woods, floodplains, and the edges of bogs

WILDLIFE VALUE The flowers are cross-pollinated by small bees, flies, and beetles, and the red berries are eaten by birds and small mammals; Canada Mayflower does not produce nectar

BUTTERFLY LARVA HOST PLANT FOR None

MOTH LARVA HOST PLANT FOR None

USDA HARDINESS ZONES 3–6

PROPAGATION Seeds are hydrophilic and cannot be allowed to dry out — best results when sown fresh. If starting indoors, at least 90 days of cold, moist stratification is needed, which typically will result in about 50 percent germination. This plant is readily propagated by dividing the rhizome in the fall, after the leaves have yellowed.

Leaves

Whole plants

ADDITIONAL INFO This plant is a good choice for acidic soils, but it will not compete well with taller species.

SAR STATUS N/A

Maianthemum racemosum
False Solomon's Seal

Flowers

Berries

FAMILY *Liliaceae* (Lily Family)

ALTERNATE COMMON NAMES False Spikenard, Feathery False Lily-of-the-valley, Feathery False Solomon's Seal, Large False Solomon's Seal, Smilacina, Solomon's Plume, Treacleberry, Wild Spikenard

PLANT DESCRIPTION False Solomon's Seal features arching, unbranched stems that are light green and slightly hairy. Sometimes they will zigzag between leaves. Leaves are attached to the stem in an alternate pattern with little-to-no leaf stem. They each measure up to 15 cm long and 7.5 cm wide, are oval with a pointed tip and toothless with wavy edges, and have a smooth upper surface. Stems terminate with 10 cm long, plume-like clusters of 20 to 80 star-like flowers. Each flower is about 0.4 cm across and is characterized by six tepals and six protruding stamens with creamy-yellow tips. Flowers give way to globular berries about 0.6 cm across that turn bright red when ripe.

IN THE GARDEN False Solomon's Seal features graceful, arching stems topped by plumes of star-like flowers. Bold foliage and bright red berries, sometimes striped brown or purple, mean that this plant extends seasonal interest well past its blooms. It will spread slowly by rhizomes over time.

SKILL LEVEL Beginner

LIFESPAN Perennial

EXPOSURE Full shade to part shade (can tolerate full sun with sufficient moisture, and its blooms will actually grow larger with more light)

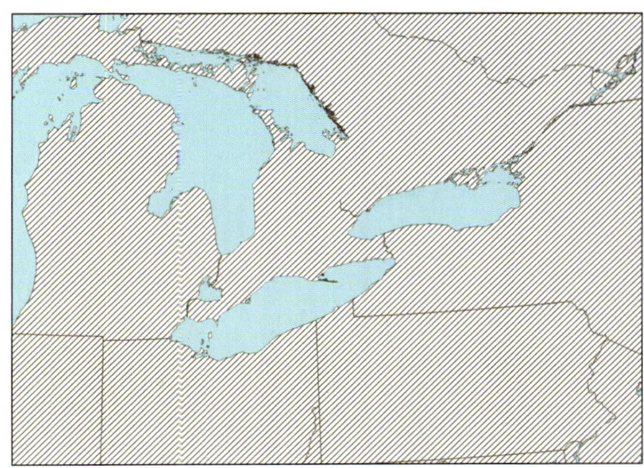

SOIL TYPE Fertile, humus-rich soil, though sandy and rocky soils are also tolerated

MOISTURE Moist to medium

HEIGHT 30–80 cm

SPREAD 60 cm

BLOOM PERIOD May, Jun

COLOR White

FRAGRANT ✓

SHOWY FRUIT ✓

CUT FLOWER ✓

PESTS No serious insect or disease problems

NATURAL HABITAT Wooded areas and clearings

WILDLIFE VALUE Flowers are pollinated by small bees, flies, and beetles, and the berries are eaten occasionally by woodland birds

BUTTERFLY LARVA HOST PLANT FOR None

MOTH LARVA HOST PLANT FOR None

USDA HARDINESS ZONE 3–8

PROPAGATION Seeds are hydrophilic and should not be allowed to dry out. If starting seeds indoors, cold, moist stratification for 90 to 120 days is needed to break dormancy. Often a second period of cold stratification is also needed. Divide rhizomes in early spring or fall but note that roots do not like to be disturbed, particularly before the plant becomes established.

ADDITIONAL INFO False Solomon's Seal is similar to Starry False Solomon's Seal (*Maianthemum*

Leaves

Whole plant

stellatum). You can tell them apart by the fact that *M. stellatum* has fewer, larger flowers, narrower leaves that clasp the stem, and a more upright form.

SAR STATUS N/A

Maianthemum stellatum
Starry False Solomon's Seal

QUICK GUIDE

PROPAGATION

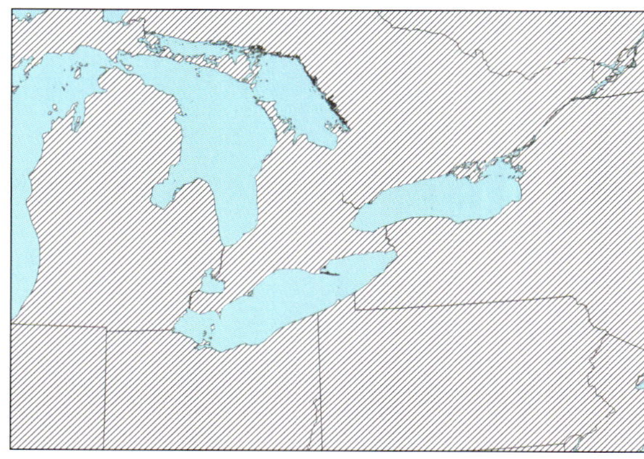

FAMILY *Liliaceae* (Lily Family)

ALTERNATE COMMON NAMES Small False Solomon's Seal, Star Flower, Star-flowered Solomon's Seal, Starry False Lily-of-the-valley, Starry Solomon's Plume, Wild Lily-of-the-valley

PLANT DESCRIPTION Starry False Solomon's Seal features smooth stems that slightly zigzag between the leaves. You may notice that the stem leans slightly to one side rather than being fully upright. Alternate leaves clasp the stem and are elliptical in shape with a pointed tip. Each leaf is toothless, finely hairy underneath, and slightly folded lengthwise and measures up to 15 cm long and 5 cm wide. At the top of the stem, you will find a 2.5 to 10 cm long flower cluster made up of about 20 small, white flowers. Each flower has six tepals, six protruding stamens and is just under 1 cm wide. Flowers turn into globular berries that start out green with purple stripes then turn to solid reddish purple at maturity. Each berry is about 0.6 cm across.

IN THE GARDEN Starry False Solomon's Seal puts on a show in the spring with its plumes of white, star-like flowers. It works well as a groundcover, readily forming small colonies, and looks especially striking in the fall when the leaves fade to hues of yellow and the berries ripen to a deep red. This is a versatile plant and will adapt to a variety of growing conditions.

SKILL LEVEL Beginner

LIFESPAN Perennial

EXPOSURE Full shade to full sun (the more sun, the more moisture required)

SOIL TYPE Often found naturally in sandy soil, but it's not very particular about soil texture

MOISTURE Moist to dry

HEIGHT 20–50 cm

SPREAD 20–45 cm

BLOOM PERIOD May, Jun

COLOR White

FRAGRANT ✓

SHOWY FRUIT ✓

CUT FLOWER ✓

PESTS No serious insect or disease problems

NATURAL HABITAT Moist woods, thickets, meadows, and savannas

WILDLIFE VALUE The flowers attract sweat bees (including green metallic bees), flower flies, and some parasitic flies (e.g., *Tachinid* flies), all of which seek nectar or pollen; the berries are eaten by woodland songbirds and mice

BUTTERFLY LARVA HOST PLANT FOR None

MOTH LARVA HOST PLANT FOR None

USDA HARDINESS ZONES 3–8

PROPAGATION The seeds of Starry False Solomon's Seal, like the other *Maianthemums*, are hydrophilic and should not be allowed to dry out before planting. The seeds require two separate periods of cold, moist stratification separated by a warm period in order to germinate. Each period should be at least 90 days. Plants may also be divided in the fall or early spring, though the roots do not like to be disturbed and flowering will be affected for a year or two when transplanted.

ADDITIONAL INFO A good choice as a groundcover under walnut trees, this plant will produce a dense mat of roots and can be an aggressive spreader in the right conditions. Although I've never seen it used as such, I have to think it would be excellent for bank stabilization.

SAR STATUS N/A

Mertensia virginica
Virginia Bluebells

FAMILY *Boraginaceae* (Borage Family)

ALTERNATE COMMON NAMES Eastern Bluebells, Lungwort, Oysterleaf, Roanoke Bells, Virginia Cowslip

PLANT DESCRIPTION Virginia Bluebells have both basal and alternate leaves. Basal leaves are borne on long stalks and are the largest leaves at 20 cm long and almost 13 cm wide. Alternate leaves are found along a smooth, light green stem and measure up to 17 cm long and 7.5 cm wide. Leaves are oval to egg shaped, toothless, and hairless and usually taper to winged leaf stalks. You will notice that the leaves have a soft, floppy texture. The stems terminate with clusters of nodding, bell-shaped flowers with long, tubular throats. Each flower is about 3 cm long and has five barely noticeable

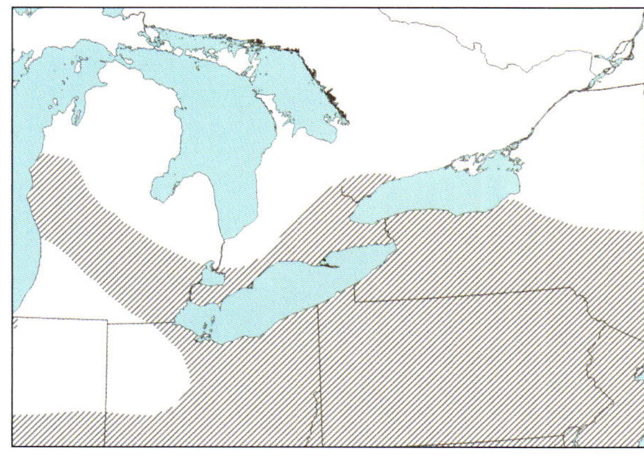

shallow lobes. The flower buds start out pinkish purple then fade to blue as the flower opens. Flowers mature to four dark brown, dry nutlets with a wrinkled texture to them.

Leaves

Whole plant

Seed pods

IN THE GARDEN Virginia Bluebells add a dynamic, ephemeral beauty to the spring garden. Its broad leaves first emerge with a deep purple hue then fade to green, while its flower buds start out with pastel hues of pink and purple then develop into vibrant blue flowers. This can be a very versatile choice in the garden if interplanted with species that will fill in the gaps it leaves behind when it goes dormant in June.

SKILL LEVEL Beginner

LIFESPAN Perennial (spring ephemeral)

EXPOSURE Full shade to part shade (tolerates morning sun)

SOIL TYPE Most soils, but does best in rich, well-drained soils

MOISTURE Moist to medium

HEIGHT 60 cm

SPREAD 30–45 cm

BLOOM PERIOD May

COLOR Starts out pink, matures to blue

FRAGRANT 

SHOWY FRUIT 

CUT FLOWER ❌

PESTS No serious insect or disease problems

NATURAL HABITAT Moist woodlands

WILDLIFE VALUE Butterflies and long-tongued bees are the most common pollinators, and White-tailed Deer (*Odocoileus virginianus*) occasionally browse the foliage

BUTTERFLY LARVA HOST PLANT FOR None

MOTH LARVA HOST PLANT FOR None

USDA HARDINESS ZONES 3–8

PROPAGATION Keep seeds moist and cool if not sowing directly when ripe. If starting indoors or in the spring, cold, moist stratify for 60 days. Plants may be divided in early fall, though once the plant is established it doesn't like to be disturbed. Taking root cuttings once the plant has gone dormant in the fall is also an option. (William Cullina suggests allowing the root segments to air dry for a week to form a callus in order to prevent root rot.)

ADDITIONAL INFO A great companion plant for ferns, which also like the dappled shade and moist soils preferred by Virginia Bluebells, and the ferns will nicely fill in the spaces when these spring ephemerals go dormant by the end of June.

SAR STATUS ON – S3/(not listed); MI – S1/E; NY – (not ranked)/EV

Mimulus ringens
Monkey Flower

  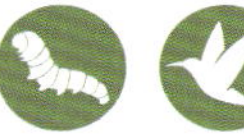

Flower

Whole plant

FAMILY *Phrymaceae* (Lopseed Family)

ALTERNATE COMMON NAMES Allegheny Monkeyflower, Sessile Monkeyflower, Square Stemmed Monkeyflower

PLANT DESCRIPTION Monkey Flower has four-sided, light green stems that are usually branched. Opposite leaves clasp the stem, each measuring up to 10 cm long and 2.5 cm wide. Leaves are lanceolate in shape with finely toothed margins, pointed tips, and rounded bases. Flowers develop on long stalks from the leaf axils (where the leaf meets the stem) and each measure up to 2.5 cm long. Each flower has two upright upper petals, while the lower lip has three rounded petals with the middle one arching into the mouth of the flower. You will notice two pale yellow marks at the base of the middle

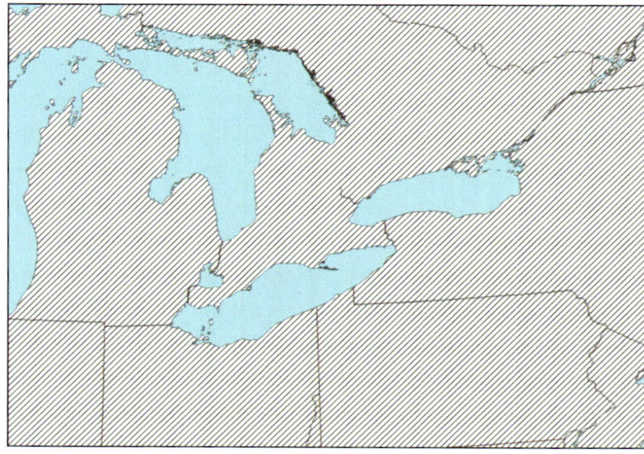

petal. All five petals fuse at their bases to form a tube that is connected to a 2.5 cm long whorl of sepals with five long teeth. Flowers mature into rounded seed capsules that contain many very tiny seeds.

Flower

Leaves and seed capsules

IN THE GARDEN Monkey Flower adds a wonderful, refined look to wet sites and will spread slowly by rhizomes to take on a bushy look once mature. Its snapdragon-like flowers have a long bloom time, which is great for bees and gardeners alike. The dried seed heads provide great textural interest over the winter months.

SKILL LEVEL Intermediate

LIFESPAN Perennial

EXPOSURE Full sun to part shade

SOIL TYPE Rich loamy soil

MOISTURE Wet to moist

HEIGHT 30–90 cm

SPREAD 20–30 cm

BLOOM PERIOD Jun, Jul, Aug, Sep

COLOR Violet

FRAGRANT ✖

SHOWY FRUIT ✖

CUT FLOWER ✖

PESTS No serious insect or disease problems

NATURAL HABITAT Wet, often muddy areas such as shorelines, swamps, drainage ditches, and wet meadows and is typically found in areas that are prone to occasional flooding or standing water

WILDLIFE VALUE Bumblebees are one of the few insects that are strong enough to force their way into the partially closed throat of the corolla as they seek nectar

BUTTERFLY LARVA HOST PLANT FOR Baltimore Checkerspot (*Euphydryas phaeton*), Common Buckeye (*Junonia coenia*)

MOTH LARVA HOST PLANT FOR Chalcedony Midget (*Elaphria chalcedona*)

USDA HARDINESS ZONES 3–8 (values range from 2, 3, or 4 at the low end to 7, 8, or 9 at the high end, depending on the source)

PROPAGATION Direct sow in late fall or cold, moist stratify for 60 days if starting indoors. The seeds need light to germinate so simply press the seed into the surface of the soil and do not cover. Be patient, as the seeds are slow to germinate. New plants may also be started from stem cuttings.

ADDITIONAL INFO Because bumblebee pollinators of Monkey Flower favor Purple Loosestrife (*Lythrum salicaria*), it has been shown that where the two plants coexist, Monkey Flower populations decline. Monkey Flower was formerly classified as *Scrophulariaceae* (Figwort Family).

SAR STATUS N/A

Monarda didyma
Bee Balm

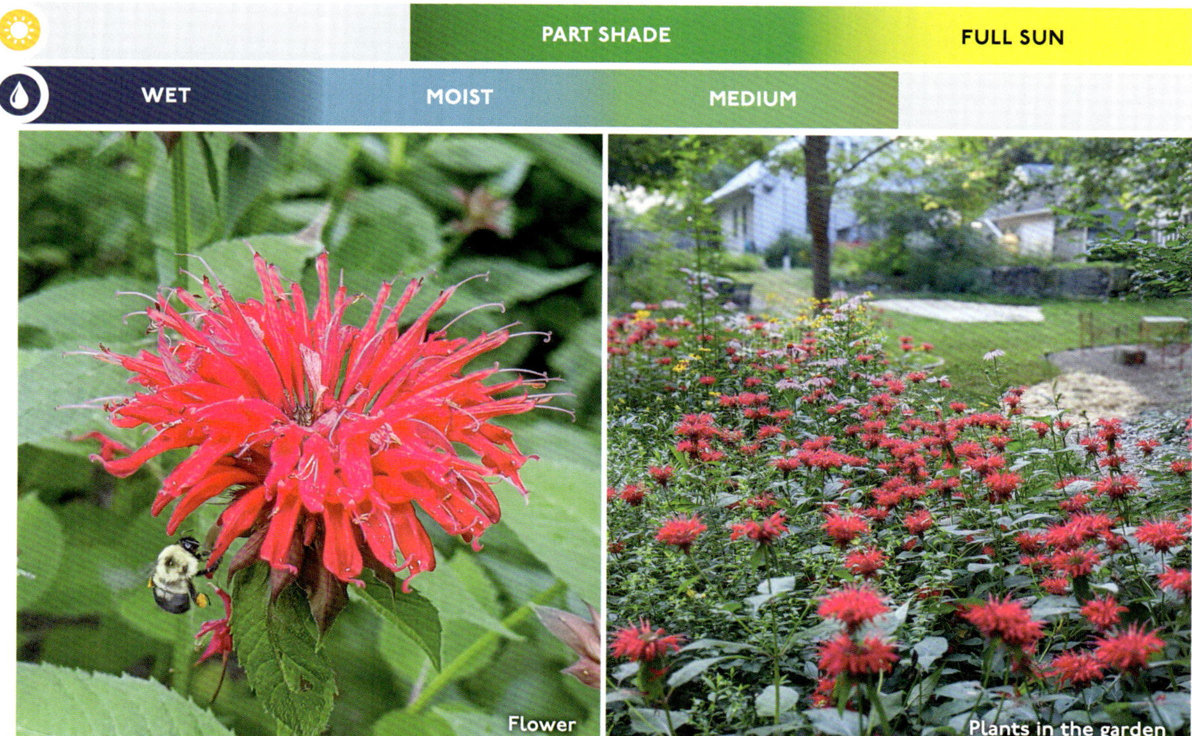

FAMILY *Lamiaceae* (Mint Family)

ALTERNATE COMMON NAMES American Bee Balm, Crimson Bee Balm, Firecracker Plant, Fragrant Balm, Oswego Tea, Red Bergamot, Scarlet Bee Balm, Scarlet Monarda, Wild Oregano

PLANT DESCRIPTION Bee Balm features square, reddish-brown stems that are slightly hairy and branch occasionally. Along the stem you will find opposite leaves that each measure up to 13 cm long and 5 cm wide with 2.5 cm long leaf stalks. Leaves are ovate to heart shaped with pointed tips and toothed edges and may have a purple/reddish tint to them. Stems are topped with a single cluster of tubular, red flowers that measures up to 10 cm wide, with each individual flower measuring just under 4 cm long. Each flower is characterized by an upper

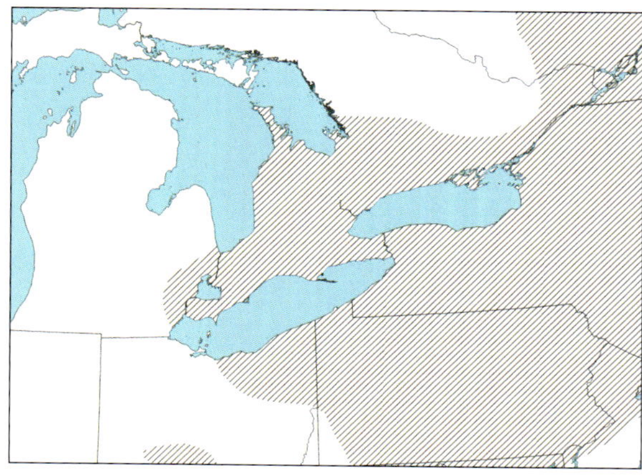

lip with two stamens protruding from it and a lower lip that arches downwards. The outer surface of the upper lip is finely hairy. Each flower head is backed

Leaves

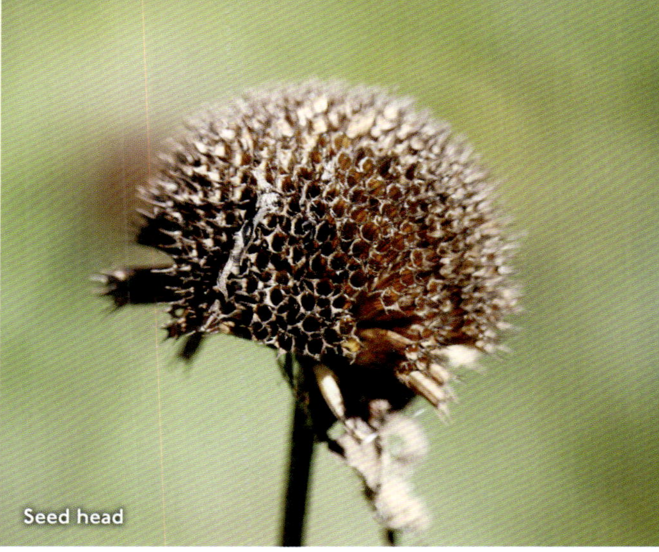

Seed head

by multiple leaf-like bracts that are tinted with red or purple. Each flower produces small, dry, ovoid seeds.

IN THE GARDEN Reminiscent of jester hats, the vivid scarlet flowers of Bee Balm contrast beautifully with surrounding foliage. Brush up against the leaves and you will notice a pleasant minty smell. The rigid stems and rounded seed heads persist into winter months to provide wonderful textural interest, especially when covered in frost or snow.

SKILL LEVEL Beginner

LIFESPAN Perennial

EXPOSURE Full sun to part shade

SOIL TYPE Well-drained sand, clay, or loam

MOISTURE Medium to wet

HEIGHT 120 cm

SPREAD 90 cm

BLOOM PERIOD Jun, Jul, Aug

COLOR Red

FRAGRANT 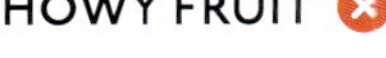 (no floral scent but foliage is aromatic)

SHOWY FRUIT ✖

CUT FLOWER ✔

PESTS Susceptible to powdery mildew (full sun and good air circulation will help)

NATURAL HABITAT Moist, open woods and meadows and along stream banks

WILDLIFE VALUE Hummingbirds and swallowtail butterflies (*Papilionidae*) are especially attracted to the red flowers. The flowers are of special value for bumblebees and other native bees, but the strongly scented leaves and stems are avoided by mammalian herbivores. *Dufourea monardae*, a small black bee, specializes in the pollination of *Monarda* flowers

BUTTERFLY LARVA HOST PLANT FOR None

MOTH LARVA HOST PLANT FOR Gray Marvel (*Anterastria teratophora*), Orange Mint Moth (*Pyrausta orphisalis*), Raspberry Pyrausta Moth (*Pyrausta signatalis*), Hermit Sphinx (*Sphinx eremitus*)

USDA HARDINESS ZONES 4–7

PROPAGATION No pretreatment is needed for seeds to germinate. Store seeds in a cool, dry environment for spring sowing. Plants may be propagated by softwood cuttings taken in late spring. Mature clumps of plants can be divided in the spring before they send up stems, although I have had great success dividing the clumps at almost any time of the year. The later the division, though, the less likely you will have blossoms in the first year.

ADDITIONAL INFO This species blooms more vigorously if divided in spring or fall every three to four years — the very shallow root system is easily lifted and divided. It will spread into a large colony in just a few years if left alone. Although this plant typically has a single flower head, it is not that uncommon for it to have a "double decker" flower or, on rare occasions, a "triple decker."

SAR STATUS ON – S3/(not listed); MI – SX/(not listed); NY – (not ranked)/EV

Monarda fistulosa
Wild Bergamot

QUICK GUIDE

PROPAGATION

	PART SHADE	FULL SUN
	MOIST	MEDIUM	DRY

Flower

Whole plant

FAMILY *Lamiaceae* (Mint Family)

ALTERNATE COMMON NAMES Bee Balm, Wild Horsemint, Mint-leaf Bee Balm, Purple Bee Balm

PLANT DESCRIPTION Wild Bergamot features multiple light green stems that are four sided and varyingly hairy. Opposite leaves are found along the stem, measuring up to 10 cm long and 3.8 cm wide and are borne on 1.5 cm long leaf stalks. Leaves are broadly lanceolate to ovate, coarsely toothed, and hairless to finely hairy and have rounded bases with pointed tips. Branching stems are topped with 7.6 cm wide clusters of tubular flowers. Each flower has a tubular upper lip, with protruding stamens and tufts of white hairs at its tips, and a curved lower lip. The outer surfaces of the lips have fine hairs. Flower heads are backed by green bracts that

may have a pinkish tinge. Flowers turn into rounded seed heads that contain small, dry, oval seeds.

IN THE GARDEN Wild Bergamot blooms profusely with pastel-purple flower heads that resemble

mini fireworks displays. The leaves have a lovely minty-oregano fragrance when rubbed. The rigid stems and rounded seed heads stand strong through the winter months to extend seasonal interest. Herbivores tend to avoid this plant.

SKILL LEVEL Beginner

LIFESPAN Perennial

EXPOSURE Full sun to part shade

SOIL TYPE Thrives in a wide range of soils — from acid to lime, rich to poor, and sand to clay

MOISTURE Dry to moist

HEIGHT 60–120 cm

SPREAD 60–90 cm

BLOOM PERIOD Jul, Aug, Sep

COLOR Pink, lavender, rarely white

FRAGRANT ✓ (foliage is aromatic)

SHOWY FRUIT ✗

CUT FLOWER ✓

PESTS Powdery mildew can be a significant problem with the *Monardas*, particularly in crowded gardens with poor air circulation; rust can also be a problem

NATURAL HABITAT Open wooded sites, prairie ditches, meadows, sunny hillsides, and rocky slopes

WILDLIFE VALUE Butterflies and many, many native bees are attracted to Wild Bergamot — it is one of the busiest flowers in my garden when in bloom. Hummingbirds may also visit occasionally. The aromatic foliage is unpalatable to most herbivores. *Dufourea monardae*, a small black bee, specializes in the pollination of *Monarda* flowers

BUTTERFLY LARVA HOST PLANT FOR None

MOTH LARVA HOST PLANT FOR Gray Marvel (*Anterastria teratophora*), Orange Mint Moth (*Pyrausta orphisalis*), Raspberry Pyrausta Moth (*Pyrausta signatalis*), Sparganothis Leafroller Moth (*Sparganothis sulfureana*), Hermit Sphinx (*Sphinx eremitus*)

USDA HARDINESS ZONES 3–9

PROPAGATION This plant is very easy to start from seed and should be surface sown because they need light to germinate. No other pretreatment is necessary. Store seeds in a cool, dry environment for spring sowing. Wild Bergamot colonizes by rhizomes so lift and divide every three years to control its spread, improve air circulation and for general plant health. Most sources recommend dividing in early spring before new growth starts, but I have successfully divided Wild Bergamot all summer long. Plants may also be propagated in the greenhouse from stem cuttings.

ADDITIONAL INFO There are conflicting reports about Wild Bergamot's salt tolerance. Some believe it's not a good choice for boulevard or verge gardens, while others have had success growing it close to the road. It prefers drier soils than *Monarda didyma*. In most years in my garden, Wild Bergamot becomes white with powdery mildew by the time the blossoms are nearly done, and I often get good regrowth and a second flush of flowers by cutting them back at this time to the lowest set of leaves.

SAR STATUS N/A

Monarda punctata
Spotted Bee Balm

QUICK GUIDE

PROPAGATION

| | PART SHADE | FULL SUN |
| | DRY |

Flower

Flower

Flower with pink bracts

FAMILY *Lamiaceae* (Mint Family)

ALTERNATE COMMON NAMES Bee Balm Horsemint, Dotted Horsemint, Dotted Mint, Horsemint

PLANT DESCRIPTION The stems of Spotted Bee Balm are brown to reddish purple, four sided, and densely hairy. Leaves are softly hairy and found in an opposite arrangement, measuring up to 7.5 cm long and 1.2 cm wide. You will notice that smaller leaves emerge from leaf axils (where leaves meet the stem). Lower leaves are serrated while upper leaves may have smooth margins. Tubular flowers are found in whorls around the upper leaf axils, with one cluster being found at the very top of each stem. Each flower is 2.5 cm long and yellow with purple spots. The upper lip of each flower is long, narrow, and arching while the lower lip is three lobed. Each

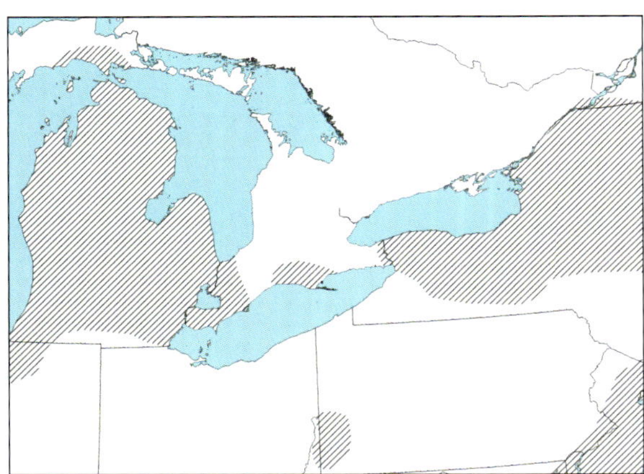

flower cluster is backed by five to 10 leaf-like bracts with pink, lavender, or white upper surfaces. Note that some leaves can take on the color of the bracts.

Whole plant | Seed heads

Flowers turn into seed heads containing small, dry, oval seeds.

IN THE GARDEN The quirky beauty of Spotted Bee Balm is sure to turn heads in your garden! It is one of the most drought tolerant of the *Monarda* species and certainly the most unique. Its foliage has a wonderful minty aroma and, despite being in the mint family, it retains a clumping habit. Herbivores rarely bother with this plant.

SKILL LEVEL Beginner

LIFESPAN Annual, biennial, or short-lived perennial

EXPOSURE Full sun to part shade

SOIL TYPE Sandy soil — requires excellent drainage

MOISTURE Dry

HEIGHT 15–60 cm (occasionally to 90 cm)

SPREAD 30 cm

BLOOM PERIOD Jul, Aug, Sep

COLOR Yellow with maroon markings; however, the bracts are showier and may be purple, pink, white, or yellow

FRAGRANT  (foliage)

SHOWY FRUIT ❌

CUT FLOWER ✔

PESTS Susceptible to powdery mildew, though because this plant prefers drier conditions than its cousins, *Monarda didyma* and *M. fistulosa*, it tends to be affected less often than the other two

NATURAL HABITAT Sandy prairies and savannas, sand dunes around the Great Lakes, and sandy fields with little grassy competition

WILDLIFE VALUE Butterflies (it is a favorite of the endangered Karner Blue, *Lycaeides melissa samuelis*), skippers, hummingbird moths, hummingbirds, honeybees, bumblebees, and other native bees sip nectar from the flowers. In my garden, if you want to see a Great Black Digger Wasp (*Sphex pensylvanicus*) just stand by the *Monarda punctata* for a few moments when it is blooming — they are almost always on the plant. The strongly scented leaves and stems are usually avoided by mammalian herbivores

BUTTERFLY LARVA HOST PLANT FOR None

MOTH LARVA HOST PLANT FOR Gray Marvel (*Anterastria teratophora*), Orange Mint Moth (*Pyrausta orphisalis*), Raspberry Pyrausta Moth (*Pyrausta signatalis*)

USDA HARDINESS ZONES 3–9

PROPAGATION No pretreatment is needed, though if holding for spring sowing, the seeds should be stored in a cool, dry environment. Seeds need light to germinate, so sow on the surface. It can also be propagated by two-to-three-node cuttings of young plants. Because they are such a short-lived perennial, root division is usually not worth the effort.

ADDITIONAL INFO Spotted Bee Balm does not tolerate grassy competition.

SAR STATUS ON – S1/(not listed); OH – S1/E

Oenothera biennis
Evening Primrose

FAMILY *Onagraceae* (Evening Primrose Family)

ALTERNATE COMMON NAMES Bastard Evening Primrose, Common Evening Primrose, Evening Star, Fever Plant, Four-o'clock, German Rampion, Golden Candlestick, Hoary Evening Primrose, Hog Weed, King's Cure-all, Night Primrose, Night Willow-herb, Sand Lily, Scabish, Scavey, Scurvish, Speckled John, Tree Primrose, Weedy Evening Primrose, Wild Beet, Wine-trap

PLANT DESCRIPTION Evening Primrose has a two-year life cycle. During the first year, it produces a low rosette of basal leaves. During the second year, it sends up a light-green-to-red central stem covered in white hairs. Lance-elliptic leaves are found in an alternate pattern along the stem and measure up to 20 cm long and 5 cm wide (they are usually smaller

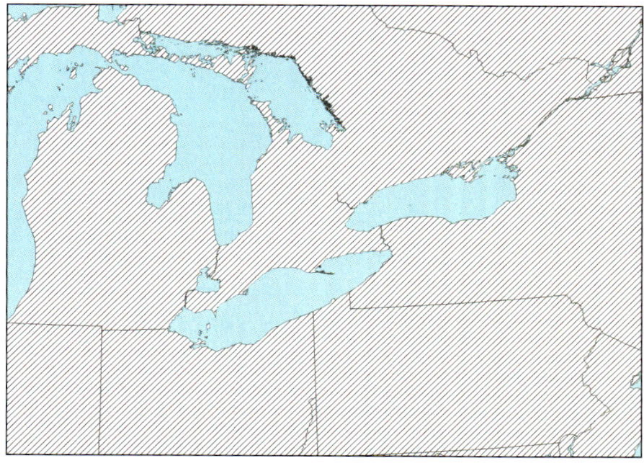

than this). They are hairless to finely hairy, toothless or with small teeth, and borne on little-to-no leaf stalk. Stems terminate with spike-like clusters of many yellow flowers, each about 5 cm across. Each

First-year basal leaves

Whole plant

Seed pods

flower has four heart-shaped petals surrounding eight yellow stamens and a cross-shaped stigma. Each flower appears to be borne on a long stalk, but this is actually an elongated calyx tube (the part that bears the sepals and stamens). Four sepals are found behind each flower, measuring about 3 cm long and bending backwards as the flower matures. As the flowers fade, they produce 3 to 4 cm long tubular seed pods that contain hundreds of tiny seeds.

IN THE GARDEN Evening Primrose features a spire of vibrant lemon-yellow flowers. These flowers open up in the evening and stay open until hit by the morning sun. They also stay open to brighten up cloudy days! This is a dependable choice for nutrient-poor soils, and the sturdy stems will persist into the winter months to extend interest.

SKILL LEVEL Beginner to intermediate

LIFESPAN Biennial

EXPOSURE Full sun to full shade

SOIL TYPE Rocky, gravelly, or sandy soils

MOISTURE Dry to medium

HEIGHT 90–150 cm (up to 210 cm in ideal conditions)

SPREAD 30–90 cm

BLOOM PERIOD Jun, Jul, Aug, Sep

COLOR Yellow

FRAGRANT ✓ (lemon scent)

SHOWY FRUIT ✗

CUT FLOWER ✓

PESTS No serious insect or disease problems, though leaf spot and powdery mildew may occur; this plant is a favorite of the invasive Japanese Beetle (*Popillia japonica*)

NATURAL HABITAT Occurs throughout the region in disturbed areas and along roadsides, lakeshores, and river valleys

WILDLIFE VALUE Moths are the main pollinators of the flowers, especially sphinx moths; the high-oil-content seeds are eaten by goldfinches and other small birds

BUTTERFLY LARVA HOST PLANT FOR None

MOTH LARVA HOST PLANT FOR Pearly Wood Nymph (*Endryas unio*), Grape Leaffolder Moth (*Desmia funeralis*), White-lined Sphinx (*Hyles lineata*), Red-streaked Momphid (*Mompha eloisella*), Primrose Moth (*Schinia florida*), Sparganothis Leafroller Moth (*Sparganothis sulfureana*)

USDA HARDINESS ZONES 4–9

PROPAGATION The very small seeds need light to germinate (surface sow) but do not need to be stratified. Store in a cool, dry environment if sowing in spring. Stem cuttings can be taken in spring.

ADDITIONAL INFO Evening Primrose often volunteers in gardens, mysteriously showing up when never seen there before, but it does not persist. However, the seeds remain viable in the soil for many years, waiting for a disturbance. Evening Primrose has a very deep taproot, and being short-lived (a biennial), the decaying root helps to aerate and take nutrients deep into the soil profile.

SAR STATUS N/A

Oenothera pilosella
Prairie Sundrops

Flowers

FAMILY *Onagraceae* (Evening Primrose Family)

ALTERNATE COMMON NAMES Meadow Evening Primrose, Meadow Sundrops, Sundrops

PLANT DESCRIPTION Prairie Sundrops have a low, bushy form with greenish-red, hairy stems. Lance-shaped leaves are found in an alternate arrangement along the stem and measure about 7.5 cm long and 2.5 cm wide. They are stalkless and covered in fine hairs and have smooth to lightly toothed edges. Flowers are found in short, spiked clusters at the top of the stems. Each flower measures about 5 cm wide when fully open and appears to be borne on a long stalk, but this is actually an elongated calyx tube (the part that bears the sepals and stamens). You will notice four yellow petals with notched tips surrounding eight stamens and a

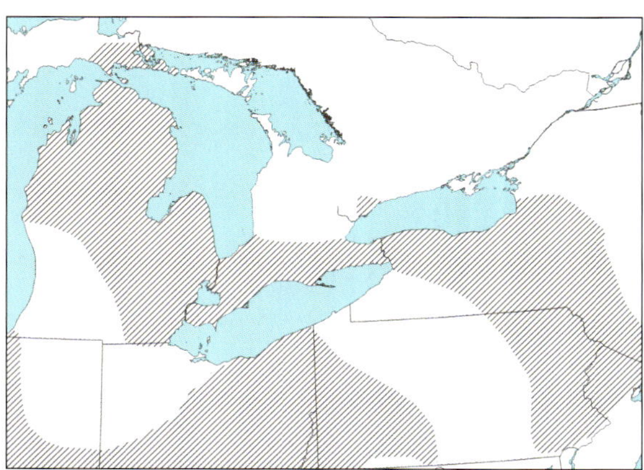

single cross-like stigma. Each petal is broad at the tip, becoming narrower at the base. Flowers mature into small, elliptical seed pods that split open at the tip to release many tiny seeds.

IN THE GARDEN Prairie Sundrops is sure to bring cheer to your garden with its abundant, dazzling yellow flowers. It makes for a compact and reliable groundcover that will spread by rhizomes to form colonies. Despite being in the Evening Primrose family, the flowers stay open during the day!

SKILL LEVEL Beginner

LIFESPAN Perennial

EXPOSURE Full sun to part shade

SOIL TYPE Most soils

MOISTURE Moist to medium

HEIGHT 30–60 cm

SPREAD 45–60 cm

BLOOM PERIOD May, Jun

COLOR Yellow

FRAGRANT ✓ (faint)

SHOWY FRUIT ✗

CUT FLOWER ✗

PESTS No serious insect or disease problems

NATURAL HABITAT Prairies, fields, wet uplands, open woods, open grassy places, railroads, and other disturbed sites

WILDLIFE VALUE Although this species attracts a large number of native bees, some of the literature suggests it may be pollinated primarily by moths

BUTTERFLY LARVA HOST PLANT FOR None

MOTH LARVA HOST PLANT FOR A variety of momphid moths (*Mompha* spp.)

USDA HARDINESS ZONES 3–8

PROPAGATION No pretreatment of seeds is necessary but, if storing, keep seeds cool and dry. May be propagated by taking root cuttings.

ADDITIONAL INFO Prairie Sundrops can be an aggressive spreader, but unwanted plants can be easily removed from the garden due to its shallow root systems. It is not a good choice for boulevard or verge plantings as this species has no salt tolerance.

SAR STATUS ON – S2/(not listed)

Packera aurea
Golden Ragwort

Flowers

Whole plant

FAMILY *Asteraceae* (Aster Family)

ALTERNATE COMMON NAMES Butterweed, Golden Groundsel, Heartleaf Ragwort, Squaw-weed

PLANT DESCRIPTION Golden Ragwort has both basal and stem leaves. Basal leaves are found in a rosette that measures up to 20 cm wide, with each leaf measuring up to 10 cm long and 5 cm wide. These leaves are oblong with a heart-shaped base, have rounded-toothed edges, and are borne on long stalks. A flower stalk rises up from the center of the rosette and this is where you will find the stem leaves. Stem leaves clasp the stem and are narrow with deep lobes and blunt teeth. The flower stalk is topped with an open, flat-topped flower cluster made up of many daisy-like flowers. Each one of these flowers measures about 2 cm wide and is

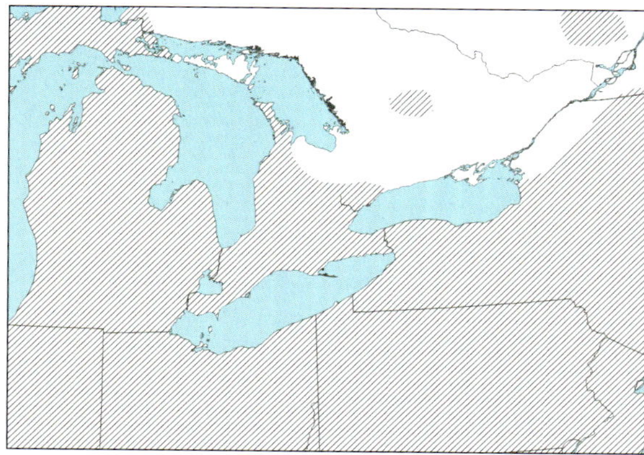

characterized by six to 16 ray florets (petals) surrounding a slightly darker center disk. Linear, green bracts surround the base of the flower heads and often have purple-tinged tips. Flowers mature into

Leaves | Whole plant | Seed heads

round, fluffy seed heads, with each seed having a tuft of hair that allows it to be carried by the wind.

IN THE GARDEN Golden Ragwort puts on a luminous and long-lasting display of golden-yellow flowers in the springtime. It is a robust, bold-textured groundcover that will spread by both rhizomes and seeds. Herbivores usually ignore this plant.

SKILL LEVEL Beginner

LIFESPAN Perennial

EXPOSURE Full sun to part shade (more sun requires more moisture)

SOIL TYPE Well-drained, average soils with sufficient organic matter to retain moisture

MOISTURE Moist (but also found growing in dryish areas; tolerates seasonal flooding)

HEIGHT 30–60 cm

SPREAD 30–45 cm

BLOOM PERIOD May, Jun, Jul

COLOR Yellow

FRAGRANT ✅ (foliage)

SHOWY FRUIT ❌

CUT FLOWER ✅

PESTS No serious insect or disease problems

NATURAL HABITAT Meadows, boggy swales, low woods, and shaded swampy areas along streams

WILDLIFE VALUE Nectar and pollen of the flowers attract small bees and flies; an andrenid bee called *Andrena gardineri* is a specialist pollinator of *Packera* species

BUTTERFLY LARVA HOST PLANT FOR None

MOTH LARVA HOST PLANT FOR Canadian Agonopterix (*Agonopterix canadensis*), The Gem (*Orthonama obstipata*)

USDA HARDINESS ZONES 3–9 (evergreen in Zone 6 and warmer)

PROPAGATION The seeds do not require pretreatment to germinate and the plant freely self-seeds, but they do need light to germinate so do not cover them — simply press them into the soil. The plant is easy to propagate by division in the spring.

ADDITIONAL INFO Golden Ragwort will form dense colonies even within the root system of mature trees and may be considered "weedy" in some circumstances.

SAR STATUS N/A

Packera paupercula
Balsam Groundsel

FAMILY *Asteraceae* (Aster Family)

ALTERNATE COMMON NAMES Balsam Ragwort, Canadian Butterweed, Common Groundsel, Northern Meadow Groundsel, Northern Ragwort

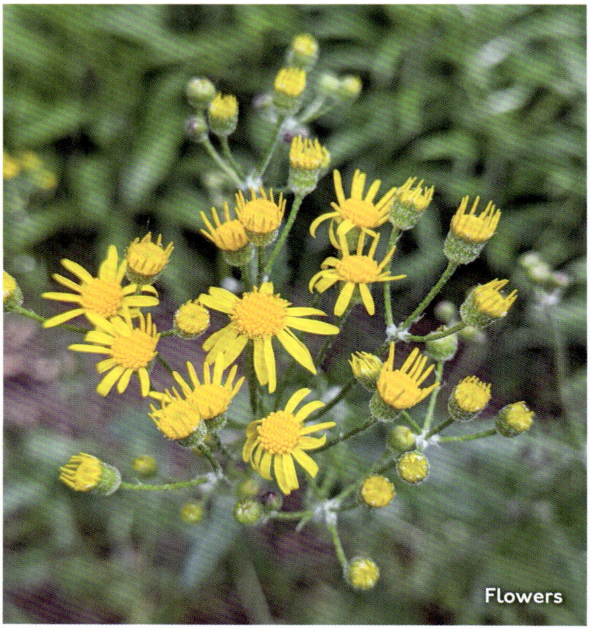

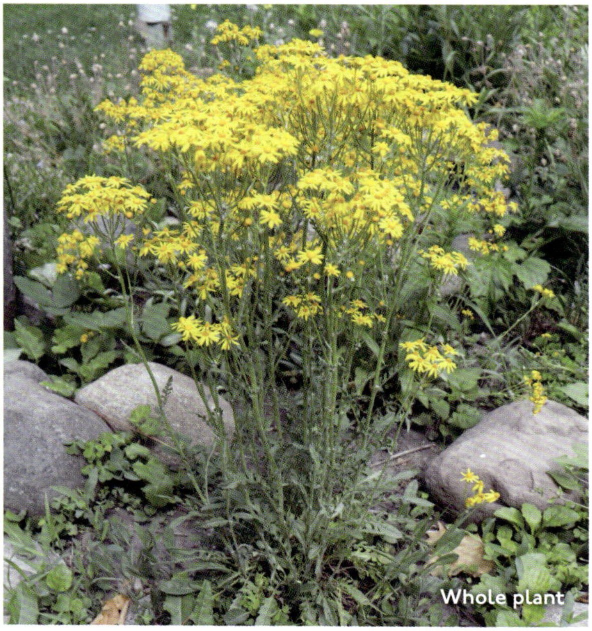

PLANT DESCRIPTION Balsam Groundsel has both basal leaves and stem leaves. Basal leaves are found in a 10 cm wide rosette, with each leaf measuring 6 cm long and 1 cm wide (they are at least twice as long as they are wide). These leaves are oblong to spoon shaped with a tapered base, stalked, and toothed. A central stem rises up from the basal leaves and this is where you will find two to four alternate stem leaves. These leaves are smaller and narrower than the basal leaves and have little-to-no leaf stalk. Stem leaves are deeply lobed with coarse teeth and become smaller as they ascend the stem.

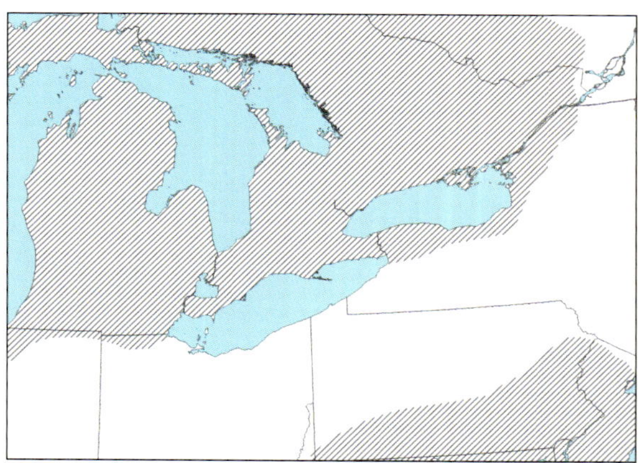

All leaves are smooth but may have fine, woolly hairs especially when young (as with the stems). The central stem terminates with a flat-topped cluster of about five to 30 slender-stalked, yellow

flowers. Each flower is just under 2 cm wide and features eight to 13 ray florets (petals) surrounding a golden-yellow center disk. Flowers are backed by smooth, linear bracts that are green with pale purple tips. Seed heads resemble dandelions and contain brown seeds with tufts of hairs that allow them to be carried by the wind.

IN THE GARDEN Balsam Groundsel features radiant clusters of golden-yellow flowers. It's a dependable choice for tough, gravelly, or rocky sites. Once the flowers fade, the basal leaves retain a lush, mounding form. Spreads slowly by rhizomes but self-seeds freely in the right conditions. Herbivores tend to ignore this plant.

SKILL LEVEL Beginner

LIFESPAN Perennial

EXPOSURE Full sun to part shade

SOIL TYPE Loam, sandy loam, or rocky, gravelly soil

MOISTURE Moist to medium

HEIGHT 45 cm

SPREAD 30 cm

BLOOM PERIOD May (mid), Jun, Jul, Aug (mid)

COLOR Yellow

FRAGRANT ✓

SHOWY FRUIT ✗

CUT FLOWER ✓

PESTS Unknown

NATURAL HABITAT Meadows, moist cliffs, and woods

WILDLIFE VALUE An important source of nectar and pollen for native bees; the foliage is toxic to most mammals; an andrenid bee called *Andrena gardineri* is a specialist pollinator of *Packera* species

BUTTERFLY LARVA HOST PLANT FOR None

MOTH LARVA HOST PLANT FOR Canadian Agonopterix (*Agonopterix canadensis*), The Gem (*Orthonama obstipata*)

USDA HARDINESS ZONES 3–8

PROPAGATION This plant is easily grown from seeds, which require about 60 days of cold, moist stratification to germinate.

ADDITIONAL INFO Contact may cause skin irritation in some people. Balsam Goundsel does not tolerate much competition in rich soils. This is a great native plant for planters and hanging baskets.

SAR STATUS NY – S3/(not listed); OH – S3/T

Pedicularis lanceolata
Swamp Betony

QUICK GUIDE

PROPAGATION

	PART SHADE	FULL SUN

WET	MOIST	

Flowers
Whole plant

FAMILY *Scrophulariaceae* (Figwort Family)

ALTERNATE COMMON NAMES Fen Betony, Marsh Betony, Swamp Lousewort, Swamp Wood Betony, Yellow Lousewort

PLANT DESCRIPTION Swamp Betony features stout, unbranched stems that have short hairs but may become smooth with age. Yellowish-green leaves are found in an opposite arrangement along the stem, and each measures up to 13 cm long and 3 cm wide. Leaves are pinnately divided into small, rounded lobes that have tiny teeth along their edges. Note how the leaves are hairless and thick textured. Tubular flowers are found in dense spikes at the top of the stems. Each flower is pale yellow to creamy white, measures 2.5 cm wide, and features an upper

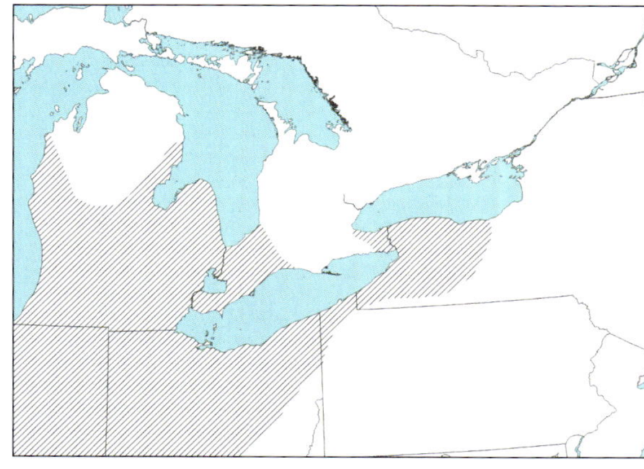

lip that curves over the end of the lower lip. Flowers mature into ovoid seed capsules that open at the top to release many small, irregularly shaped seeds.

IN THE GARDEN Swamp Betony is an interesting-looking plant with fern-like leaves and an attractive form.

SKILL LEVEL Beginner to intermediate

LIFESPAN Short-lived perennial

EXPOSURE Full sun to part shade

SOIL TYPE Rich, calcareous soils, but will tolerate most soils

MOISTURE Wet to moist

HEIGHT 30–90 cm

SPREAD 30 cm

BLOOM PERIOD Aug, Sep

COLOR Cream

FRAGRANT ❌

SHOWY FRUIT ❌

CUT FLOWER ✅

PESTS No serious insect or disease problems

NATURAL HABITAT Swamps, fens, wet sand prairies, and other wet areas, in full sun or wooded areas (this plant is usually only found in higher-quality wetlands)

WILDLIFE VALUE The flowers are cross-pollinated primarily by bumblebees, which obtain both nectar and pollen

BUTTERFLY LARVA HOST PLANT FOR Baltimore Checkerspots (*Euphydryas phaeton*) lay their eggs on White Turtlehead (*Chelone glabra*), but often the later instars of the caterpillar will move to feed on the foliage of Swamp Betony and other plants

MOTH LARVA HOST PLANT FOR Pink-patched Looper (*Eosphoropteryx thyatiroides*)

USDA HARDINESS ZONES 3–8

PROPAGATION The seeds of the semi-parasitic Swamp Betony, unlike its cousin Wood Betony (*Pedicularis canadensis*), germinate readily, but without a host plant the seedlings will be stunted and chlorotic (see Additional Info). If starting indoors, wheat may be used as a short-term host. In the garden, Swamp Betony should be planted where other plants are present or with the seed of another plant such as grass. The tiny seeds need light to germinate,

so press seeds lightly into the surface of the soil and keep the soil consistently moist until germination. For spring planting, mix the seeds with moist sand and store in a cool location. It then should have cold, moist stratification for 30 days before sowing.

ADDITIONAL INFO *Pedicularis lanceolata* is a hemiparasite. It produces chlorophyll but it also obtains some of its resources by tapping into the roots of other plants, often asters and native grasses. This is a good choice to include in wetland restoration projects.

SAR STATUS NY – S2/T; PA – S1/(not listed)

Penstemon digitalis
Foxglove Beardtongue

QUICK GUIDE

PROPAGATION

C(60) L D LYR CS

	PART SHADE	FULL SUN

WET	MOIST	MEDIUM	DRY

Flowers

Whole plant

FAMILY *Scrophulariaceae* (Figwort Family)

ALTERNATE COMMON NAMES Foxglove Penstemon, Mississippi Beardtongue, Mississippi Penstemon, Smooth Penstemon, Smooth White Beardtongue, Smooth White Penstemon, Tall Beardtongue, Talus Slope Beardtongue, Talus Slope Penstemon, White Beardtongue

PLANT DESCRIPTION Foxglove Beardtongue has both basal and stem leaves. Basal leaves are variable in shape but are generally spatula shaped, stalked, and toothless and measure up to 15 cm long and 6 cm wide. In mature plants, flowering stems will emerge from the basal rosette, and this is where you will find the stem leaves. Stem leaves are lance shaped with pointed tips and finely toothed margins. They clasp the stem and are found in an

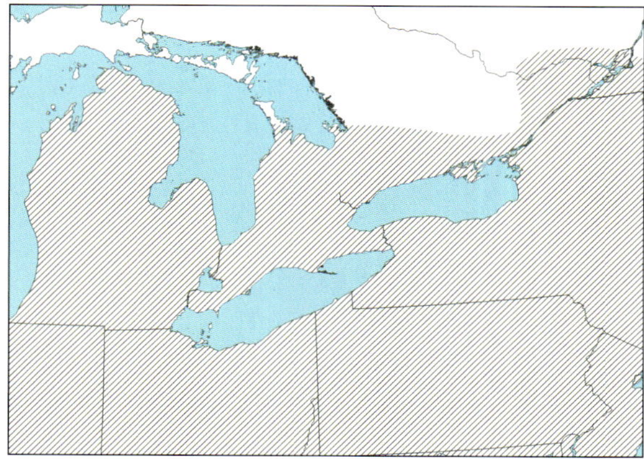

opposite arrangement. All leaves have smooth surfaces and are green but may display varying levels of red/maroon. Clusters of long-stalked, tubular flowers arise from upper leaf axils (where the leaf meets the stem). Each flower is 2.5 cm long and consists of a lower lip with three lobes and upper lip with two lobes. Inside the flower, you will notice purplish lines and four black-tipped stamens. The flower and flower stalks are covered in short hairs. Flowers turn into brown, teardrop-shaped seed pods. They split open when ripe to release many small, irregularly angled seeds.

IN THE GARDEN Foxglove Beardtongue helps transition the garden from spring into summer with an elegant vertical display of tubular, white flowers. In the fall months, the foliage takes on wonderful burgundy hues and in the winter months the seed heads provide reliable winter interest. This is a well-behaved clumping plant. Herbivores usually avoid it unless food is scarce.

SKILL LEVEL Beginner

LIFESPAN Perennial

EXPOSURE Full sun to part shade

SOIL TYPE Fertile, well-drained loams, clay loams, and sand

MOISTURE Dry to wet

HEIGHT 75–90 cm (occasionally to 150 cm)

SPREAD 45–60 cm

BLOOM PERIOD May, Jun

COLOR White

FRAGRANT ✖

SHOWY FRUIT ✖

CUT FLOWER ✔

PESTS No serious insect or disease problems, though root rot can occur in wet, poorly drained soils, and leaf spots are occasional problems

NATURAL HABITAT Low, moist-to-mesic areas in prairies and open woodlands

WILDLIFE VALUE Long-tongued bees, including

honeybees, bumblebees, miner bees, and mason bees, and hummingbirds

BUTTERFLY LARVA HOST PLANT FOR There are reports that caterpillars of the Baltimore Checkerspot (*Euphydryes phaeton*) feed on the foliage of various beardtongues

MOTH LARVA HOST PLANT FOR Chalcedony Midget (*Elaphria chalcedonia*), Verbena Bud Moth (*Endothenia hebesana*), Sparganothis Leafroller Moth (*Sparganothis sulfureana*)

USDA HARDINESS ZONES 3–8

PROPAGATION Very easy to grow from seed, it requires at least 30 days of cold, moist stratification if starting indoors or in the spring. Seeds need light to germinate, so simply press the seeds into the soil but do not cover. It may also be propagated from stem cuttings, by layering (pinning down a stem and covering with soil, which will then root where the leaf nodes contact the soil), or by root division.

ADDITIONAL INFO Foxglove Beardtongue is one of the few late spring/early summer bloomers that will tolerate dry shade. It also does well in clay-loam areas with poor drainage. It has some tolerance to salt, so could be a good choice for boulevard or verge plantings.

SAR STATUS N/A

Penstemon hirsutus
Hairy Beardtongue

FULL SHADE	PART SHADE	FULL SUN
	MOIST	MEDIUM	DRY

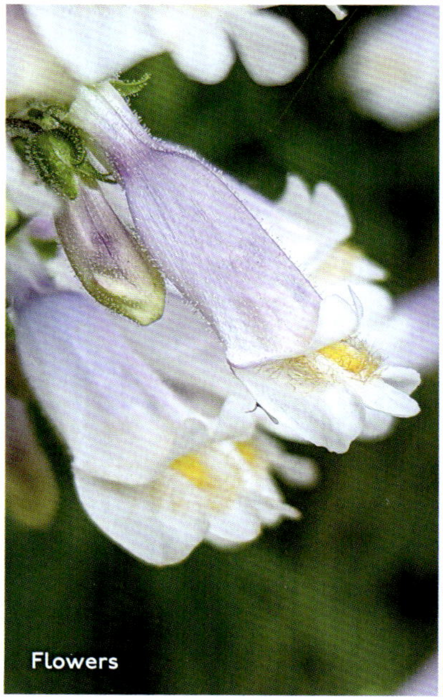

Flowers

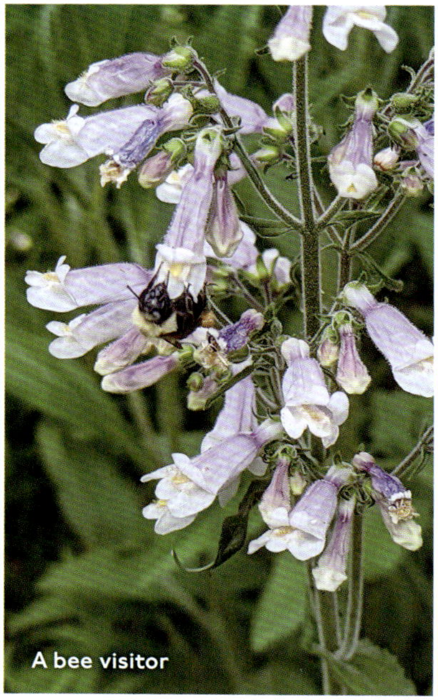

A bee visitor

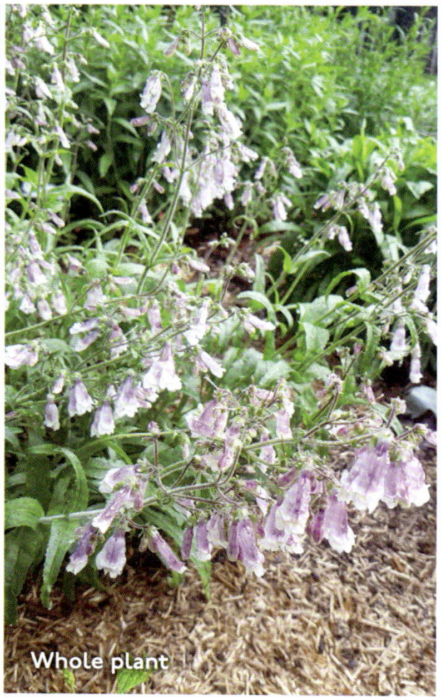

Whole plant

FAMILY *Scrophulariaceae* (Figwort Family)

ALTERNATE COMMON NAMES Dwarf Hairy Penstemon, Eastern Penstemon, Hairy Penstemon, Northeastern Beardtongue, Pride of the Mountain

PLANT DESCRIPTION Hairy Beardtongue features green-to-reddish-brown, hairy stems. Along the stems are opposite, lance-shaped leaves with pointed tips and toothed edges. Leaves clasp the stem. Stems terminate with clusters of slender, tubular flowers. Flower stalks also emerge from upper leaf axils. Each flower is purple to violet with white tips and has five petals with a protruding lower lip. You will notice that the flowers are hairy inside. Flowers turn into brown, teardrop-shaped seed pods. They split open when ripe to release many small, irregularly angled seeds.

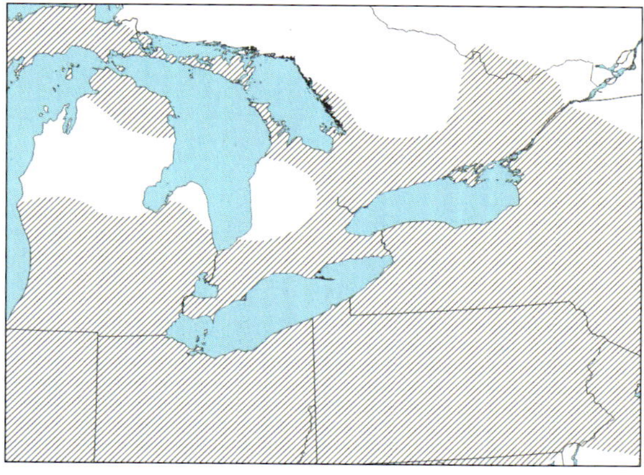

IN THE GARDEN Hairy Beardtongue transitions the garden into early summer with a fantastic display of tubular, pink flowers. It maintains a

well-behaved, clumping habit and adapts to a wide range of growing conditions. The seed heads provide reliable winter interest. Plants may suffer from a reduced lifespan if grown in rich soil.

SKILL LEVEL Beginner

LIFESPAN Perennial

EXPOSURE Full sun to full shade

SOIL TYPE Thin, well-drained soils

MOISTURE Moist to dry

HEIGHT 45 cm

SPREAD 25–30 cm

BLOOM PERIOD May, Jun

COLOR Pink, purple

FRAGRANT ❌

SHOWY FRUIT ❌

CUT FLOWER ✅

PESTS No serious insect or disease problems

NATURAL HABITAT Fields and open areas

WILDLIFE VALUE Attracts hummingbirds, butterflies, and native bees

BUTTERFLY LARVA HOST PLANT FOR Baltimore Checkerspot (*Euphydryas phaeton*)

MOTH LARVA HOST PLANT FOR Chalcedony Midget (*Elaphria chalcedonia*), Verbena Bud Moth (*Endothenia hebesana*), Sparganothis Leafroller Moth (*Sparganothis sulfureana*)

USDA HARDINESS ZONES 3–9

PROPAGATION Direct sow in late fall through to late winter, pressing seeds into the surface of the soil without covering as the seeds need light to germinate. For starting indoors, germination is enhanced with 60 days of cold, moist stratification. The seeds are slow to germinate, and the plant will spend its first year getting established then flower in year two. Plants seem to germinate best in cool soil. *Penstemon* readily self-seeds. Mature clumps of plants can be divided in early spring or late fall. Plants may also be propagated by layering (pinning the stem down and covering the leaf nodes with soil, which will then root) or by stem cuttings using sections of stem with two to four nodes.

ADDITIONAL INFO This low-maintenance plant blooms in my southwestern Ontario garden in the early summer, after the spring ephemerals are done and before the bulk of my other natives begin — providing a wonderful splash of color during the "quiet" period before the blooms of summer reach their peak. As a result, it has become one of my favorites.

SAR STATUS N/A

Phlox divaricata
Wild Blue Phlox

QUICK GUIDE

Flower

Flowers

Leaves

FAMILY *Polemoniaceae* (Phlox Family)

ALTERNATE COMMON NAMES Blue Phlox, Blue Woodland Phlox, Eastern Blue Phlox, Forest Phlox, Louisiana Phlox, Sweet William, Timber Phlox, Wild Sweet William, Wood Phlox

PLANT DESCRIPTION Wild Blue Phlox features light-green-to-reddish-brown stems that are covered in fine hairs. They are unbranched except at the top, where the flowers are. Stalkless leaves are found in an opposite arrangement along the stem and are rotated 90 degrees from the leaves above and below them. Each leaf is 5 cm long, about 1 cm wide, toothless, and finely hairy and has a pointed tip with a rounded base. Upper stems are topped with loosely branched clusters of nine to 30 flat-faced flowers. Each flower is about 2.5 cm wide and has five petals

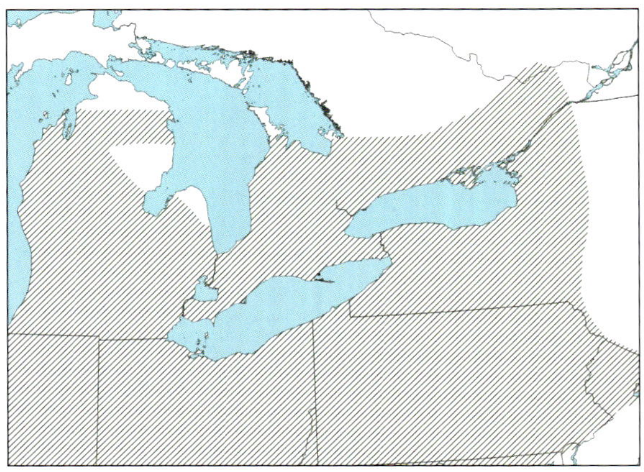

that join at the base to form a slender tube with a narrow opening. Each petal has a narrow base with a wider tip. The tips are shallowly notched. Flower

color is variable and may be tones of violet, purple, blue, or sometimes white. Each flower matures into an ovoid seed capsule with several small seeds inside.

IN THE GARDEN Wild Blue Phlox puts on a show of dainty, pastel-blue flowers in the spring. It makes a fine groundcover and spreads non-aggressively by stolons to form small colonies. The leaves are semi-evergreen.

SKILL LEVEL Beginner

LIFESPAN Perennial

EXPOSURE Full shade to part shade (will tolerate full sun in cooler climates)

SOIL TYPE Not very fussy — will grow in most soils but prefers fertile, humus-rich, well-drained soils

MOISTURE Moist to medium

HEIGHT 30 cm

SPREAD 20–30 cm

BLOOM PERIOD May, Jun

COLOR Violet-blue but may also be white or pink

FRAGRANT ✅ (lightly)

SHOWY FRUIT ❌

CUT FLOWER ❌

PESTS This plant has few pest problems other than powdery mildew, although spider mites can also be a problem, particularly in hot, dry conditions; rabbits will browse

NATURAL HABITAT Moist, rich, deciduous woods and bluffs

WILDLIFE VALUE Pollinated by long-tongued insects, including butterflies (especially swallowtails and Gray Hairstreaks [*Strymon melinus*]), skippers, Hummingbird Clearwing (*Hemaris thysbe*) and sphinx moths, and bumblebees, which can reach the nectar; foliage is eaten by rabbits and roots are consumed by voles

BUTTERFLY LARVA HOST PLANT FOR None

MOTH LARVA HOST PLANT FOR Spotted Straw (*Heliothis phloxiphagus*), Olive Arches (*Lacinipolia olivacea*)

USDA HARDINESS ZONES 3–8

PROPAGATION Seeds should be planted just below the soil surface when fresh (late spring) for germination the following spring, or they can be cold, moist stratified for 60 days if starting in the winter or spring. Plants may also be propagated by root division, layering, or stem cuttings taken in late spring.

ADDITIONAL INFO Wild Blue Phlox typically has blue flowers, but occasionally they're white or even pink. In my garden, they often start blue and turn whiter as the flowers mature.

SAR STATUS N/A

Physostegia virginiana
Obedient Plant

FAMILY *Lamiaceae* (Mint Family)

ALTERNATE COMMON NAMES Fall Obedient Plant, False Dragonhead, Virginia Lions-heart

PLANT DESCRIPTION Obedient Plant features four-sided, hairless stems that are unbranched, except for at the top where flowers occur. Opposite leaves are found along the stem. They are lance-elliptic with pointed tips, sharply toothed, and hairless and measure up to 13 cm long and just under 4 cm wide. Lower leaves are stalked but upper leaves are stalkless. Stems terminate with 25 cm tall flower spikes that are densely packed with four rows of tubular flowers. Each flower is about 2.5 cm long and has two lips — the upper lip is broad while the lower lip is divided into three lobes, the biggest of which is in the middle. The insides of the flowers are often dotted or striped with a slightly darker color. Flowers mature into capsules that contain sharply angled, dry, brown seeds.

IN THE GARDEN Obedient Plant puts on a floriferous late-summer display of snapdragon-like, pink flowers found along spikes that bloom from the bottom up. The seeds heads persist well into the winter months and provide excellent seasonal interest. A great choice for filling in large areas quickly.

SKILL LEVEL Beginner

LIFESPAN Perennial

EXPOSURE Full sun (preferred) to part shade

SOIL TYPE Sand, loam, clay

MOISTURE Wet to medium (but drought tolerant once established)

HEIGHT 120 cm

SPREAD 30–60 cm

BLOOM PERIOD Aug, Sep

COLOR Pink

FRAGRANT 

SHOWY FRUIT 

CUT FLOWER 

PESTS No serious insect or disease problems, though in some seasons rust can be a problem; deer resistant

NATURAL HABITAT Moist or mesic prairies, woodland edges, moist meadows, and moist roadside or railroad rights-of-way

WILDLIFE VALUE Flowers are generally pollinated by bumblebees, while other long-tongued bees, hummingbirds, and the occasional butterfly also sip nectar

BUTTERFLY LARVA HOST PLANT FOR None

MOTH LARVA HOST PLANT FOR None

USDA HARDINESS ZONES 3–9

PROPAGATION Cold, moist stratify for 60 days or fall sow for natural stratification. Plants spread rapidly by rhizomes, and the clumps that form can be divided in early spring or late fall. New plants can also be started from stem cuttings taken before flowering begins.

ADDITIONAL INFO The Obedient Plant gets its name from the fact that if you reposition an individual flower, it will stay where you put it. Many consider this species to be too aggressive for small gardens, especially in moist, rich soil. If exposure is shaded, plants are likely to flop.

SAR STATUS N/A

Podophyllum peltatum
Mayapple

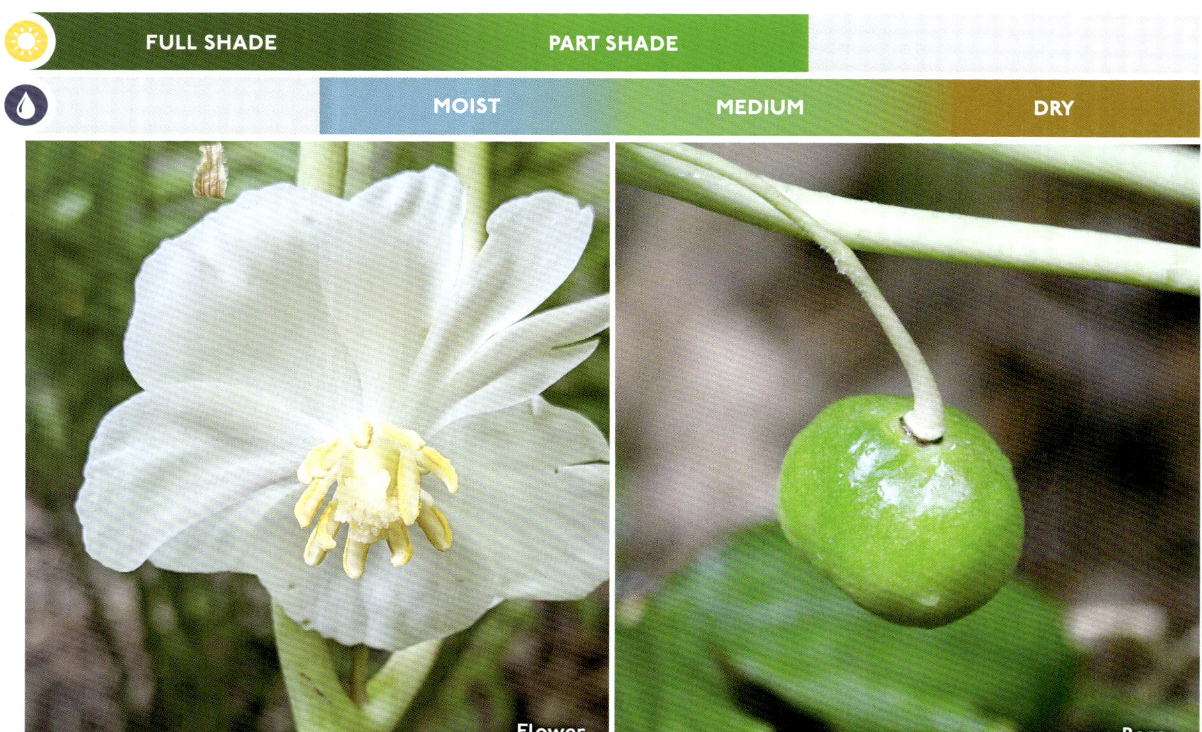

Flower

Berry

FAMILY *Berberidaceae* (Barberry Family)

ALTERNATE COMMON NAMES American Mandrake, Devil's Apple, Duck's Foot, Ground Lemon, Hog Apple, Indian Apple, Raccoon Berry, Umbrella Leaf, Wild Jalap, Wild Lemon, Wild Mandrake

PLANT DESCRIPTION Mayapple features a light green, smooth stem. In flowering plants, two long-stalked leaves fork out from the top of the stem. Non-flowering, immature plants produce only a single leaf on a long stalk. Leaves are palmately divided (meaning the lobes span out from a central point) into five to nine lobes, often with notched tips. They measure up to 40 cm wide at maturity, but younger plants usually have smaller leaves. A solitary, nodding flower is found at the fork of the stem and is just under 4 cm wide. This flower has

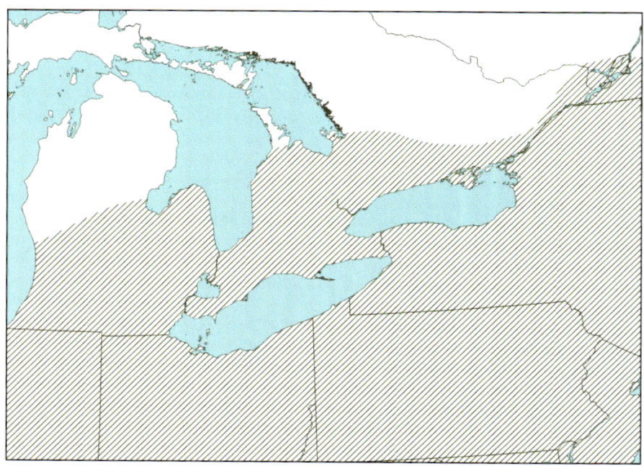

six to nine white, oval petals surrounding 12 to 18 stamens and a columnar ovary. The flower turns into an oval berry that matures to a pale yellow color and contains several seeds.

IN THE GARDEN Mayapple emerges early in the spring with bold, umbrella-like leaves. It is an excellent mid-height groundcover and will spread slowly by rhizomes to form colonies. The single flower on each plant is very showy but is usually hidden by the leaves. Herbivores avoid eating the foliage.

SKILL LEVEL Beginner

LIFESPAN Perennial (spring ephemeral)

EXPOSURE Full shade to part shade

SOIL TYPE Almost any humus-rich, well-drained soil

MOISTURE Dry to moist

HEIGHT 30–45 cm

SPREAD 20–30 cm

BLOOM PERIOD May

COLOR White

FRAGRANT ❌

SHOWY FRUIT ✅

CUT FLOWER ❌

PESTS No serious insect or disease problems

NATURAL HABITAT Mixed deciduous forests, shaded fields, shaded moist road banks, and shaded riverbanks

WILDLIFE VALUE The flowers are cross-pollinated by bumblebees and other long-tongued bees, and the berries are eaten by box turtles and possibly by such mammals as opossums, raccoons, and skunks

BUTTERFLY LARVA HOST PLANT FOR Variegated Fritillary (*Euptoieta claudia*)

MOTH LARVA HOST PLANT FOR Golden Borer Moth (*Papaipema cerina*)

USDA HARDINESS ZONES 3–8

PROPAGATION Seeds are hydrophilic and should not be allowed to dry out. Plant directly in the late summer when harvested (once the fruit has turned yellow), or cold, moist stratify for 90 days. Seedlings take several years to mature. The easiest way to propagate is to divide the rhizomes in fall once the plant has gone dormant, with at least one bud per section.

ADDITIONAL INFO Mayapple will not grow under coniferous trees and mowing will kill the plants. Young foliage is vulnerable to late-frost damage. The foliage dies back by midsummer, so consider interplanting it with something that can fill the space it leaves behind.

SAR STATUS N/A

Polygonatum biflorum
Solomon's Seal

QUICK GUIDE

PROPAGATION

FULL SHADE	PART SHADE	
	MOIST	MEDIUM

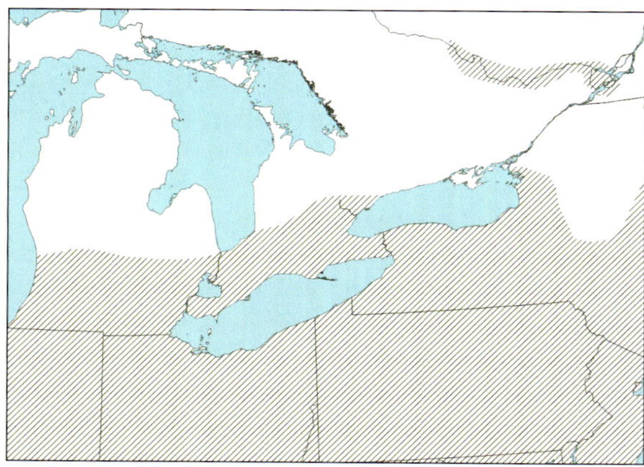

FAMILY *Liliaceae* (Lily Family)

ALTERNATE COMMON NAMES Giant Solomon's Seal, Great Solomon's Seal, King Solomon's Seal, Seal-wort, Small Solomon's Seal, Smooth Solomon's Seal

PLANT DESCRIPTION Solomon's Seal has smooth, unbranched stems that arch as if they are top heavy. Alternate leaves clasp the stem and measure up to 15 cm long and 7.5 cm wide. They are oval, toothless, and smooth and have prominent veins. The pale, yellowish-green flowers are tubular in shape, and you will find about two to 10 of them hanging down from the leaf axils (where the leaf meets the stem). Each flower is about 2.5 cm long and has six flaring lobes. Flowers each mature into a small berry that ripens from green to dark blue.

This species is differentiated from Starry False Solomon's Seal (*Maianthemum stellatum*) and False Solomon's Seal (*Maianthemum racemosum*), which both have flowers at the tip of the stem rather than hanging underneath.

Berries

IN THE GARDEN Solomon's Seal is an excellent structural plant for shady landscapes with its arching stems. Delightful, bell-shaped flowers dangle beneath the leaves then mature into dark blue berries that contrast well with its yellow fall foliage. Expect it to spread slowly by rhizomes to form colonies.

SKILL LEVEL Beginner

LIFESPAN Perennial

EXPOSURE Full shade to part shade (will tolerate sun in cooler climates if moist enough)

SOIL TYPE Almost any fertile, humus-rich soil

MOISTURE Moist to medium

HEIGHT 60–90 cm

SPREAD 45 cm

BLOOM PERIOD May, Jun

COLOR White/green

FRAGRANT ❌

SHOWY FRUIT ✅

CUT FLOWER ❌

PESTS No serious insect or disease problems

NATURAL HABITAT In both high-quality and degraded, moist to slightly dry, deciduous woodlands

WILDLIFE VALUE Various bees and even hummingbirds visit the plant, and the berries are eaten by woodland birds while deer enjoy the foliage

BUTTERFLY LARVA HOST PLANT FOR None

MOTH LARVA HOST PLANT FOR Black-patched Clepsis (*Clepsis melaleucanus*)

USDA HARDINESS ZONES 3–9

PROPAGATION This plant requires patience to grow from seed. First, the seeds need to be cleaned of any pulp from the fruit, which contains a germination inhibitor. They must then be kept moist, and they need a double dormancy. After the first cold, moist period followed by a warm, moist period, they will put down a root. Then, after a second cold, moist period they will send up a shoot. Or sow outdoors in the fall and allow two years for germination.

Plants may be divided in spring or started from root cuttings (with at least one eye per cutting). Rhizome cuttings are apt to lie dormant for a year before starting to grow.

ADDITIONAL INFO The plant will slowly expand via rhizomes into an open clone of which all the stems will arch in the same direction. In the wild, new plants usually appear beneath places where birds perch.

SAR STATUS N/A

Pycnanthemum tenuifolium
Slender Mountain Mint

FAMILY *Lamiaceae* (Mint Family)

ALTERNATE COMMON NAMES Common Horse-mint, Narrowleaf Mountain Mint

PLANT DESCRIPTION Slender Mountain Mint has four-sided, hairless stems that branch frequently to create a bushy effect. Opposite leaves are attached to the stem with no leaf stalk and are very narrow, at about 7.5 cm long and 0.3 cm wide. Leaves are hairless and toothless and feature a prominent central vein. Numerous, densely packed, rounded heads of flowers are found at the top of the stems. Individual flowers measure just under 1 cm long, are white with small purple dots, and are divided into a single upper lip with a three-lobed lower lip. Flowers mature into small, dry seeds.

IN THE GARDEN Slender Mountain Mint has a compact, bushy form and blooms with a profusion of long-blooming, white flowers. The leaves give off a minty smell when crushed, meaning herbivores won't touch this plant. The attractive seed heads persist through the winter months to extend seasonal interest.

SKILL LEVEL Beginner

LIFESPAN Perennial

EXPOSURE Full sun to part shade

SOIL TYPE Sand, clay, or loam

MOISTURE Moist to medium

HEIGHT 60–90 cm

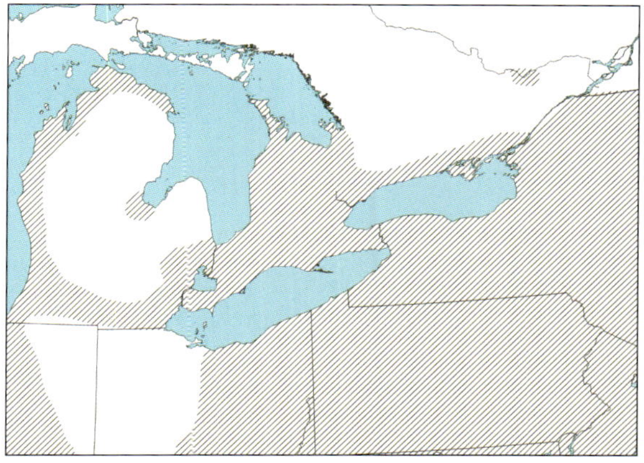

SPREAD 60–90 cm

BLOOM PERIOD Jul, Aug, Sep

COLOR White

FRAGRANT ✔ (foliage, very mild)

SHOWY FRUIT ✖

CUT FLOWER ✔

PESTS No serious insect or disease problems, and deer resistant

NATURAL HABITAT Open woods, dry prairies, and fields

WILDLIFE VALUE Typical visitors include honeybees, butterflies, skippers, and a wide variety of native bees and flies

BUTTERFLY LARVA HOST PLANT FOR None

MOTH LARVA HOST PLANT FOR None

USDA HARDINESS ZONES 4–8

PROPAGATION No pretreatment of seeds is necessary, but the tiny seeds need light to germinate, so simply press them into the soil in spring. Easily propagated by tip cuttings taken in June, or by lifting the clump in late fall or early spring and dividing the shallow root system.

ADDITIONAL INFO Some have reported this mint to be mildly aggressive in some gardens. This is the least aromatic of the mountain mints in this region.

SAR STATUS ON – S3/(not listed)

Leaves

Seed heads

Pycnanthemum virginianum
Virginia Mountain Mint

QUICK GUIDE

PROPAGATION

PART SHADE　　FULL SUN

WET　　MOIST　　MEDIUM

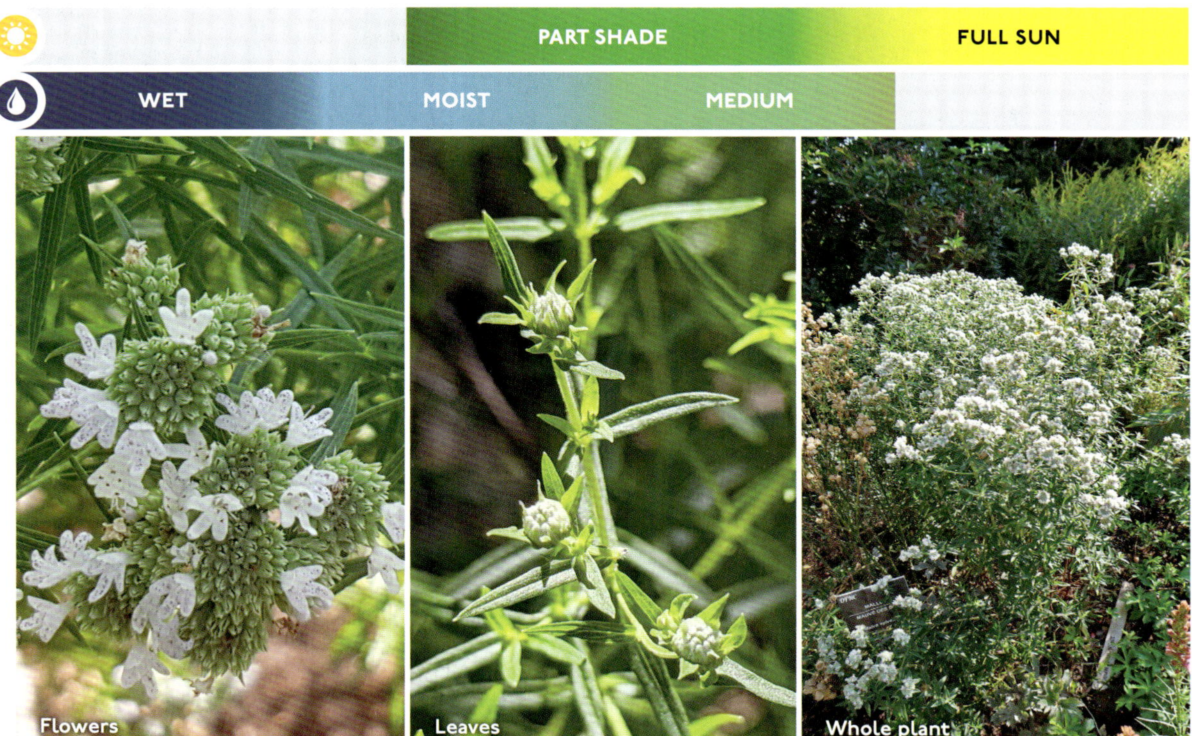

Flowers

Leaves

Whole plant

FAMILY *Lamiaceae* (Mint Family)

ALTERNATE COMMON NAMES American Mountain Mint, Common Mountain Mint, Mountain Mint, Mountain Thyme, Pennyroyal, Prairie Hyssop, Virginia Thyme, Wild Basil, Wild Hyssop

PLANT DESCRIPTION Virginia Mountain Mint is a bushy plant with frequently branching stems that are four sided and green to reddish and have scattered hairs along the edges. Opposite, stalkless leaves are found along the stem, the largest of which measure up to 6 cm long and 1 cm wide. Each leaf is toothless and hairless and has a pointed tip and rounded base. Stems terminate with numerous flat clusters of densely packed, tubular flowers. Each flower is small, at about 0.6 cm wide, and features an upper lip with two lobes and a lower lip with three lobes. The upper lips often look like one lip. Both lips are white with purple spots. Flowers each mature into a dry capsule that holds four tiny, black seeds.

This species is similar to Slender Mountain Mint (*Pycnanthemum tenuifolium*), which has smooth stems and narrower leaves.

IN THE GARDEN There are many reasons to love Virginia Mountain Mint, including its copious, long-lasting blooms and its persistent seed heads that provide excellent winter interest. The minty foliage not only smells delightful but is rarely, if ever, bothered by browsing herbivores.

SKILL LEVEL Beginner

LIFESPAN Perennial

EXPOSURE Full sun to part shade

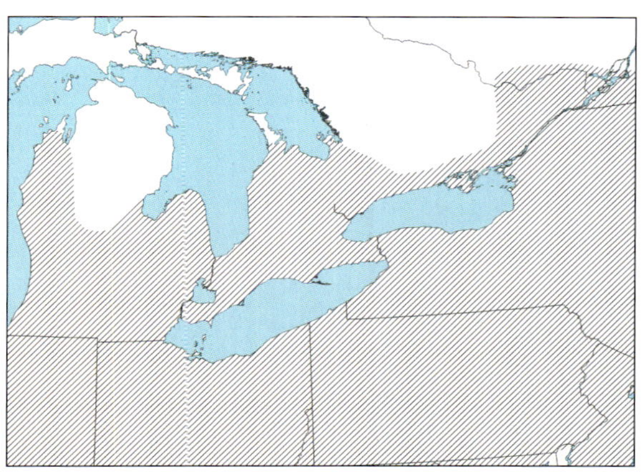

SOIL TYPE Sand, clay, or loam

MOISTURE Medium to wet

HEIGHT 75–90 cm

SPREAD 30–45 cm

BLOOM PERIOD Jul, Aug, Sep

COLOR White with purple

FRAGRANT ✓ (foliage)

SHOWY FRUIT ✗

CUT FLOWER ✓

PESTS No serious insect or disease problems, though stressed plants are susceptible to rust

NATURAL HABITAT Mesic to wet prairies, edges of streams, marshes, and sedge meadows

WILDLIFE VALUE Typical visitors include honeybees and a wide variety of native bees and beetles; its seems to be a favorite of Pearl Crescent (*Phyciodes tharos*) butterflies

BUTTERFLY LARVA HOST PLANT FOR None

MOTH LARVA HOST PLANT FOR None

USDA HARDINESS ZONES 3–7

PROPAGATION No pretreatment of seeds is necessary, but the tiny seeds need light to germinate, so simply press into the soil in spring. Easily propagated by tip cuttings taken in June or by lifting the clump in late fall or early spring and dividing.

ADDITIONAL INFO Virginia Mountain Mint tolerates flooding early in the growing season only. It is drought tolerant. It can be an aggressive spreader but is less so in drier soil.

SAR STATUS N/A

Ranunculus fascicularis
Early Buttercup

FULL SHADE	PART SHADE	FULL SUN
	MEDIUM	DRY

Flowers

Whole plant

FAMILY *Ranunculaceae* (Buttercup Family)

ALTERNATE COMMON NAMES Acrid Crowfoot, Bundle-root Buttercup, Cowslip, Dwarf Buttercup, Early Crowfoot, Low Buttercup, Prairie Buttercup, Thick-root Buttercup, Tufted Buttercup

PLANT DESCRIPTION Early Buttercup features both basal leaves and stems. Basal leaves are borne on long, hairy leaf stalks and divided into three to five leaflets that each measure about 2.5 cm long. Each leaflet is lobed into three to five parts and has rounded tips that sometimes have a sharp point. Leaf surfaces are silky-hairy. Flowering stalks rise up from the basal leaves, bearing one to two stalk-less leaves themselves. These stalks are green to purplish brown and hairy. At the top of each stalk is a single flower characterized by five shiny, yellow, oblong petals surrounding a yellow center that turns green with age. Each flower is about 2.5 cm wide. Behind the flowers are five yellowish-green bracts that are shorter than the petals and covered in spreading hairs. The center of the flower matures into an oval-shaped cluster of beaked seeds.

IN THE GARDEN The radiant yellow flowers of Early Buttercup are some of the first blooms you will see in the spring. This plant is great for rock gardens and borders, where it can be paired with plants of similar height. Herbivores leave this plant alone.

SKILL LEVEL Beginner

LIFESPAN Perennial (spring ephemeral)

EXPOSURE Full sun to full shade

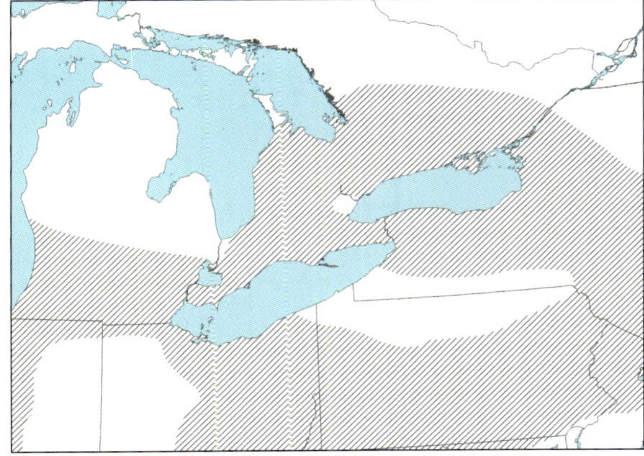

SOIL TYPE Any well-drained soil, prefers a rather poor soil containing rocky material or sand

MOISTURE Dry to medium

HEIGHT 15–20 cm

SPREAD 15 cm

BLOOM PERIOD Apr (mid), May

COLOR Yellow

FRAGRANT ✖

SHOWY FRUIT ✖

CUT FLOWER ✖

PESTS No serious insect or disease problems

NATURAL HABITAT Dry, open woods and prairies

WILDLIFE VALUE A source of pollen for native bees, and the seeds are eaten to a limited extent by various game birds and small mammals

BUTTERFLY LARVA HOST PLANT FOR None

MOTH LARVA HOST PLANT FOR Sparganothis Leafroller Moth (*Sparganothis sulfureana*)

USDA HARDINESS ZONES 3–9

PROPAGATION The seeds are viable for a relatively short period only. Ideally propagate by sowing just before seeds ripen and keep the soil moist. Germination may take some time. Seeds need 60 days of cold, moist stratification if starting indoors.

ADDITIONAL INFO This spring ephemeral goes dormant in the summer. Caution for some people: contact with cell sap can result in skin redness, a burning sensation, and blisters. It does not like competition from taller plants.

SAR STATUS OH – S2/T; PA – S1/E

Ratibida pinnata
Gray-headed Coneflower

QUICK GUIDE

PROPAGATION

 FULL SUN

 MOIST | MEDIUM | DRY

Flowers

Whole plant

FAMILY *Asteraceae* (Aster Family)

ALTERNATE COMMON NAMES Drooping Coneflower, Globular Coneflower, Globular Prairie Coneflower, Gray Headed Mexican Hat, Pinnate Prairie Coneflower, Weary Susan, Yellow Coneflower

PLANT DESCRIPTION Gray-headed Coneflower features slender, ridged stems that feel rough from small hairs. Basal leaves are found at the base of the plant and measure 20 cm long and 12 cm wide. They are pinnately divided into three to seven irregular lobes and sometimes further divided into one to two secondary lobes. You will also notice some leaves along the stem that have fewer lobes than the basal leaves and become smaller as they ascend the stem. Leaves are smooth to sparsely toothed and feel rough from small hairs. One to 12 long-stalked flowers are found at the top of the plant. Each flower has 13 droopy, yellow ray florets (petals) surrounding a gray-brown, 2 cm tall cone. This cone is made up of numerous small, brown disk flowers that bloom from the bottom up. Each disk will mature into a compact head of small, brown seeds.

IN THE GARDEN Gray-headed Coneflower has a lot going for it: drought tolerance, showy flowers, wildlife value, and excellent winter interest. However, the trait that I like best is how the flower heads and droopy petals sway in the breeze to add a new level of visual interest to the landscape.

SKILL LEVEL Beginner

LIFESPAN Perennial

EXPOSURE Full sun

SOIL TYPE Sandy loam to clay loam

MOISTURE Moist to dry

HEIGHT 50–150 cm

SPREAD 60 cm

BLOOM PERIOD Jun, Jul, Aug, Sep

COLOR Yellow

FRAGRANT ✓ (faintly sweet to none; cones emit an anise scent when crushed)

SHOWY FRUIT ✗

CUT FLOWER ✓

PESTS No serious insect or disease problems

NATURAL HABITAT Mesic prairies and savannas, woodland openings and edges, and limestone outcrops

WILDLIFE VALUE A valuable food source for native bees, and goldfinches occasionally eat the seeds, while some mammalian herbivores (especially groundhogs) eat the foliage and flowering stems

BUTTERFLY LARVA HOST PLANT FOR Silvery Checkerspot (*Chlosyne nycteis*)

MOTH LARVA HOST PLANT FOR Wavy-Lined Emerald (*Eynchlora acida*), Common Eupithecia (*Eupithecia miserulata*)

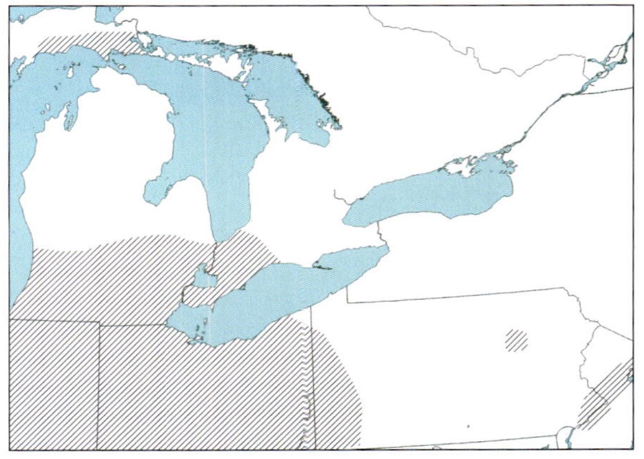

USDA HARDINESS ZONES 3–9

PROPAGATION Surface sow outdoors in the fall, or cold, moist stratify for 30 days if starting indoors or in the spring. Seeds should not be covered with soil. This plant usually blooms in the second year from seed. Clumps may be divided.

ADDITIONAL INFO You may read in some places that this is a "rough" or "messy-looking" plant, but I couldn't disagree more. If overwatered or fertilized, this plant can be leggy and flop over, but under its preferred growing conditions of dry, poor soils, the long-lasting, soft-yellow flowers are an asset to the garden. They are especially magnificent in mass plantings. Deadhead to encourage a second flowering.

SAR STATUS ON – S3/(not listed); PA – S1/E

Rudbeckia fulgida
Orange Coneflower

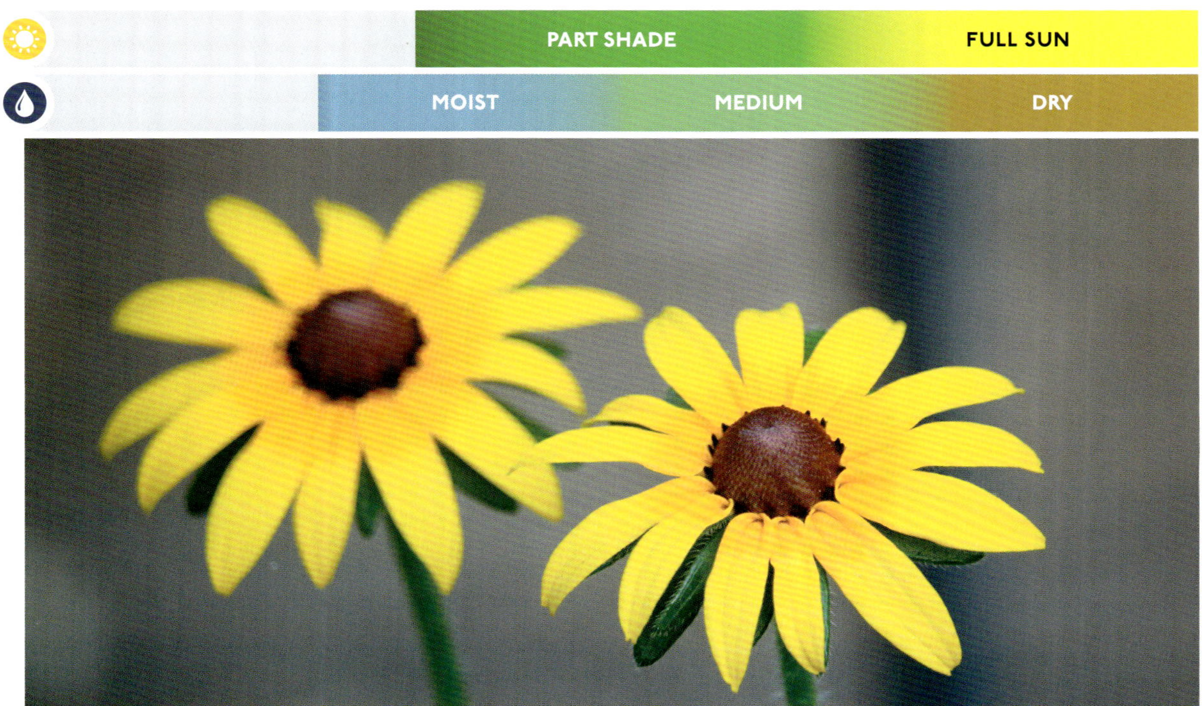

FAMILY *Asteraceae* (Aster Family)

ALTERNATE COMMON NAMES Black-eyed Susan, Brilliant Coneflower, Brown-eyed Susan, Orange Rudbeckia, Perennial Black-eyed Susan, Showy Black-eyed Susan, Showy Coneflower

PLANT DESCRIPTION Orange Coneflower is a heavily branched plant with rigid, sparsely hairy stems. Leaves measure up to 15 cm long, are sparsely haired and ovate to oblong and have sparsely toothed or smooth edges. Stems are topped by daisy-like flowers that measure up to 7.5 cm wide. Flowers are characterized by 10 to 15 yellow ray florets (petals) surrounding a domed, brown center disk. Flowers mature into domed seed heads that contain many small, black seeds.

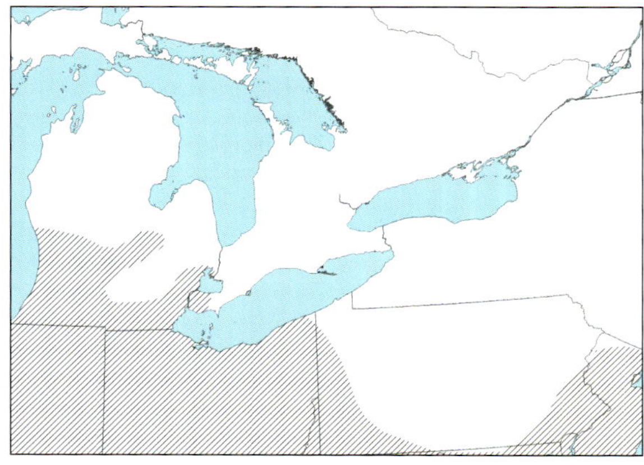

IN THE GARDEN Orange Coneflower is a vigorous plant that blooms with a profusion of long-lasting, yellow flowers. The seed heads persist throughout the winter months. Herbivores usually ignore this plant.

Whole plant

Seed head

Leaves

SKILL LEVEL Beginner

LIFESPAN Perennial

EXPOSURE Full sun to part shade

SOIL TYPE Well drained, clay to sandy

MOISTURE Dry to moist

HEIGHT 60–90 cm

SPREAD 60–75 cm

BLOOM PERIOD Jul, Aug, Sep, Oct (to frost)

COLOR Yellow

FRAGRANT ✖

SHOWY FRUIT ✖

CUT FLOWER ✔

PESTS No serious insect or disease problems

NATURAL HABITAT Moist woods, meadows, savannas, and swamps

WILDLIFE VALUE Visited by butterflies and other pollinating insects for its excellent source of nectar

BUTTERFLY LARVA HOST PLANT FOR Silvery Checkerspot (*Chlosyne nycteis*)

MOTH LARVA HOST PLANT FOR Blackberry Looper Moth (*Chlorochlamys chloroleuca*), Common Eupithecia (*Eupithecia miserulata*), Sunflower Moth (*Homoeosoma electellum*), Meadowrue Borer (*Papaipema unimoda*), Wavy-lined Emerald (*Synchlora aerata*)

USDA HARDINESS ZONES 4–8

PROPAGATION Sow seeds in fall or provide 60 days of cold, moist stratification. Plants slowly spread in the garden by rhizomes, which can be divided every four years in very early spring.

ADDITIONAL INFO Longer bloom times can be achieved by deadheading. The commonly grown cultivar "Goldsturm" originated from this species.

SAR STATUS ON – S1/(not listed); IN (*R. fulgida* var. *fulgida*) – S3/WL; PA – S3/(not listed)

Rudbeckia hirta
Black-eyed Susan

PART SHADE | FULL SUN

MOIST | MEDIUM | DRY

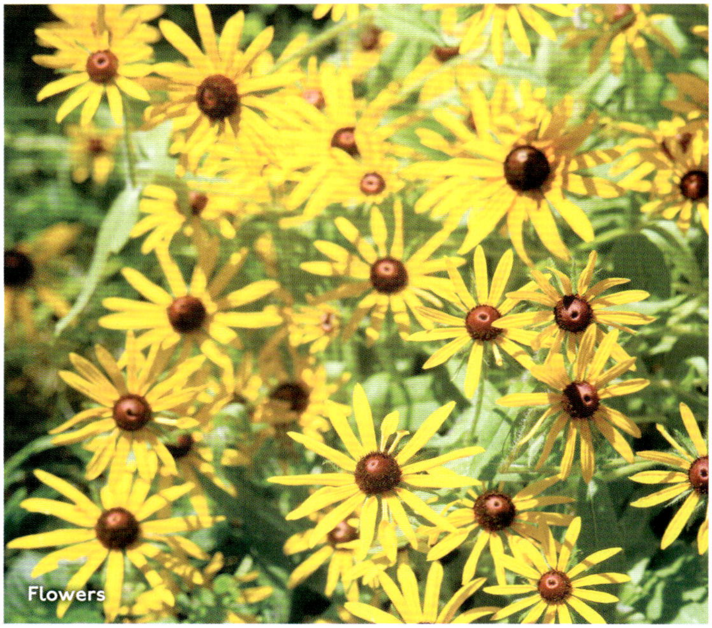

Flowers

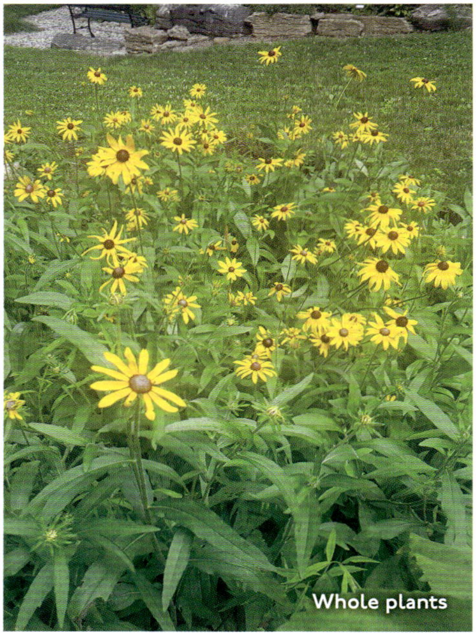

Whole plants

FAMILY *Asteraceae* (Aster Family)

ALTERNATE COMMON NAMES Bristly Cone-flower, Brown Betty, Brown-eyed Susan, Common Black-eyed Susan, English Bull's Eye, Gloriosa Daisy, Golden Jerusalem, Poor-land Daisy, Yellow Daisy, Yellow Ox-eye Daisy

PLANT DESCRIPTION Black-eyed Susan features upright, green-to-reddish, occasionally branching stems that are densely hairy with long, white hairs. Leaf shape is variable, but they range from oblan-ceolate or spatula shaped to ovate. Most leaves are basal and measure up to 18 cm long and 5 cm wide with long stalks. Basal leaves usually die back by flowering time. Stem leaves are found in an alter-nate arrangement and become shorter stalked as they climb the stem. All leaves are densely hairy

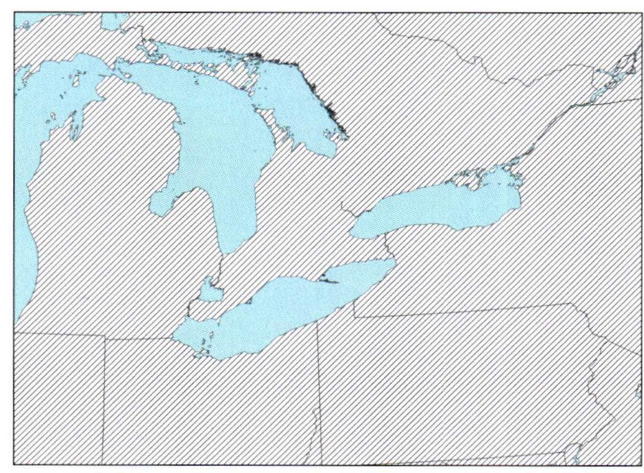

with smooth margins or a few blunt teeth. Each stem terminates with a single flower that measures up to 7.6 cm wide. They are characterized by eight to

"

Leaf

Flower with a Harris Checkerspot

Seed head

20 yellow ray florets (petals) surrounding a domed, brown center disk. This center disk matures into a head of small, black seeds.

IN THE GARDEN Black-eyed Susan is a classic wildflower with long-lasting, yellow flowers. Being a short-lived plant, it persists in the garden by self-seeding where it pleases; whether or not this is a good thing is up to you, but I like how it adds a dynamic effect to a space. The attractive seed heads persist through the winter months. Herbivores rarely touch this plant due to its dense hairs.

SKILL LEVEL Beginner

LIFESPAN Annual, biennial, or short-lived perennial

EXPOSURE Full sun to part shade

SOIL TYPE Any reasonably fertile, well-drained soil

MOISTURE Dry to moist

HEIGHT 100 cm

SPREAD 30–60 cm

BLOOM PERIOD Jul, Aug, Sep

COLOR Yellow

FRAGRANT ❌

SHOWY FRUIT ❌

CUT FLOWER ✅

PESTS No serious insect or disease problems, though stands can be affected by powdery mildew

NATURAL HABITAT Fields, open woods, and roadsides

WILDLIFE VALUE A wide range of insects, particularly bees and flies, as well as some wasps, butterflies, and beetles, are attracted to the flowers, and the seeds are eaten occasionally by goldfinches

BUTTERFLY LARVA HOST PLANT FOR Gorgone Checkerspot (*Chlosyne gorgone*), Silvery Checkerspot (*Chlosyne nycteis*), Pearl Crescent (*Phyciodes tharos*)

MOTH LARVA HOST PLANT FOR Common Pug (*Euphithecia miserulata*)

USDA HARDINESS ZONES 3–9

PROPAGATION Black-eyed Susan propagates very easily from seed sown in the fall or spring. Spring-sown seed should be cold, moist stratified for 30 days. Surface sow as the seeds need light to germinate and should germinate in one to two weeks with adequate moisture.

ADDITIONAL INFO Deadhead to prolong the bloom season or to prevent volunteers — those plants that pop up in your garden without any help from you. There is a lot of controversy over the native status of *Rudbeckia hirta* in many of the northeastern states, with opponents claiming there are few historical records to substantiate the claim, and proponents arguing that given the ubiquity of the plants throughout the region, it must be native. I'll let you decide.

SAR STATUS N/A

Rudbeckia laciniata
Green-headed Coneflower

Flowers

Whole plant

FAMILY *Asteraceae* (Aster Family)

ALTERNATE COMMON NAMES Cutleaf Coneflower, Outhouse Plant, Sochan, Tall Coneflower, Wild Goldenglow

PLANT DESCRIPTION Green-headed Coneflower has light green, smooth stems that sometimes have a waxy bloom on them. They branch occasionally in the upper part of the plant. Basal and lower stem leaves measure up to 25 cm long and wide. They are borne on 10 cm long leaf stalks and are deeply lobed into three to seven segments, with each lobe being irregularly toothed. Leaves become smaller stalked and fewer lobed as they climb the stem, becoming stalkless and unlobed towards the top. Basal leaves often wither away by flowering time. All leaves are smooth but may have sparse hairs. Stems

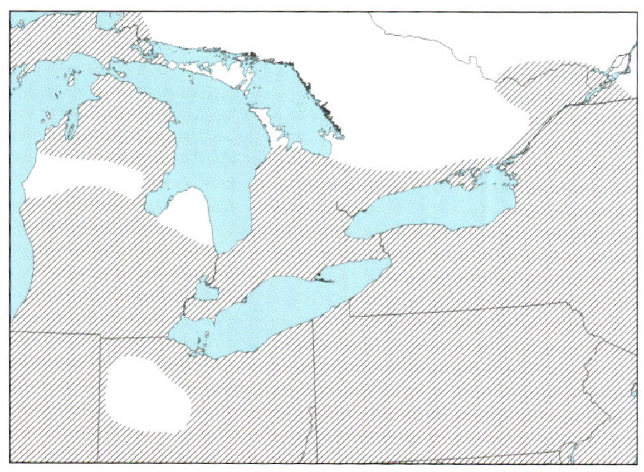

are topped with two to 25 daisy-like flowers borne singly on long stalks. Each flower is about 7.5 cm wide and features six to 12 yellow ray florets (petals)

surrounding a greenish-yellow center cone. The center cone is made of many small disk flowers that bloom from the bottom up. The center cone matures into a head of long, dry, black, four-sided seeds.

IN THE GARDEN The gorgeous, long-blooming yellow flowers of Green-headed Coneflower rise tall above lush, green foliage. This plant has an impressive shade tolerance. The stems and seed heads persist well into the winter months to extend seasonal interest.

SKILL LEVEL Beginner

LIFESPAN Perennial

EXPOSURE Full sun to full shade

SOIL TYPE Sand to clay

MOISTURE Medium to moist

HEIGHT 90–360 cm

SPREAD 45–90 cm

BLOOM PERIOD Jul, Aug, Sep, Oct

COLOR Yellow

FRAGRANT ✔

SHOWY FRUIT ✘

CUT FLOWER ✔

PESTS No serious insect or disease problems

NATURAL HABITAT Low, rich woods and wet fields, fens, and floodplain thickets

WILDLIFE VALUE Many native bees and butterflies are attracted to the flowers for nectar and/or pollen, and goldfinches are attracted to the seeds

BUTTERFLY LARVA HOST PLANT FOR Gorgone Checkerspot (*Chlosyne gorgone*), Silvery Checkerspot (*Chlosyne nycteis*)

MOTH LARVA HOST PLANT FOR Common Pug (*Eupithecia miserulata*), Meadowrue Borer (*Papaipema unimoda*), Wavy-lined Emerald (*Synchlora aerata*), some tortrix moths

USDA HARDINESS ZONES 3–9

PROPAGATION If starting seeds indoors, they require 30 days of cold, moist stratification, otherwise direct sow outdoors in the fall. Plants may be divided in spring before they get too tall.

ADDITIONAL INFO Because it spreads aggressively by underground stems, Green-headed Coneflower is only suitable for larger garden beds. A double-flowered heirloom cultivar from the Victorian era, *Rudbeckia laciniata* "Hortensia," is commonly known as the Outhouse Plant because it was often planted to beautify (and perhaps give some privacy to) that important structure. However, these double blooms don't provide much benefit to pollinators. Green-headed Coneflower may need staking in the average garden but otherwise is very hardy.

SAR STATUS N/A

Sanguinaria canadensis
Bloodroot

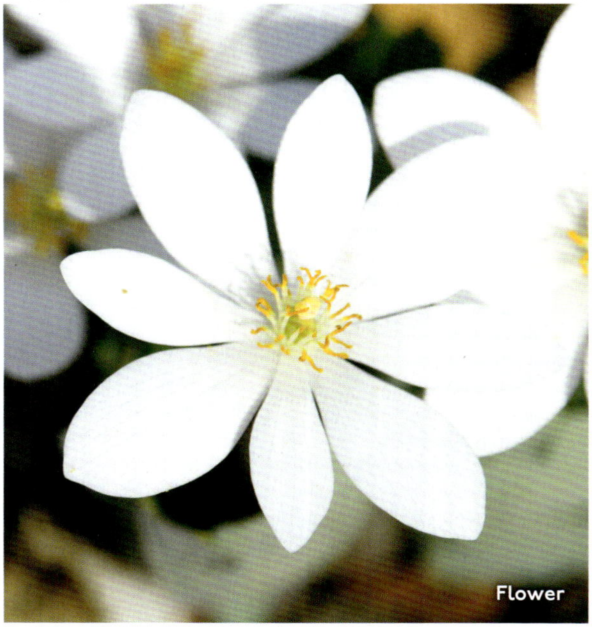

Flower

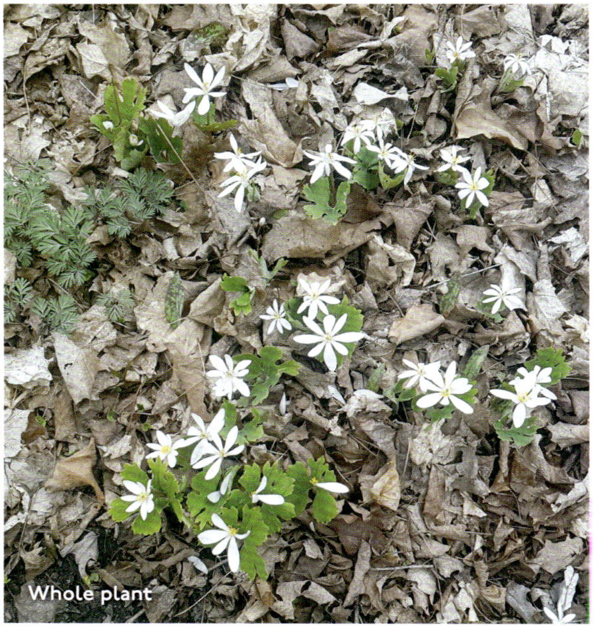

Whole plant

FAMILY *Papaveraceae* (Poppy Family)

ALTERNATE COMMON NAMES Bloodwort, Indian Paint, Puccoon, Red Puccoon

PLANT DESCRIPTION Bloodroot only has basal leaves. They measure up to 13 cm wide, are lobed into three to nine parts, and have a deep indent at the base. The leaf edges have shallow, rounded teeth and the leaf surfaces are smooth. Flowers open before the leaves fully unfurl in the spring. A single flower is borne at the top of each naked, 10 cm tall, reddish stem. Each flower measures 7.6 cm wide and is characterized by eight to 16 white petals surrounding numerous yellow stamens. Each flower matures into a long, tapered seed capsule that splits open to release 10 to 15 dark red seeds.

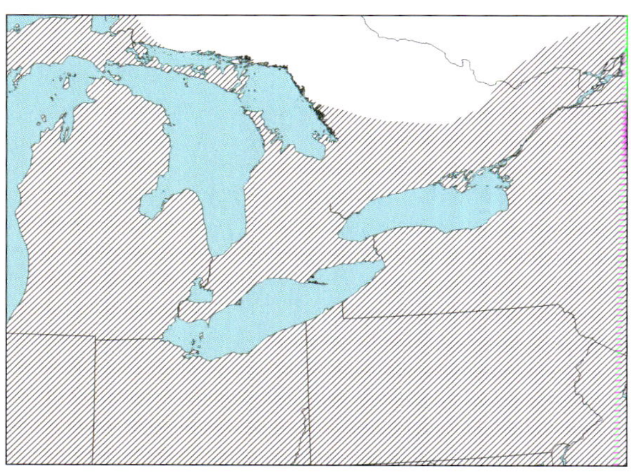

IN THE GARDEN The early, fleeting beauty of Bloodroot flowers is a springtime show you don't want to miss! Each delicate flower blooms for only

Leaf

Seed capsule

Open seed capsule

Seeds with elaiosomes attached, with a Canadian dime for scale

one to two days, but the bold leaves that emerge shortly after will persist well into late summer and make an excellent groundcover. The foliage is not often eaten by herbivores.

SKILL LEVEL Beginner

LIFESPAN Perennial

EXPOSURE Full shade to part shade (during early to midspring, this plant should have access to some sunlight, otherwise the flowers may fail to open)

SOIL TYPE Well-drained, humus-rich soils

MOISTURE Medium

HEIGHT 10–20 cm

SPREAD 7.5–15 cm

BLOOM PERIOD Apr, May

COLOR White

FRAGRANT ✔

SHOWY FRUIT ✘

CUT FLOWER ✘

PESTS No serious insect or disease problems

NATURAL HABITAT Rich deciduous woods and forests

WILDLIFE VALUE Pollen of the flowers attracts various kinds of bees and other insects

BUTTERFLY LARVA HOST PLANT FOR None

MOTH LARVA HOST PLANT FOR Southern Armyworm (*Spodoptera eridania*), Tufted Apple Bud Moth (*Platynota idaeusalis*)

USDA HARDINESS ZONES 3–8

PROPAGATION The most reliable method of propagation is by sowing seeds, which have a double dormancy requiring two 30-day periods of cold separated by a 30-day mild period. Seeds must not be allowed to dry out and are best planted immediately following harvest. In nature they take two years to sprout, and some seeds may not sprout for two years even with artificial stratification. Plants can be propagated by rhizome division in either fall or early spring, but wear gloves and wash your hands after handling the roots as the sap is potentially toxic. Bloodroot is a challenge to germinate and grow to maturity. I have had considerable success growing new Bloodroot plants from pieces that break off when being dug up in my garden — as long as there is a piece of root still attached.

ADDITIONAL INFO Bloodroot seeds are dispersed by ants, which take the seeds back to their nest to consume the energy-rich appendage called the elaiosome before discarding the seed.

SAR STATUS N/A

Silphium laciniatum
Compass Plant

FULL SUN

DRY

Flower

Flowers from below

FAMILY *Asteraceae* (Aster Family)

ALTERNATE COMMON NAMES None

PLANT DESCRIPTION Compass Plant has thick, light green stems with long hairs. They are unbranching except for around the flowers. At the base of the plant, you will find stalked or stalk-less basal leaves that measure up to 60 cm long and 30 cm wide. They are finely hairy, broadly lanceolate, deeply lobed, and toothless (or have a few irregular teeth). Leaves are also found along the stems. These leaves become smaller and shorter stalked as they ascend the stem and are unlobed towards the top. Clusters of stalked, sunflower-like flowers are found at the top of the stem sild arise from upper leaf axils (where the leaf meets the stem). Each flower is 10 cm wide with 17 to 35 yellow ray florets surrounding a yellow center disk. Flowers are backed by densely hairy, lance-shaped bracts. Ray florets mature into flat, dark seeds that can be carried by the wind.

IN THE GARDEN The sunflower-like bloom of Compass Plant towers over its large, distinctive, deeply cut foliage. The sturdy stems make it an excellent structural plant, and they persist well into the winter months.

SKILL LEVEL Beginner

LIFESPAN Perennial

EXPOSURE Full sun

SOIL TYPE Various well-drained soils

MOISTURE Dry

Leaves

Whole plant

Seed head

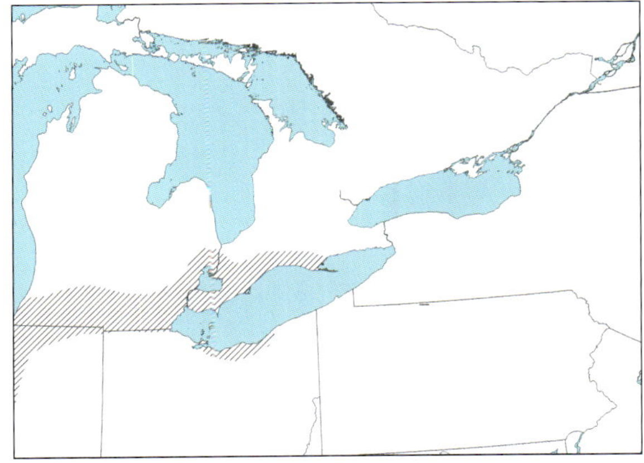

HEIGHT 180–300 cm

SPREAD 60–90 cm

BLOOM PERIOD Jul, Aug, Sep

COLOR Yellow

FRAGRANT ✗

SHOWY FRUIT ✗

CUT FLOWER ✓

PESTS No serious insect or disease problems

NATURAL HABITAT Tall-grass prairie

WILDLIFE VALUE Several native bees and occasionally sulphur butterflies (*Phoebis* spp.) and Monarch Butterflies (*Danaus plexippus*) visit the flowers for nectar, while the large seeds are eaten by birds and small mammals

BUTTERFLY LARVA HOST PLANT FOR None

MOTH LARVA HOST PLANT FOR None

USDA HARDINESS ZONES 4–8

PROPAGATION Seeds germinate after 60 days of cold, moist stratification. Some photodormancy has been reported in seeds, so only cover lightly with soil. Seedlings can take two to four years before they flower. The taproot, which can reach down 4.5 m, precludes division.

ADDITIONAL INFO The name Compass Plant comes from the observation that the leaves tend to orient themselves on a north-south axis. Some specimens have been known to live for 100 years!

SAR STATUS ON – S1/(not listed); MI – S1/T; OH – S1/E

Silphium perfoliatum
Cup Plant

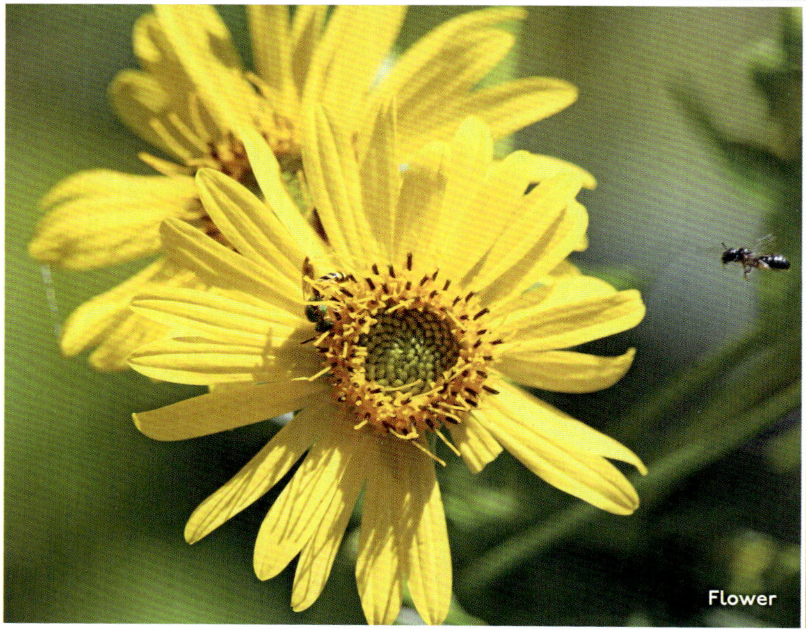

Flower

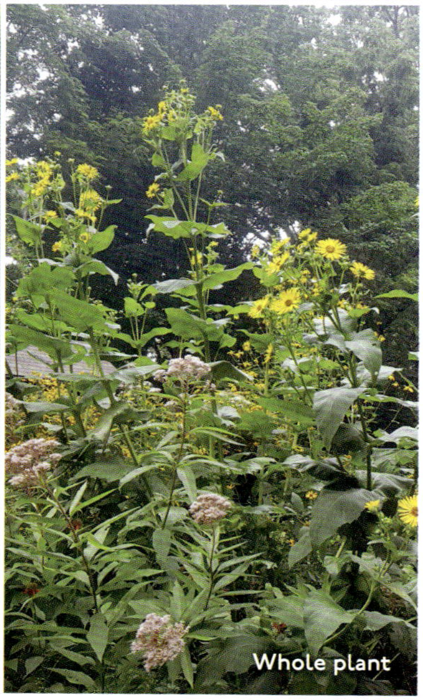

Whole plant

FAMILY *Asteraceae* (Aster Family)

ALTERNATE COMMON NAMES Carpenter's Weed, Cup Rosin Weed, Indian Cup, Ragged Cup

PLANT DESCRIPTION Cup Plant features thick, four-sided, smooth stems that are green to reddish. Along the stem you will find large, opposite leaves that measure up to 20 cm long and just under 13 cm wide. These leaves are fused together around the stem and form a distinctive "cup" that holds rainwater. Leaves are coarsely toothed with pointed tips and rough surfaces. Stalked basal leaves also occur, but they usually die back by flowering time. Upwards of 30 sunflower-like flowers are found in branching clusters at the tops of the stems. Each flower is up to 9 cm wide with 17 to 40 yellow ray florets (petals) surrounding a green to yellow center

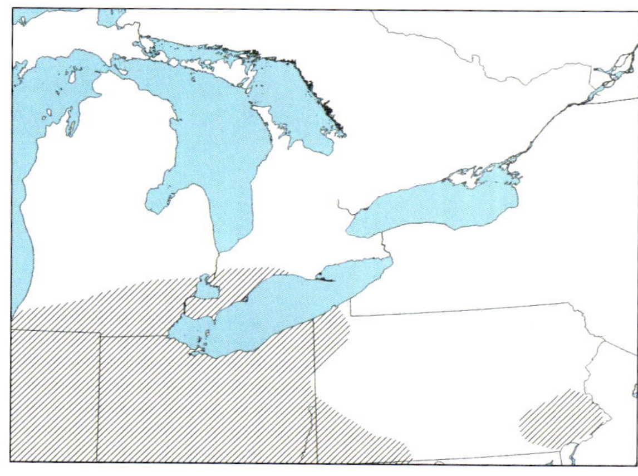

disk. Behind the flowers are smooth, egg-shaped bracts with flaring tips. Ray florets mature into flat, black, winged seeds that can be carried by the wind.

Distinctive cup

Leaf

Seed head

IN THE GARDEN Cup Plant towers above surrounding plants to boast its bright yellow flowers. It makes for a dramatic structural plant, and the rigid stems persist well into the winter months. Great for the back of a border or used as a privacy "hedge." Make sure to give it space!

SKILL LEVEL Beginner

LIFESPAN Perennial

EXPOSURE Full sun to part shade

SOIL TYPE Clay to sand — prefers loamy soil

MOISTURE Wet to dry (may lose lower leaves if dry for too long)

HEIGHT 75–270 cm

SPREAD 60–90 cm

BLOOM PERIOD Jul, Aug, Sep

COLOR Yellow

FRAGRANT ❌

SHOWY FRUIT ❌

CUT FLOWER ✅

PESTS No serious insect or disease problems; however, Cup Plant is sometimes attacked by red aphids, which gather on the undersides of the leaves

NATURAL HABITAT Moist woods, prairies, lowlands, river floodplains, forest openings, and forest edges

WILDLIFE VALUE The flowers attract many native bees and several butterflies — it seems to be a favorite of the Eastern Tiger Swallowtail (*Papilio glaucus*) in my garden — and goldfinches are very fond of the seeds. I have often witnessed a variety of insects and small birds drinking water from the "cup" formed by the leaves

BUTTERFLY LARVA HOST PLANT FOR None

MOTH LARVA HOST PLANT FOR None

USDA HARDINESS ZONES 3–9

PROPAGATION The seeds need to be cold, moist stratified for 60 to 90 days for successful germination, or you can sow them outdoors in the fall.

ADDITIONAL INFO Cup Plant is reportedly sensitive to herbicide drift. Note that this plant is a very prolific spreader and is considered an invasive species in New York, where it is illegal to grow it for sale or distribution.

SAR STATUS ON – S2/(not listed); MI – S2/T

Sisyrinchium angustifolium
Blue-eyed Grass

Flowers

Whole plant

FAMILY *Iridaceae* (Iris Family)

ALTERNATE COMMON NAMES Bermuda Blue-eyed Grass, Common Blue-eyed Grass, Narrowleaf Blue-eyed Grass, Pointed Blue-eyed Grass, Stout Blue-eyed Grass

PLANT DESCRIPTION Blue-eyed Grass features a tuft of basal leaves that emerge straight from the ground. Leaves are very narrow, at 6 mm wide, and have a bluish-green color. Narrow flowering stalks also emerge straight from the ground and reach the same height or just a little bit higher than the leaves. Flowering stalks are winged, sometimes branched, and each ends with a long, leaf-like bract from which a sheath develops. It is from within this sheath that the flowers will emerge. This species has one sheath per stem. The flowers themselves are just over 1 cm wide and feature three petals and three sepals — both of which are bluish purple with pointed tips and dark veins. Each petal and sepal tapers to a narrow base with a yellow mark on it. Flowers mature into three-celled seed capsules that split open to reveal small, black seeds.

IN THE GARDEN Blue-eyed Grass has all the beauty that one would expect from a plant in the Iris family, but in a compact package! It is recognized for its grass-like leaves adorned with star-like, violet-blue flowers. This plant has a clumping habit.

SKILL LEVEL Intermediate

LIFESPAN Perennial

EXPOSURE Full sun to part shade (more sun equals more flowers)

SOIL TYPE Poor-to-average sand, clay, or loam

MOISTURE Medium to wet

HEIGHT 15–30 cm

SPREAD 10–15 cm

BLOOM PERIOD May, Jun

COLOR Blue

FRAGRANT ❌

SHOWY FRUIT ❌

CUT FLOWER ❌

PESTS No serious insect or disease problems

NATURAL HABITAT Meadows, damp fields, and low, open woods

WILDLIFE VALUE Native bees, especially bumble-bees, are attracted to the flowers for its pollen

BUTTERFLY LARVA HOST PLANT FOR None

MOTH LARVA HOST PLANT FOR Verbena Bud Moth (*Endothenia hebesana*)

USDA HARDINESS ZONES 4–9

PROPAGATION The seeds need 60 days of cold, moist stratification for germination, or fall sow outdoors. The small seeds need light to germinate, so merely press them into the soil and do not cover. This plant needs to be divided at least every two to three years or it tends to die out. Several dozen divisions can be expected from a mature, healthy specimen.

ADDITIONAL INFO Another "Blue-eyed Grass," *Sisyrinchium montanum* (sometimes called Mountain Blue-eyed Grass), can be quite challenging to tell apart from *S. angustifolium*. *S. montanum* is a more northern species than its cousin, though their ranges overlap in the southern Great Lakes region. They differ primarily in that *S. montanum* has unbranched flower stems and slightly wider leaves than *S. angustifolium*. Both species are short-lived perennials and are a tad fussy about their growing conditions. They will decline if allowed to dry out, while heavy mulch can cause crown rot and rich, organic soils encourage excessive vegetative growth and few flowers.

SAR STATUS N/A (*Sisyrinchium montanum* is ranked as S1/E in Indiana and S1/T in Ohio)

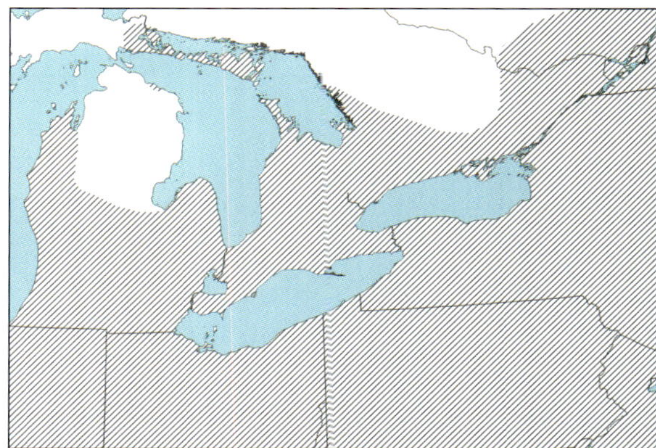

Range map for *Sisyrinchium angustifolium*

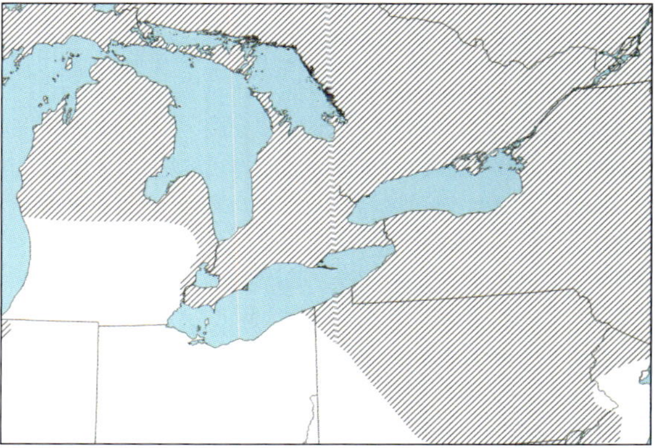

Range map for *Sisyrinchium montanum*

Solidago bicolor
Silverrod

	FULL SHADE	PART SHADE	FULL SUN
		MEDIUM	DRY

FAMILY *Asteraceae* (Aster Family)

ALTERNATE COMMON NAMES White Goldenrod

PLANT DESCRIPTION Silverrod features erect stems that are finely hairy as to appear grayish green. Leaves are alternate, stalked, elliptical to obovate (egg-shaped leaf with the narrower end at the base) with tapered bottoms, and toothed. Each leaf measures up to 10 cm long, but they become smaller as they ascend the stem. The smaller leaves at the top may also lack teeth and leaf stems. Flowers are less than 1 cm wide and found in cylindrical clusters emerging from the uppermost leaf axils (where the leaf meets the stem). Each flower is borne on a short stalk and features seven to nine white ray florets (petals) surrounding a center of yellow stamens. Flowers mature into tiny seeds with tufts of hairs that allow them to be carried by the wind.

IN THE GARDEN Silverrod stands out among other goldenrods with its unique spires of creamy-white flowers. It is adaptable to a variety of conditions, including nutrient-poor soils. This is a clump-forming species, so it won't take over your garden. Herbivores rarely eat this plant.

SKILL LEVEL Beginner

LIFESPAN Perennial

EXPOSURE Full sun to full shade

SOIL TYPE Not fussy about soil type — clay to sandy

MOISTURE Medium to dry (drought tolerant)

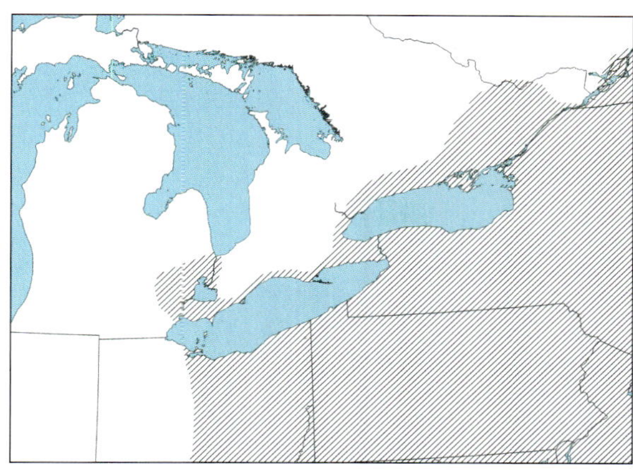

HEIGHT 100 cm

SPREAD 30–60 cm

BLOOM PERIOD Jul, Aug, Sep

COLOR White

FRAGRANT ✗

SHOWY FRUIT ✗

CUT FLOWER ✓

PESTS No serious insect or disease problems

NATURAL HABITAT Sandy and clay soils, dry open woods, rocky slopes, and disturbed soils

WILDLIFE VALUE Flowers are attractive to butterflies, native bees, honeybees, and other pollinators, while songbirds eat the seeds

BUTTERFLY LARVA HOST PLANT FOR None

MOTH LARVA HOST PLANT FOR At least 45 species of moths, including members of the tiger moths, ribbed cocoon-maker moths, case-bearer moths, twirler moths, geometer moths, leaf-blotch miner moths, owlet moths, pyralid moths, trumpet leafminer moths, and tortrix moths

USDA HARDINESS ZONES 3–9

PROPAGATION Seeds germinate after a 60-day period of moist, cold stratification. Seeds are very small and need light to naturally break dormancy and germinate.

ADDITIONAL INFO This very early blooming goldenrod is one of only two white-flowered *Solidago* species (the other, *S. ptarmicoides* [Upland White Goldenrod], was only recently added to the goldenrods) and, except for the white flowers, it is almost identical to *S. hispida* (Hairy Goldenrod), which has yellow blossoms. Some authors consider *S. hispida* to simply be a variety of *S. bicolor*, thus you may see the former plant listed as *Solidago bicolor* var. *hispida*.

SAR STATUS MI – S1/E

Seed heads

Whole plant

Solidago caesia
Blue-stemmed Goldenrod

Flowers

Whole plant

FAMILY *Asteraceae* (Aster Family)

ALTERNATE COMMON NAMES Woodland Goldenrod, Wreath Goldenrod

PLANT DESCRIPTION Blue-stemmed Goldenrod has sparingly branched, arching stems that are green when young but turn bluish purple with age. Leaves are alternate and elliptic-oblong (long and rounded) in shape, and they measure about 12 cm long and 2 cm wide and become smaller as they ascend the stem. They are stalkless and hairless and have serrated edges. Small clusters of one to 12 yellow flowers develop from upper leaf axils (where the leaf meets the stem) and at the very tip of the stem. Each flower is 3 mm wide and has four to five ray florets (petals) surrounding four to five disk flowers. At the base of each flower, you will

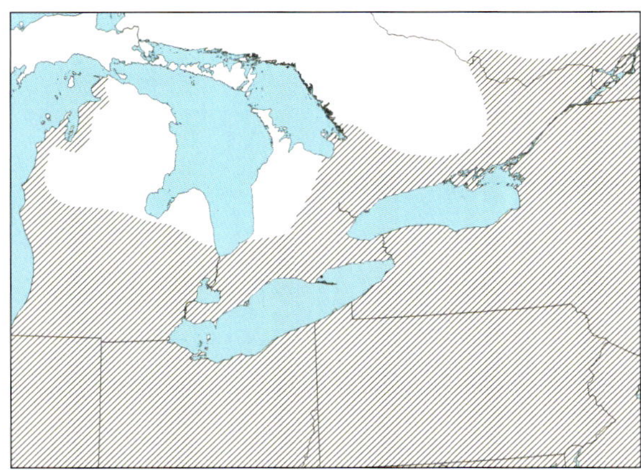

notice small bracts that are overlapping, smooth, and oblong. Flowers mature into small, finely hairy seeds that have small tufts of hairs attached to them.

IN THE GARDEN Blue-stemmed Goldenrod is a shade-loving goldenrod with a well-behaved clumping habit. As if its graceful, arching stems and dark green leaves aren't reason enough to love this plant, it is adorned with bright yellow flowers late in the season that make it a real crowd-pleaser.

SKILL LEVEL Beginner

LIFESPAN Perennial

EXPOSURE Part shade to full sun

SOIL TYPE Most well-drained soils — tolerates poor soil

MOISTURE Medium to dry

HEIGHT 45–90 cm

SPREAD 30–60 cm

BLOOM PERIOD Aug (late), Sep

COLOR Yellow

FRAGRANT ❌

SHOWY FRUIT ❌

CUT FLOWER ✅

PESTS No serious insect or disease problems, though rust may be an occasional problem

NATURAL HABITAT Rich, deciduous or open woods, the edges of woods, and clearings

WILDLIFE VALUE Attracts native and honeybees, wasps, flies, and butterflies, and the seeds are occasionally eaten by several small songbirds

BUTTERFLY LARVA HOST PLANT FOR None

MOTH LARVA HOST PLANT FOR At least 45 species of moths, including members of the tiger moths, ribbed cocoon-maker moths, case-bearer moths, twirler moths, geometer moths, leaf-blotch miner moths, owlet moths, pyralid moths, trumpet leafminer moths, and tortrix moths

USDA HARDINESS ZONES 4–8

PROPAGATION Small seeds need light to germinate and benefit from 90 days of cold, moist stratification. Mature plants may be divided, and new plants can be started from stem cuttings.

ADDITIONAL INFO This species is primarily clump forming and does not spread aggressively as some of the other goldenrod species and hybrids do.

SAR STATUS N/A

Solidago flexicaulis
Zigzag Goldenrod

QUICK GUIDE

PROPAGATION

FULL SHADE · PART SHADE

MEDIUM

Flowers

Whole plant

FAMILY *Asteraceae* (Aster Family)

ALTERNATE COMMON NAMES Broad Leaf Goldenrod

PLANT DESCRIPTION Zigzag Goldenrod has smooth to slightly hairy, green stems. You will notice that the stem usually zigzags between the leaves, but it may also be more on the straight side. The leaves are found in an alternate pattern and measure about 12 cm long and 7.5 cm wide. They are softly hairy and coarsely toothed, have a rounded base with a pointed tip, and are borne on winged leaf stalks. They become smaller as they ascend the stem. Yellow flowers are found in clusters at the top of the plant and from upper leaf axils. Each flower head is about 0.6 cm wide and has three to four yellow ray florets (petals) surrounding four to eight disk flowers. Flowers mature into small, finely hairy seeds with tufts of hairs that allow them to be carried by the wind.

IN THE GARDEN Zigzag Goldenrod fills shady gardens with bold, deeply toothed foliage and spikes of abundant yellow flowers. It spreads nicely to form patches but is not aggressive as other goldenrods can be. Adaptable to a variety of growing conditions.

SKILL LEVEL Beginner to intermediate

LIFESPAN Perennial

EXPOSURE Part shade to full shade (this is one of the most shade-tolerant goldenrods)

SOIL TYPE Most well-drained soils, will tolerate heavy clay

MOISTURE Medium

HEIGHT 90 cm

SPREAD 30–60 cm

BLOOM PERIOD Aug, Sep, Oct

COLOR Yellow

FRAGRANT ❌

SHOWY FRUIT ❌

CUT FLOWER ✅

PESTS No serious insect or disease problems, though rust, powdery mildew, and leaf spot may occur during years with a lot of humidity

NATURAL HABITAT Rich woods and sandy streambanks

WILDLIFE VALUE An important source of pollen and nectar for many kinds of insects, including long-tongued bees, small-tongued bees, wasps, flies, and butterflies, and the seeds are eaten by sparrows and mice

BUTTERFLY LARVA HOST PLANT FOR None

MOTH LARVA HOST PLANT FOR At least 45 species of moths, including members of the tiger moths, ribbed cocoon-maker moths, case-bearer moths, twirler moths, geometer moths, leaf-blotch miner moths, owlet moths, pyralid moths, trumpet leafminer moths, and tortrix moths

USDA HARDINESS ZONES 3–8

PROPAGATION Seeds need light to germinate. Cold, moist stratify for 60 days or direct sow in the fall. Clumps may be divided.

Leaves

Seed heads

ADDITIONAL INFO Zigzag Goldenrod tolerates dry shade.

SAR STATUS N/A

Solidago hispida
Hairy Goldenrod

FULL SHADE	PART SHADE	FULL SUN
	MEDIUM	DRY

Flowers

Spike of flowers

FAMILY *Asteraceae* (Aster Family)

ALTERNATE COMMON NAMES Upland Goldenrod

PLANT DESCRIPTION Hairy Goldenrod is an upright plant with ridged, green-to-purplish, densely hairy stems that are usually unbranched but can be heavily branched in open settings. Most of its leaves are basal, but there are some alternate stem leaves. Basal leaves are obovate (shaped like the longitudinal section of an egg with the narrower end at the base), taper to a winged leaf stalk, have shallow teeth, and are densely hairy above and below. The largest leaves measure up to 20 cm long, but usually closer to 11 cm, and 5 cm wide. As the leaves ascend the stem, they become smaller, more lance-shaped, stalkless, and less toothed. Clusters of flowers arise from the upper leaf axils to form a spike-like column of flowers. Individual flowers are 0.6 cm long and are characterized by seven to 14 ray florets (petals) surrounding a yellow center disk. Behind the flowers you will find three to four layers of flattened, green-tipped bracts that can be hairless or slightly hairy. Flowers mature into small, finely hairy seeds with tufts of hairs that allow the wind to carry them.

IN THE GARDEN Hairy Goldenrod is a versatile, drought-tolerant choice for your garden. It has a slender, upright form and puts on a stunning columnar display of yellow flowers. This is a well-behaved, clumping goldenrod, so it won't take over your garden.

SKILL LEVEL Beginner

LIFESPAN Perennial

EXPOSURE Full sun to full shade

SOIL TYPE Sandy to clay soils and disturbed soils

MOISTURE Medium to dry (drought tolerant)

HEIGHT 45–90 cm

SPREAD 30–45 cm

BLOOM PERIOD Jul, Aug

COLOR Yellow

FRAGRANT ✗

SHOWY FRUIT ✗

CUT FLOWER ✓

PESTS No serious insect or disease problems, though rust can be an occasional problem

NATURAL HABITAT Dry woods and wooded banks

WILDLIFE VALUE Flowers are attractive to butterflies, native bees, honeybees, and other pollinators, and songbirds eat the seeds

BUTTERFLY LARVA HOST PLANT FOR None

MOTH LARVA HOST PLANT FOR At least 45 species of moths, including members of the tiger moths, ribbed cocoon-maker moths, case-bearer moths, twirler moths, geometer moths, leaf-blotch miner moths, owlet moths, pyralid moths, trumpet leafminer moths, and tortrix moths

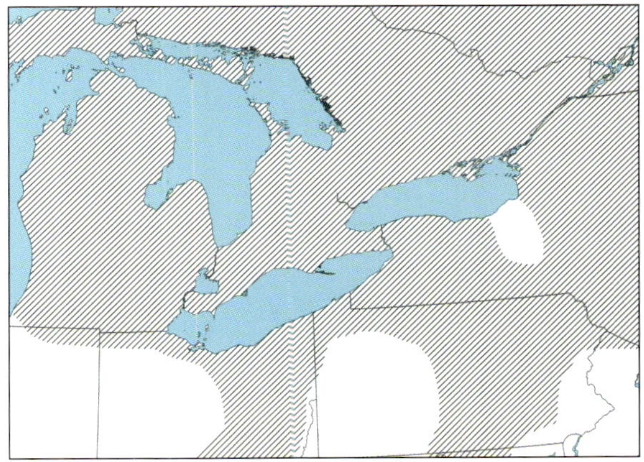

USDA HARDINESS ZONES 2–8

PROPAGATION Seeds need light to germinate and 60 days of cold, moist stratification, or you can scatter seeds outdoors in the fall. Plants may also be divided in early spring.

ADDITIONAL INFO Other than the yellow flowers, this species looks very similar to the white-flowered *Solidago bicolor* (Silverrod), and some authors suggest it is simply a variety of that species. Most, however, treat it as a separate species.

SAR STATUS IN – S3/WL

Solidago juncea
Early Goldenrod

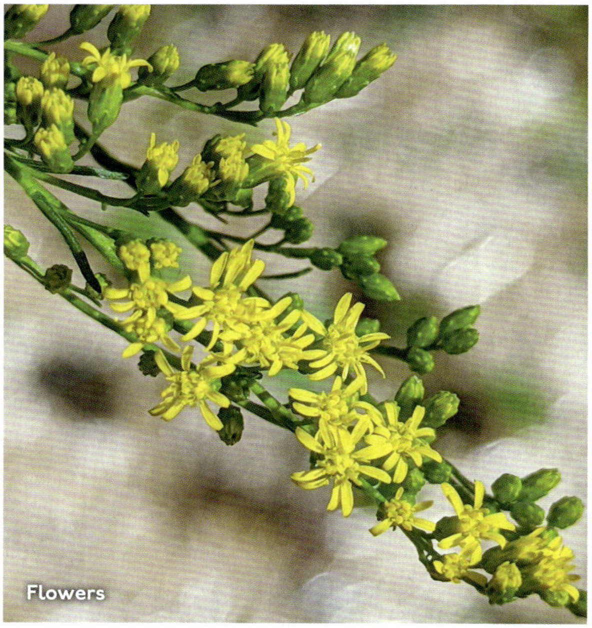

Flowers

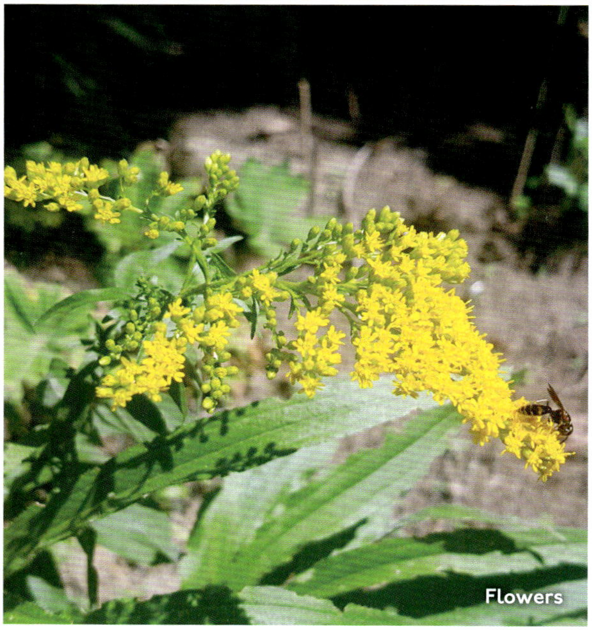

Flowers

FAMILY *Asteraceae* (Aster Family)

ALTERNATE COMMON NAMES Sharp-toothed Goldenrod

PLANT DESCRIPTION Early Goldenrod stems are unbranched, ridged, hairless, and green to reddish. It has both basal leaves and alternate stem leaves. Basal and lower stem leaves are lance-elliptic to spatula shaped, taper to a pointed tip, are borne on long, winged leaf stalks, and measure up to 30 cm long and 5 cm wide. They have toothed edges and are hairless, except for fine hairs around the leaf edges. As the leaves ascend the stem, they become smaller, more lance-shaped, stalkless, and toothless. You will also notice small leaves emerging from the upper leaf axils. Flowers are found in clusters of broadly spreading branches that arch outwards.

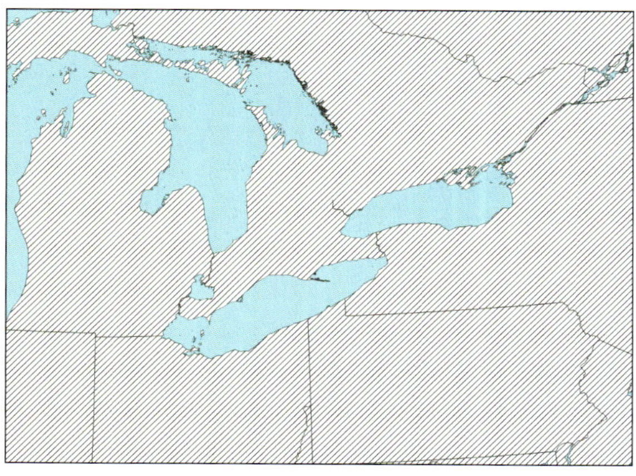

The flowers all emerge from one side of the branch on short stalks. Each flower measures about 0.6 cm wide and has four to 12 unevenly spaced, yellow ray

Leaf

Whole plant

Seed heads

florets (petals) surrounding a yellow center with eight to 15 disk flowers. Flowers mature into finely hairy, narrow seeds each with a tuft of hair that allows them to be carried by the wind.

Early Goldenrod can be distinguished from other goldenrods by its overall lack of hairs (except for the leaf edges) and by the winged leaf stalks that emerge from upper leaf axils. In addition, its flower clusters arch outwards rather than upwards.

IN THE GARDEN Early Goldenrod puts on a fireworks-like display of bright yellow flowers before most other goldenrods have started blooming.

SKILL LEVEL Beginner

LIFESPAN Perennial

EXPOSURE Full sun to part shade

SOIL TYPE Sandy to clay soils

MOISTURE Medium to dry (drought tolerant)

HEIGHT 200 cm

SPREAD 60–90 cm

BLOOM PERIOD Jun (late), Jul, Aug, Sep, Oct

COLOR Yellow

FRAGRANT ✘

SHOWY FRUIT ✘

CUT FLOWER ✔

PESTS No serious insect or disease problems, though leaf rust can be an occasional problem

NATURAL HABITAT Open areas, dry old fields, and other sunny, disturbed ground

WILDLIFE VALUE Flowers are attractive to butterflies, native bees, honeybees, and other pollinators, and songbirds eat the seeds

BUTTERFLY LARVA HOST PLANT FOR None

MOTH LARVA HOST PLANT FOR At least 45 species of moths, including members of the tiger moths, ribbed cocoon-maker moths, case-bearer moths, twirler moths, geometer moths, leaf-blotch miner moths, owlet moths, pyralid moths, trumpet leafminer moths, and tortrix moths

USDA HARDINESS ZONES 3–8

PROPAGATION Seeds need light and 60 days of cold, moist stratification to germinate if starting indoors, or you can scatter seeds outdoors in the fall. Mature clumps may be divided.

ADDITIONAL INFO Early Goldenrod tolerates acidic soil.

SAR STATUS N/A

253

Solidago nemoralis
Gray Goldenrod

Flowers

Whole plant

FAMILY *Asteraceae* (Aster Family)

ALTERNATE COMMON NAMES Dwarf Gold-enrod, Dyersweed Goldenrod, Field Goldenrod, Gray-stemmed Goldenrod, Old Field Goldenrod, Prairie Goldenrod

PLANT DESCRIPTION Gray Goldenrod features unbranched stems that are reddish to gray green and covered in dense hairs. It has both basal and stem leaves. Basal leaves are variable in shape but are generally spoon shaped with a gradual taper to a winged leaf stalk. Each leaf measures about 10 cm long and 1.2 cm wide and has smooth to slightly toothed edges. Stem leaves are narrowly lance shaped and stalkless and become smaller as they ascend the stem. You will notice that small leaflets emerge from upper leaf axils. All leaves are gray

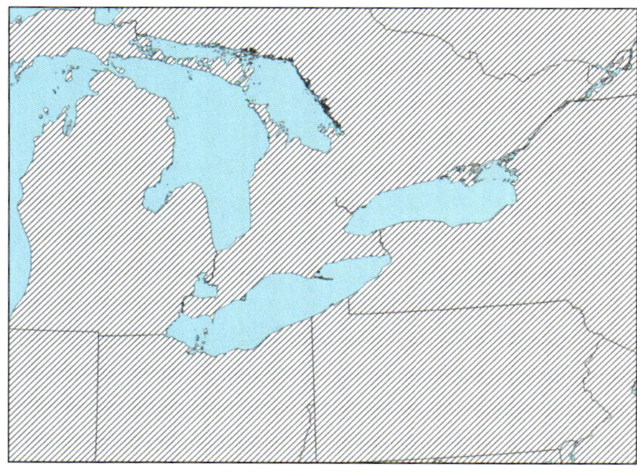

green due to short, white hairs. Flowers are found in a 20 cm long, arching cluster at the top of the stem. Individual flowers are all arranged on one side of

the stem on short stalks and feature four to 10 ray florets (petals) surrounding a yellow center. Flowers mature into small seeds that develop tufts of hairs that allow them to be carried by the wind.

IN THE GARDEN Gray Goldenrod has a compact form and beautiful arching flower clusters. It is one of the later-blooming goldenrods, so it is great for extending bloom time in your yard. It is also a good choice for difficult locations, such as sunny slopes or open areas with poor soil, where little else will grow.

SKILL LEVEL Beginner

LIFESPAN Perennial

EXPOSURE Full sun (tolerates light shade)

SOIL TYPE Sand, clay, or gravel

MOISTURE Dry

HEIGHT 30–75 cm

SPREAD 30–60 cm

BLOOM PERIOD Aug, Sep, Oct

COLOR Yellow

FRAGRANT ❌

SHOWY FRUIT ❌

CUT FLOWER ✅

PESTS No serious insect or disease problems in its natural setting, though it is quite susceptible to verticillium wilt if planted where plant density is high and morning dew persists long into the day

NATURAL HABITAT Disturbed sand and gravel soils of roadsides, fields, and openings in deciduous forests

WILDLIFE VALUE A wide variety of insects, including native bees and butterflies, visit the flowers for pollen and nectar, and seeds are eaten by songbirds

BUTTERFLY LARVA HOST PLANT FOR None

MOTH LARVA HOST PLANT FOR At least 45 species of moths, including members of the tiger moths, ribbed cocoon-maker moths, case-bearer moths, twirler moths, geometer moths, leaf-blotch miner moths, owlet moths, pyralid moths, trumpet leafminer moths, and tortrix moths

USDA HARDINESS ZONES 2–9

PROPAGATION Seeds need light and at least 60 days of cold, moist stratification to germinate. New plants can also be started by taking four-to-six-node softwood stem cuttings in the late spring, which will readily root. In addition, mature clumps may be divided before growth begins in the spring.

ADDITIONAL INFO Gray Goldenrod is a somewhat weedy plant that can colonize an area by creeping rhizomes and self-seeding. To prevent seed dispersal, remove flower heads before seeds ripen. *Solidago nemoralis* is one of the shortest goldenrods — only *S. ptarmicoides* (Upland White Goldenrod) is shorter. Note that Gray Goldenrod does not like competition from taller plants.

SAR STATUS N/A

Solidago ohioensis
Ohio Goldenrod

QUICK GUIDE

PROPAGATION

FULL SUN

MOIST · **MEDIUM**

Flowers

Whole plant

FAMILY *Asteraceae* (Aster Family)

ALTERNATE COMMON NAMES None

PLANT DESCRIPTION Ohio Goldenrod features smooth, erect stems. It has both basal leaves and alternate stem leaves. Basal leaves are lance shaped with a blunt tip, smooth with a prominent central vein, and borne on long, winged leaf stalks and measure up to 25 cm long (some describe them as looking like rabbit ears). Stem leaves point upwards and become smaller and shorter stalked as they ascend the stem. Numerous small, yellow flowers are found in a flat-topped, branching cluster at the top of the plant. Each flower is about 0.6 cm wide and features six to eight ray florets (petals) surrounding eight to 20 disk flowers. Flowers mature into angled, smooth, dry seeds that have tufts of hairs that allow them to be carried by the wind.

Ohio Goldenrod is similar to Riddell's Goldenrod (*Solidago riddellii*), but the latter has folded leaves with multiple veins.

IN THE GARDEN Ohio Goldenrod has a compact, upright form that is topped by striking, flat-topped flower clusters. Great for moist, sunny sites. It will self-seed but not aggressively.

SKILL LEVEL Beginner

LIFESPAN Perennial

EXPOSURE Full sun (though some literature suggests it will tolerate shade, even full shade)

SOIL TYPE Most well-drained soils

MOISTURE Moist to medium

Leaves

Seed heads

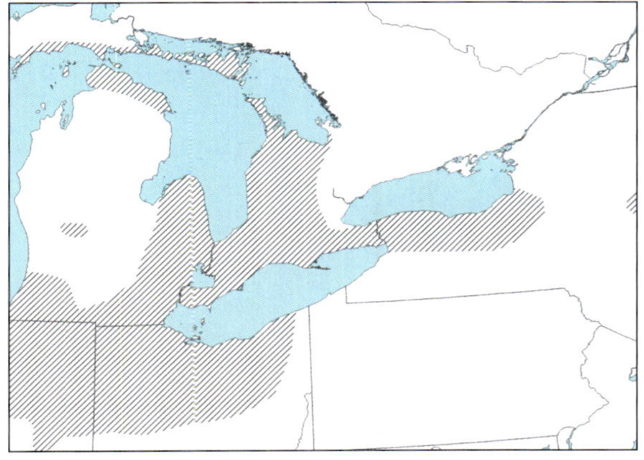

Flowers

HEIGHT 100 cm

SPREAD 30–60 cm

BLOOM PERIOD Jul, Aug, Sep

COLOR Yellow

FRAGRANT ❌

SHOWY FRUIT ❌

CUT FLOWER ✓

PESTS No serious insect or disease problems, though leaf rust can be an occasional problem

NATURAL HABITAT Wet depressions in sand dunes, marshes, fens, and riverbanks

WILDLIFE VALUE Attractive to bees and butterflies

BUTTERFLY LARVA HOST PLANT FOR None

MOTH LARVA HOST PLANT FOR At least 45 species of moths, including members of the tiger moths, ribbed cocoon-maker moths, case-bearer moths, twirler moths, geometer moths, leaf-blotch miner moths, owlet moths, pyralid moths, trumpet leafminer moths, and tortrix moths

USDA HARDINESS ZONES 4–9

PROPAGATION Seeds need light to germinate and will benefit from 30 days of cold, moist stratification (though stratification is not required for adequate germination), or you can surface sow outdoors in the fall. Plants do not tolerate dry soils, so keep moist.

ADDITIONAL INFO This species is sometimes placed in the genus *Oligoneuron* (*Oligoneuron ohioense*) as a member of the flat-topped goldenrods, most of which are found in the Great Lakes region. Most authorities, however, claim there is no real compelling argument for separating this species from *Solidago*.

SAR STATUS NY – S2/T; OH – S3/(not listed)

257

Solidago ptarmicoides
Upland White Goldenrod

QUICK GUIDE

PROPAGATION

 FULL SUN

 MEDIUM   DRY

Flowers

Flowers

Whole plant

FAMILY *Asteraceae* (Aster Family)

ALTERNATE COMMON NAMES Prairie Aster, Prairie Flat-top Goldenrod, Prairie Goldenrod, Sneezewort Aster, Snowy Aster, Stiff Aster, Upland White Aster, Upland Whiterod, White Flat-top Goldenrod, White Prairie Goldenrod

PLANT DESCRIPTION Upland White Goldenrod features rigid, upright stems that are unbranched, except for around the flower clusters. The lower parts of the stems are smooth while the upper parts may be roughly hairy. Basal and lower leaves are lanceolate-oval in shape, taper to a stalked base, and measure up to 20 cm long and 1 cm wide. They are smooth to rough textured (especially on the underside of the leaf) and mostly toothless, except for a few sharp, shallow teeth towards the tip of the leaf. As the leaves ascend the stem, they become smaller and stalkless. Flowers are found in flat clusters at the top of the plant and are made up of up to 60 individual, daisy-like flowers. Each flower measures just over 1 cm wide, is borne on a 2.5 cm long leaf stalk, and features 10 to 20 white ray florets (petals) surrounding a pale yellow center. Flowers mature into small seeds with tufts of hairs that allow them to be carried by the wind.

IN THE GARDEN Upland White Goldenrod has earned a place in gardens with its stylish white blooms and low, mounding habit. It grows very well in poor and shallow soils. It may self-seed in open areas but is not aggressive and can be used in smaller spaces.

SKILL LEVEL Beginner

Leaves

Cluster of flowers

Seed heads

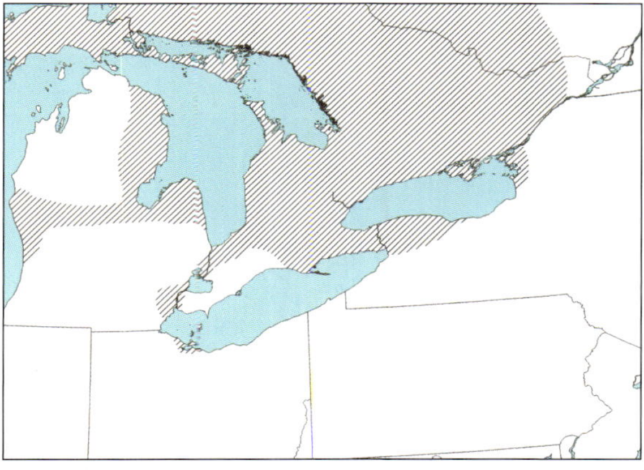

LIFESPAN Perennial

EXPOSURE Full sun (tolerates some light shade)

SOIL TYPE Neutral, sandy loams

MOISTURE Dry to medium (tolerates moist loams as long as they're well-drained)

HEIGHT 30–60 cm

SPREAD 30–45 cm

BLOOM PERIOD Jul, Aug, Sep

COLOR White

FRAGRANT ✗

SHOWY FRUIT ✗

CUT FLOWER ✓

PESTS No serious insect or disease problems, though it may be susceptible to rust, powdery mildew and leaf spot in some years, and root rot may occur in overly moist or poorly drained soils

NATURAL HABITAT Calcareous barren ground that is sandy, gravelly, or rocky

WILDLIFE VALUE As with all goldenrods, this is an important source of pollen and nectar for bees and other insects

BUTTERFLY LARVA HOST PLANT FOR None

MOTH LARVA HOST PLANT FOR At least 45 species of moths, including members of the tiger moths, ribbed cocoon-maker moths, case-bearer moths, twirler moths, geometer moths, leaf-blotch miner moths, owlet moths, pyralid moths, trumpet leafminer moths, and tortrix moths

USDA HARDINESS ZONES 3–8

PROPAGATION No pretreatment of seeds is needed for germination, but seeds need light to germinate so surface sow or cover very lightly. Mature clumps may be divided.

ADDITIONAL INFO As one of only two white goldenrods (the other is *Solidago bicolor*, or Silverrod), it is also the shortest of the goldenrods. For many years, this aster lookalike was considered to be a member of the genus *Aster*. However, it was observed to hybridize with *Solidago rigida*, *S. ohioensis*, and *S. riddellii* but not with any species of aster, and thus it is now placed in the *Solidago* genus.

SAR STATUS IN – S3/T; NY – S3/R; OH – SH/X

Solidago riddellii
Riddell's Goldenrod

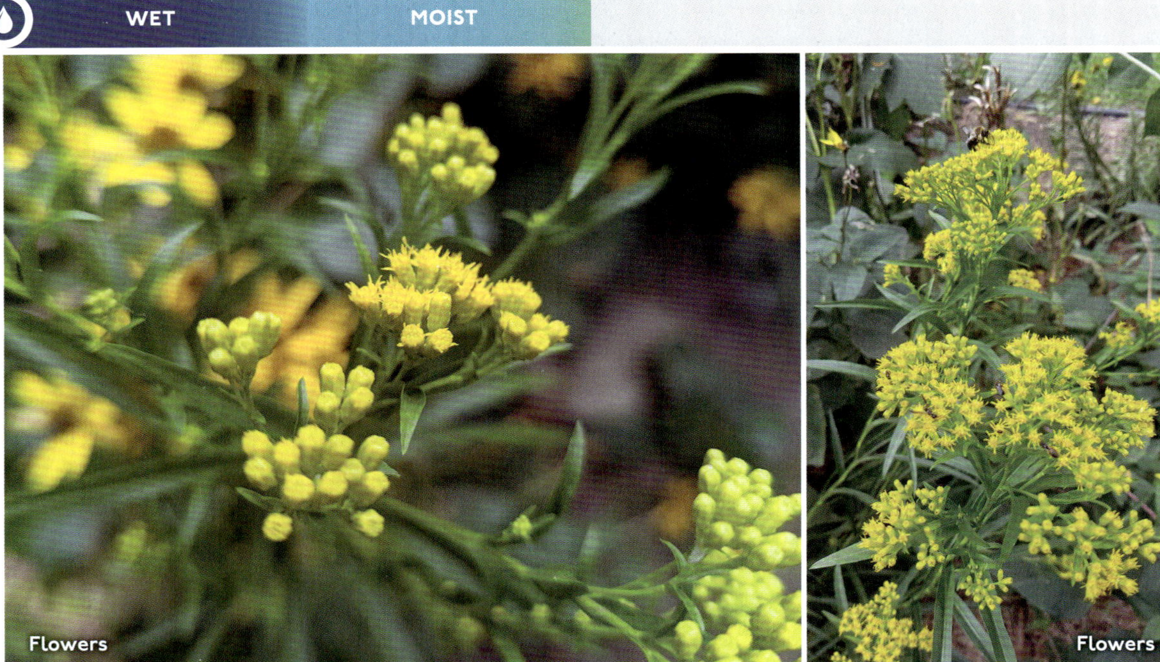
Flowers

Flowers

FAMILY *Asteraceae* (Aster Family)

ALTERNATE COMMON NAMES None

PLANT DESCRIPTION Riddell's Goldenrod features upright, smooth stems that are unbranched. Basal and lower stem leaves are oblanceolate in shape, tapering to a winged leaf stalk and a pointed tip. They measure up to 25 cm long and just over 1 cm wide, and the leaves often wither away by flowering time. Stem leaves are found in an alternate arrangement and are somewhat arching. As they ascend the stem they become smaller, more lance-linear, and stalkless to the point where they clasp the stem. All leaves are toothless, smooth with a leathery feel to them, and folded along the center vein (forming a V-shaped cross section). At the top of the plant you will find flat-topped-to-rounded

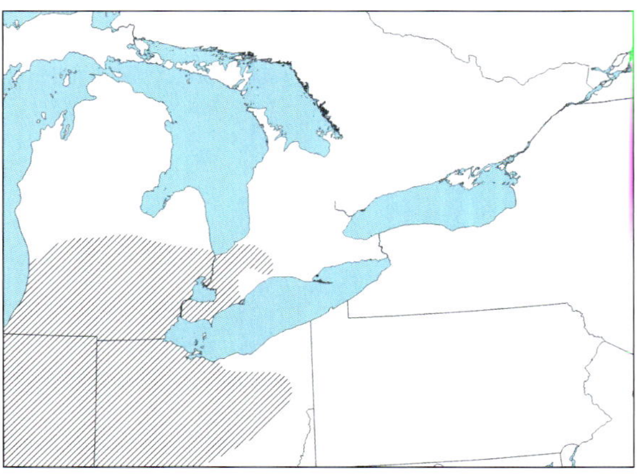

clusters of small, yellow flowers. Each flower has seven to nine yellow ray florets (petals) surrounding six to 10 yellow disk flowers. Flowers mature

into small, dry seeds with tufts of hairs that allow them to be carried in the wind.

IN THE GARDEN Riddell's Goldenrod brightens up the landscape with flat-topped clusters of yellow flowers that rise above its grass-like leaves. It is a great choice for adding a splash of yellow to moist areas or rain gardens and is not an aggressive spreader.

SKILL LEVEL Beginner

LIFESPAN Perennial

EXPOSURE Full sun

SOIL TYPE Calcareous soil containing some sand or gravel

MOISTURE Wet to moist

HEIGHT 40–100 cm

SPREAD 45–60 cm

BLOOM PERIOD Sep

COLOR Yellow

FRAGRANT ❌

SHOWY FRUIT ❌

CUT FLOWER ✅

PESTS No serious insect or disease problems, though leaf rust can be an occasional problem

NATURAL HABITAT Wet prairie-like sites and marshy ground

WILDLIFE VALUE An important source of pollen for bees, wasps, and flies

BUTTERFLY LARVA HOST PLANT FOR None

MOTH LARVA HOST PLANT FOR At least 45 species of moths, including members of the tiger moths, ribbed cocoon-maker moths, case-bearer moths, twirler moths, geometer moths, leaf-blotch miner moths, owlet moths, pyralid moths, trumpet leafminer moths, and tortrix moths

USDA HARDINESS ZONES 3–7

PROPAGATION Seeds need light and 60 days of cold, moist stratification to germinate. New plants can be started by dividing the rhizomes.

ADDITIONAL INFO This moisture-loving goldenrod is an excellent choice for wetland restoration projects. Riddell's Goldenrod is easily distinguished from other goldenrods by its folded leaves that clasp the stem.

SAR STATUS ON – S3/SC

Solidago rigida
Stiff Goldenrod

QUICK GUIDE

PROPAGATION

FAMILY *Asteraceae* (Aster Family)

ALTERNATE COMMON NAMES Bold Goldenrod, Hard-leaved Goldenrod, Prairie Goldenrod, Rigid Goldenrod, Stiff-leaved Goldenrod

PLANT DESCRIPTION Stiff Goldenrod has rigid, upright, finely hairy stems that are unbranched, except for near the flower cluster. Basal and lower leaves are up to 25 cm long and 12 cm wide and vary in shape from lanceolate to oblong or oval. They are grayish green from fine hairs (which give it the texture of felt) and toothless or slightly toothed with wavy edges. They abruptly taper to a long leaf stalk with blunt tips. Stem leaves clasp the stem in an alternate arrangement and shrink in size as they ascend. Yellow flowers are found in flat-topped-to-rounded clusters at the top of the stem. Each flower measures just over 1 cm wide and features six to 13 yellow ray florets (petals) surrounding a yellow center. Flowers mature into small, pale seeds with tufts of hairs that allow them to be carried by the wind.

IN THE GARDEN Stiff Goldenrod makes a dramatic statement in the garden with large leaves and flat-topped flower clusters that provide a perfect landing spot for butterflies. It has a clumping habit but does like to self-seed into open spots. The large, fluffy seed heads make a charming addition to the fall landscape.

SKILL LEVEL Beginner

LIFESPAN Perennial

EXPOSURE Full sun

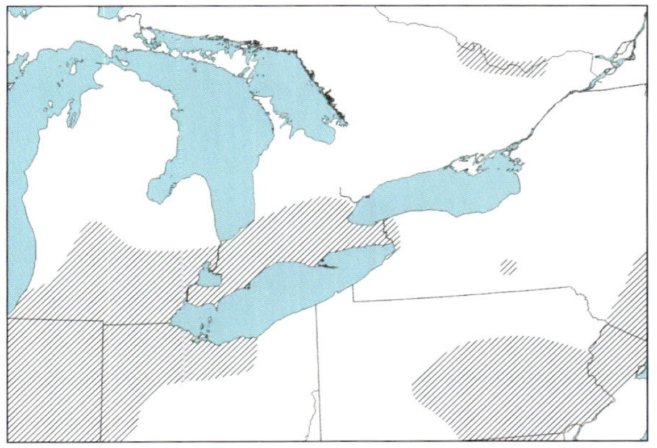

SOIL TYPE Well-drained, gravelly sandy-loam or clay-loam soils

MOISTURE Moist to medium

HEIGHT 30–120 cm

SPREAD 45–60 cm

BLOOM PERIOD Aug, Sep

COLOR Yellow

FRAGRANT ✖

SHOWY FRUIT ✖

CUT FLOWER ✔

PESTS No serious insect or disease problems, though leaf rust may be an occasional problem

NATURAL HABITAT Mesic to dry, often degraded prairies and other open spaces — often with sandy soils

WILDLIFE VALUE The flowers are an important source of pollen and attract many kinds of insects, including a variety of native bees, butterflies, and others, and goldfinches are known to eat the seeds to a limited extent

BUTTERFLY LARVA HOST PLANT FOR None

MOTH LARVA HOST PLANT FOR At least 45 species of moths, including members of the tiger moths, ribbed cocoon-maker moths, case-bearer moths, twirler moths, geometer moths, leaf-blotch miner moths, owlet moths, pyralid moths, trumpet leafminer moths, and tortrix moths

USDA HARDINESS ZONES 3–9

PROPAGATION Seed germination may be increased with 30 to 60 days of cold, moist stratification, but this pretreatment is not absolutely necessary. Seeds need light to germinate so surface sow. Plants may also be propagated by dividing mature plants in early spring or late fall.

ADDITIONAL INFO Remove spent flower clusters to encourage additional blooms.

SAR STATUS ON – S3/SC; NY – S2/T; PA – S1/TU

Solidago rugosa
Rough-stemmed Goldenrod

FULL SUN

WET **MOIST**

Flowers

Whole plant

FAMILY *Asteraceae* (Aster Family)

ALTERNATE COMMON NAMES Bitterweed, Early Wrinkle-leaf, Rough Goldenrod, Rough-leaved Goldenrod, Tall Hairy Goldenrod, Wrinkle-leaf Goldenrod

PLANT DESCRIPTION Rough-stemmed Goldenrod is an upright plant with light-green-to-reddish stems that may be densely to finely hairy. Leaves are found in an alternate arrangement and measure up to 10 cm long and 4 cm wide. They vary in shape from ovate to ovate-lanceolate, with shallow teeth that point towards the tip of the leaf. Upper leaf surfaces are hairy to hairless and have deeply indented veins that give them a wrinkly look. Stems terminate in multi-branched, arching clusters of yellow flowers. These flowers are concentrated on

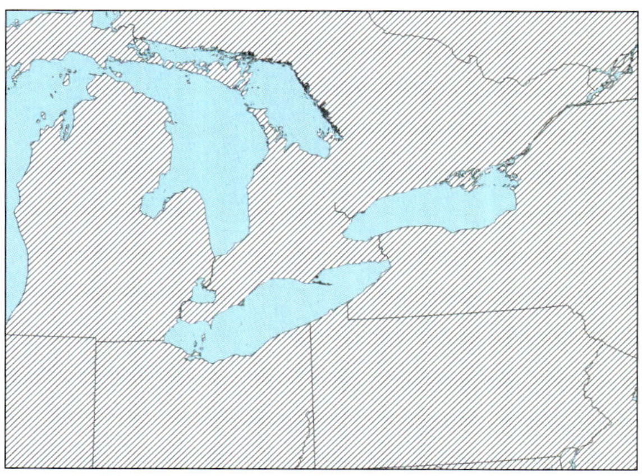

the upper side of the branches, measure about 3 mm wide, and feature four to eight ray florets (petals) surrounding an equal number of disk florets.

Note that this plant can be highly variable in size, hairiness, and flower cluster shape.

IN THE GARDEN Rough-stemmed Goldenrod brightens up the landscape with delightful sprays of yellow flowers.

SKILL LEVEL Beginner

LIFESPAN Perennial

EXPOSURE Full sun (tolerates light shade)

SOIL TYPE Most well-drained soils

MOISTURE Wet to moist

HEIGHT 60–150 cm

SPREAD 45–75 cm

BLOOM PERIOD Aug, Sep

COLOR Yellow

FRAGRANT ❌

SHOWY FRUIT ❌

CUT FLOWER ✅

PESTS No serious insect or disease problems, though rust may occur, and watch for powdery mildew and leaf spot

NATURAL HABITAT Open, disturbed habitats, meadows, old fields, pine barrens, thickets surrounding marshes and near bogs, and sterile, acidic habitats

WILDLIFE VALUE Flowers attract a variety of native bees, butterflies, and other insects

BUTTERFLY LARVA HOST PLANT FOR None

MOTH LARVA HOST PLANT FOR At least 45 species of moths, including members of the tiger moths, ribbed cocoon-maker moths, case-bearer moths, twirler moths, geometer moths, leaf-blotch miner moths, owlet moths, pyralid moths, trumpet leafminer moths, and tortrix moths

USDA HARDINESS ZONES 4–9

PROPAGATION Seeds need light to germinate, but do not need to be stratified. However, germination is often poor, so sow thickly. Stem tip cuttings can be taken in May or June, or plants may be grown from late winter division of the basal rosettes, which is the easiest method to propagate new plants.

ADDITIONAL INFO Plants are said to be aggressive spreaders in optimum growing conditions, but they have been quite well behaved in my southwestern Ontario garden in moist, well-drained, sandy-loam soil under full sun — as ideal conditions as you can get. The cultivar "Fireworks" is often available in nurseries. Plants can be cut back to half in June to keep them compact and eliminate the need to stake.

SAR STATUS N/A

Solidago speciosa
Showy Goldenrod

QUICK GUIDE

PROPAGATION

	PART SHADE	FULL SUN
	MEDIUM	DRY

Flowers

Flower cluster

FAMILY *Asteraceae* (Aster Family)

ALTERNATE COMMON NAMES Noble Goldenrod

PLANT DESCRIPTION Showy Goldenrod has smooth, unbranched stems that are green to reddish. Basal and lower stem leaves are lance-elliptic to spoon shaped with a pointed tip and a base that tapers to a winged leaf stalk. Leaves are 30 cm long, 7.5 cm wide, shallowly to coarsely toothed, and smooth to slightly rough in texture. Basal leaves wither away by flowering time. Stem leaves are found in an alternate arrangement and become more lance-elliptic, stalkless, and toothless as they climb the stem. You will notice that small leaves also develop from upper leaf axils. Flowers are found in a densely packed, erect, cylindrical cluster at the top of the plant, with additional clusters arising from

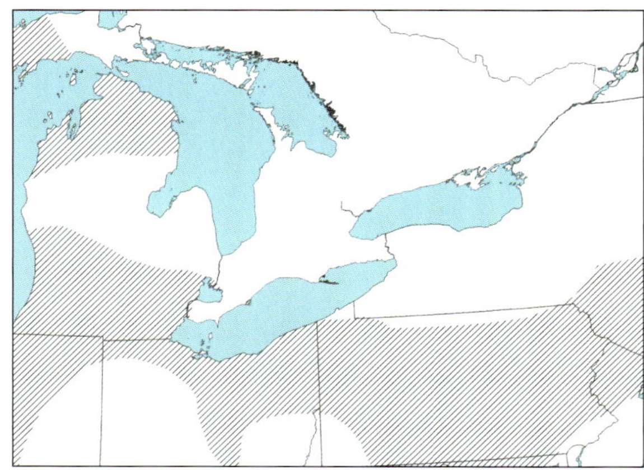

upper leaf axils. Each flower is about 6 mm wide and features four to 10 yellow ray florets (petals) surrounding yellow disk florets. Flowers mature

into small, hairless, dry seeds with tufts of hairs that allow them to be distributed by the wind.

A distinguishing feature of Showy Goldenrod is its ascending flower branches and the overall cylindrical shape of its flower cluster.

IN THE GARDEN Showy Goldenrod blooms in profusion with spires of densely packed, yellow flowers. It also blooms relatively late into the growing season, making it a good choice for extending bloom time.

SKILL LEVEL Beginner

LIFESPAN Perennial

EXPOSURE Full sun to part shade (light shade)

SOIL TYPE Sandy, well-drained soil

MOISTURE Dry to medium

HEIGHT 60–90 cm

SPREAD 60–90 cm

BLOOM PERIOD Sep (late), Oct (early)

COLOR Yellow

FRAGRANT ❌

SHOWY FRUIT ❌

CUT FLOWER ✅

PESTS No serious insect or disease problems

NATURAL HABITAT Sandy soils in open areas or under partial shade. In southwestern Ontario, it grows in prairie grasslands and oak savannas on fine sandy-loam soils

WILDLIFE VALUE The flowers attract primarily honeybees, bumblebees, ants, and beetles, while songbirds, especially goldfinches, are known to eat the seeds

BUTTERFLY LARVA HOST PLANT FOR None

MOTH LARVA HOST PLANT FOR At least 45 species of moths, including members of the tiger moths, ribbed cocoon-maker moths, case-bearer moths, twirler moths, geometer moths, leaf-blotch miner moths, owlet moths, pyralid moths, trumpet leafminer moths, and tortrix moths

USDA HARDINESS ZONES 3–8

PROPAGATION Seeds germinate after about 60 days of cold, moist stratification, and they will germinate most successfully in cool soil without covering (seeds need light to germinate). Showy Goldenrod will self-sow in open, sandy soils. Mature clumps can be divided.

ADDITIONAL INFO This very showy goldenrod can become aggressive in moist soils.

SAR STATUS ON – (not ranked)/E; IN – SU/WL; OH – S2/T; PA – S2/(not listed)

Solidago uliginosa
Bog Goldenrod

FAMILY *Asteraceae* (Aster Family)

ALTERNATE COMMON NAMES Marsh Golden-rod, Northern Bog Goldenrod, Swamp Goldenrod

PLANT DESCRIPTION Bog Goldenrod has reddish stems that are unbranched and smooth. Basal and lower stem leaves measure up to 22 cm long, just under 4 cm wide and are oblanceolate in shape. They have smooth surfaces and small teeth and taper to a long leaf stalk that sheathes the stem. As leaves ascend the stem, they become smaller, more lance shaped, stalkless (but not sheathing), and toothless. At the top of the plant, you will find densely packed, narrow flower clusters with branches that are angled closely to the stem. Each flower measures about 0.6 cm long and features one to eight yellow ray florets (petals) and six to

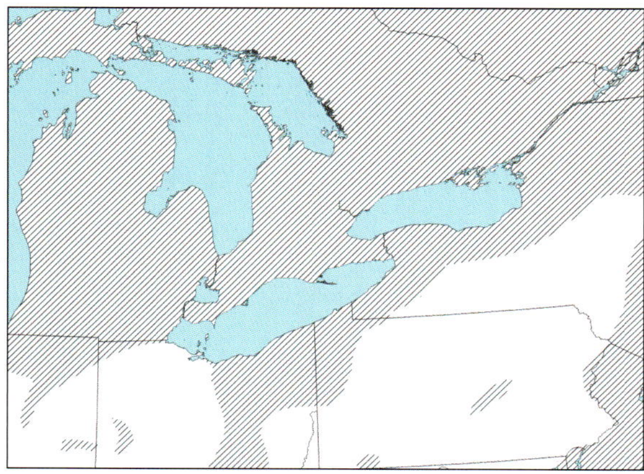

eight yellow disk florets. Flowers mature into small, dry seeds with tufts of hairs that allow them to be carried by the wind.

Leaves

Seed heads

IN THE GARDEN Bog Goldenrod makes a stylish addition to wet spots or rain gardens with its slender plumes of showy, yellow flowers. It is one of the earlier-blooming goldenrods, making it valuable for extending the beauty and wildlife value of the "goldenrod season" in your garden.

SKILL LEVEL Beginner

LIFESPAN Perennial

EXPOSURE Full sun to part shade (light shade)

SOIL TYPE Slightly acidic, well-drained soils (also tolerates average soils, but prefers boggy ones)

MOISTURE Wet to medium

HEIGHT 30–120 cm

SPREAD 30–90 cm

BLOOM PERIOD Jul, Aug

COLOR Yellow

FRAGRANT ✗

SHOWY FRUIT ✗

CUT FLOWER ✓

PESTS No serious insect or disease problems, though rust, powdery mildew, and leaf spot may be problems in some years

NATURAL HABITAT Bogs, marshes, and wet depressions

WILDLIFE VALUE Attracts butterflies and several native bee species

BUTTERFLY LARVA HOST PLANT FOR None

MOTH LARVA HOST PLANT FOR At least 45 species of moths, including members of the tiger moths, ribbed cocoon-maker moths, case-bearer moths, twirler moths, geometer moths, leaf-blotch miner moths, owlet moths, pyralid moths, trumpet leafminer moths, and tortrix moths

USDA HARDINESS ZONES 4–8

PROPAGATION Although some sources suggest Bog Goldenrod germination requires 60 days of cold, moist stratification, it seems to germinate just fine within a couple of weeks, as long as the seeds are uncovered (they need light to break dormancy) and the soil is cool. Mature clumps may also be divided in the spring, just as new growth is starting.

ADDITIONAL INFO Bog Goldenrod flowers consistently begin to open in the second week of July in my garden, two to three weeks earlier than Grass-leaved Goldenrod (*Euthamia graminifolia*) or Early Goldenrod (*Solidago juncea*), which are invariably the next two goldenrods to open.

SAR STATUS PA – S2/T

Symphyotrichum cordifolium

Blue Wood Aster

Flowers

Flower bracts

Whole plant

FAMILY *Asteraceae* (Aster Family)

ALTERNATE COMMON NAMES Broad-leaved Aster, Common Blue Wood Aster, Heartleaf Aster, Heart-leaved American Aster, Lowrie's Blue Wood Aster

PLANT DESCRIPTION Blue Wood Aster features a few light-green-to-reddish-brown stems that are smooth, upright, and branching. The basal and lower leaves are alternate and measure about 15 cm long and 8 cm wide. They are heart shaped, coarsely toothed, mostly hairless (occasionally hairy along the veins), and borne on long, winged leaf stalks. As the leaves ascend the stem, their stalks become shorter and their wings become broader, while their overall shape becomes less heart shaped and more rounded at the base. The uppermost leaves, including those in the flower heads, become stalk-less, lance shaped, and toothless and measure up to 2.5 cm long. Numerous branching flower clusters are found at the top of the plant and contain upwards of 300 individual flowers that each measure 1.2 cm across. They are characterized by seven to 15 petals surrounding a yellow center that turns a reddish color with age. The base of each flower is surrounded by four to six layers of overlapping bracts. These bracts are narrow, sharply pointed, and light green except for the tips, which are red to dark green. Flowers give way to small, dry, light brown seeds with tufts of light-brown-to-white hairs that allow them to be carried by the wind.

IN THE GARDEN Blue Wood Aster adds a calming, soft blue color to gardens and is living proof that you can have a successful pollinator garden in the shade. The lower leaves sometimes wither away before the flowers bloom, so consider pairing it with a companion that can hide the scraggly base. Plants grown in part shade will have more abundant flower clusters than plants grown in full shade. Blue Wood Aster has long-lasting blooms that often persist well into late fall.

SKILL LEVEL Beginner

LIFESPAN Perennial

EXPOSURE Part shade to full shade

SOIL TYPE Prefers soil that contains loam, clay loam, or some rocky material

MOISTURE Medium

HEIGHT 20–120 cm

SPREAD 45–60 cm

BLOOM PERIOD Aug, Sep, Oct

COLOR Blue

FRAGRANT ❌

SHOWY FRUIT ❌

CUT FLOWER ✅

PESTS No serious insect or disease problems, though there is some susceptibility to powdery mildew, leaf spots and rust; aster wilt can also be an occasional problem, particularly if plants are grown in poorly drained clay soils

NATURAL HABITAT Upland meadows and upland forests

WILDLIFE VALUE The nectar and pollen attract many insects, including native bees, wasps, flies, butterflies, skippers, and beetles, and Wild Turkey (*Meleagris gallopavo*) and Ruffed Grouse (*Bonasa umbellus*) are known to eat both the foliage and the seeds

BUTTERFLY LARVA HOST PLANT FOR Silvery Checkerspot (*Chlosyne nycteis*), Tawny Crescent (*Phyciodes batesii*), Northern Crescent (*Phyciodes cocyta*), Pearl Crescent (*Phyciodes tharos*), Painted Lady (*Vanessa cardui*)

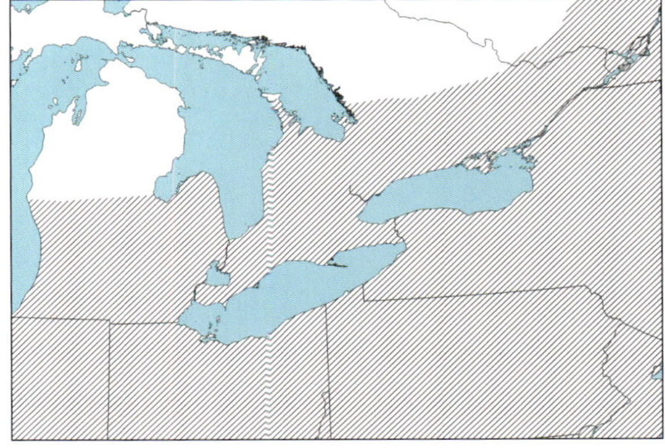

MOTH LARVA HOST PLANT FOR At least 40 species of moths, including members of the tiger moths, ribbed cocoon-maker moths, case-bearer moths, twirler moths, geometer moths, leaf-blotch miner moths, slug caterpillar moths, owlet moths, clearwing moths, flower moths, trumpet leafminer moths, and tortrix moths

USDA HARDINESS ZONES 3–9

PROPAGATION Blue Wood Aster is one of those species that likes to play mind games with gardeners. About 50 percent of the time, spring sowing works just fine; the other half of the time the seeds won't germinate without 60 days of cold, moist stratification. Plants may also be divided in spring or fall or by taking basal cuttings in late spring. Harvest the shoots when they are about 10 to 15 cm tall with plenty of underground stem.

ADDITIONAL INFO Blue Wood Aster prefers moist, rich soils, but avoid consistent moisture. Good air circulation helps reduce incidence of foliar diseases. Pinch back stems several times before mid-July to help control plant height and promote bushiness.

SAR STATUS N/A

Symphyotrichum ericoides
White Heath Aster

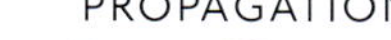

FAMILY *Asteraceae* (Aster Family)

ALTERNATE COMMON NAMES Heath Aster, Many-flowered Aster, Squarrose White Aster, Tufted White Prairie Aster, White Prairie Aster

PLANT DESCRIPTION White Heath Aster features one to a few erect stems that start out green but often turn brown and woody with age. Stems are covered in appressed hairs (hairs that lie flat against the stem) that point upwards. The lance-shaped leaves are stalkless and found in an alternate pattern along the stem. They are toothless with smooth edges and have fine hairs on both the upper and lower surfaces. Lower leaves measure 5 cm long and 6 mm wide, while upper leaves are much smaller at 2.5 cm long and 3 mm wide. A key feature of this species is the numerous needle-like bracts on the

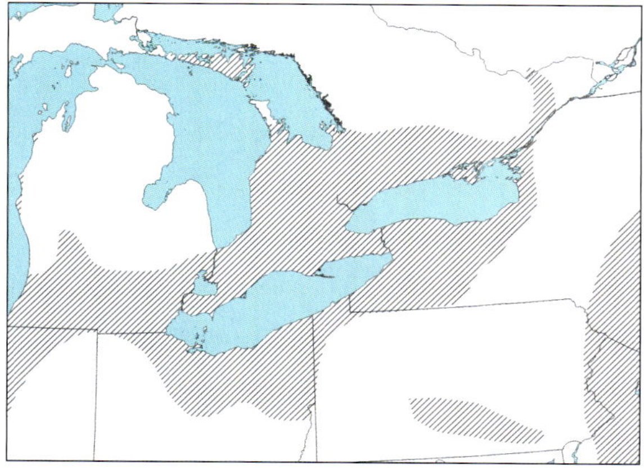

flowering stems that give the plant a heath-like look. Tightly packed, branching flower clusters are borne at the top of the plant. Individual flowers measure

less than 1 cm across and are characterized by 18 to 20 petals surrounding a yellow center that turns a reddish color with age. The base of each flower is surrounded by bracts that are whitish at the base and narrow with blunt tips and that often peel back from the base of the flower. Flowers give way to dry seeds with tufts of white hairs that allow them to be carried by the wind.

IN THE GARDEN White Heath Aster is an elegant, drought-tolerant plant with copious white blooms. The heavily branching upper stems give it a bushy appearance. It readily spreads by rhizomes and seed, so keep this in mind when choosing its location. Note that the lower leaves often fall off before flowers bloom, so it is best paired with companions that can hide the bare lower stems.

SKILL LEVEL Beginner

LIFESPAN Perennial

EXPOSURE Full sun

SOIL TYPE Average, well-drained soil

MOISTURE Dry to medium

HEIGHT 20–80 cm

SPREAD 30–45 cm

BLOOM PERIOD Aug, Sep, Oct

COLOR White

FRAGRANT 

SHOWY FRUIT 

CUT FLOWER 

PESTS No serious insect or disease problems, and it is mildew resistant

NATURAL HABITAT Fields and open areas, meadows, and along roadsides

WILDLIFE VALUE A wide variety of insects are attracted to the flowers, including many native bees and butterflies, and Wild Turkeys (*Meleagris gallopavo*) are known to nibble on the seeds and foliage to a limited extent

BUTTERFLY LARVA HOST PLANT FOR Silvery Checkerspot (*Chlosyne nycteis*), Tawny Crescent (*Phyciodes batesii*), Northern Crescent (*Phyciodes cocyta*), Pearl Crescent (*Phyciodes tharos*), Painted Lady (*Vanessa cardui*)

MOTH LARVA HOST PLANT FOR At least 40 species of moths, including members of the tiger moths, ribbed cocoon-maker moths, case-bearer moths, twirler moths, geometer moths, leaf-blotch miner moths, slug caterpillar moths, owlet moths, clearwing moths, flower moths, trumpet leafminer moths, and tortrix moths

USDA HARDINESS ZONES 3–10

PROPAGATION No pretreatment of the seeds is necessary for successful germination, indoors or out. Seeds are quite small and need light to germinate, so leave uncovered or cover very lightly and keep moist. They usually take two to three weeks to sprout. Mature plants may be divided every two to three years.

ADDITIONAL INFO Heath Aster is one of the last flowers to bloom in the year, sometimes producing little seed due to a killing frost before the seeds ripen. It is sometimes confused with Frost Aster (*Symphyotrichum pilosum*), but that plant has larger flower heads with more petaloid rays, and its leaves are larger.

SAR STATUS PA – S3/TU

Symphyotrichum laeve
Smooth Aster

Flowers

Whole plant

FAMILY *Asteraceae* (Aster Family)

ALTERNATE COMMON NAMES Glaucous Aster, Purple Aster, Smooth Blue American Aster, Smooth Blue Aster, Smooth-leaved Aster

PLANT DESCRIPTION Smooth Aster features one to a few erect, hairless stems that are usually green but can be a reddish color. Leaves clasp the stems in an alternate pattern and measure about 10 cm long and 4 cm wide. They are smooth (almost waxy), shiny, toothless, and greenish blue on top and light green underneath. The basal leaves are toothed with winged petioles and oblanceolate. Open, branching flower clusters are found at the top of the plant. Flowers can also arise from upper leaf axils (where the leaves meet the stem). Individual flowers measure up to 2.5 cm across and feature 15 to 30 oblong

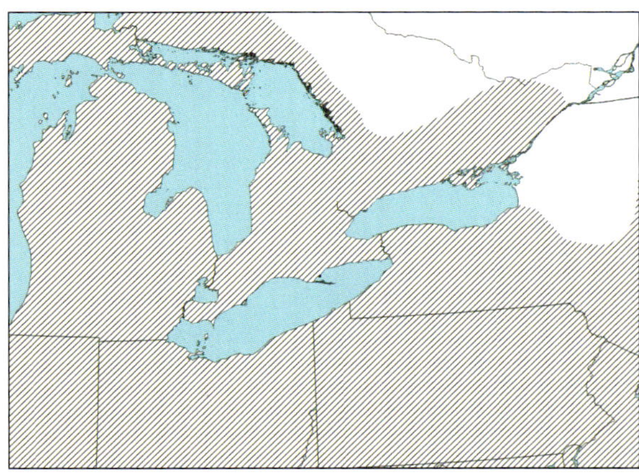

ray florets (petals) surrounding yellow centers that turn purplish red with age. Petal color can vary from light blue to light purple. Four to six layers of bracts surround the base of each flower. They are smooth, appressed (flattened), and light green to bluish green and have diamond-shaped ends with dark tips. Flowers give way to dry, brown, narrowly cone-shaped seeds, each with a tuft of light brown hair that allows them to be carried by the wind.

IN THE GARDEN Smooth Aster is valued by gardeners for its copious blue blooms and non-aggressive growth habit. The flowers are frost hardy and bloom late into fall while the tough stems will persist through the winter months. Smooth Aster is easy to grow but doesn't like being shaded by taller plants. A favorite food of rabbits, possibly due to the smooth leaves.

SKILL LEVEL Beginner

LIFESPAN Perennial

EXPOSURE Full sun

SOIL TYPE Any well-drained soil

MOISTURE Dry

HEIGHT 20–70 cm (occasionally to 120 cm)

SPREAD 30–60 cm

BLOOM PERIOD Sep, Oct (to frost)

COLOR Blue

FRAGRANT ✖

SHOWY FRUIT ✖

CUT FLOWER ✔

PESTS No serious insect or disease problems, though powdery mildew can affect the plant in some years

NATURAL HABITAT Fields, open woods, and roadsides

WILDLIFE VALUE The nectar and pollen of the flower heads attract many species of native bees, butterflies, and other insects, and Ruffed Grouse (*Bonasa umbellus*) and Wild Turkey (*Meleagris gallopavo*) feed on both the leaves and seeds of asters; the seeds are also eaten by mice and American Tree Sparrows (*Spizelloides arborea*)

BUTTERFLY LARVA HOST PLANT FOR Silvery Checkerspot (*Chlosyne nycteis*), Tawny Crescent (*Phyciodes batesii*), Northern Crescent (*Phyciodes cocyta*), Pearl Crescent (*Phyciodes tharos*), Painted Lady (*Vanessa cardui*)

Leaves

Seed heads

MOTH LARVA HOST PLANT FOR At least 40 species of moths, including members of the tiger moths, ribbed cocoon-maker moths, case-bearer moths, twirler moths, geometer moths, leaf-blotch miner moths, slug caterpillar moths, owlet moths, clearwing moths, flower moths, trumpet leafminer moths, and tortrix moths

USDA HARDINESS ZONES 3–9

PROPAGATION Direct sow the seeds in late fall or early spring. No pretreatment is necessary even when starting indoors, but seeds need light to germinate. Germination is said to be slow. Transplanted seedlings will likely bloom in their first year. You can also multiply plants from root cuttings.

ADDITIONAL INFO *Symphyotrichum laeve* will tolerate short durations of seasonal flooding. It also self-sows strongly in open areas that are burned and mowed and is walnut (juglone) tolerant.

SAR STATUS N/A

275

Symphyotrichum lanceolatum
Panicled Aster

Flowers | Whole plant | Leaves

FAMILY *Asteraceae* (Aster Family)

ALTERNATE COMMON NAMES Eastern Line Aster, Lanceleaf Aster, Lance Leaved Aster, Marsh Aster, Narrow-leaf Michaelmas Daisy, Narrow Leaved Aster, Smooth Aster, Tall White Aster, White Field Aster, White Panicle Aster

PLANT DESCRIPTION Panicled Aster features erect, light-green-to-red stems that have vertical lines of hairs along them and may be shallowly grooved. Leaves are attached directly to the stem in an alternate pattern and measure up to 13 cm long and 2 cm wide. They are hairless, lance shaped with pointed tips, and toothless, except for the lower leaves, which may have a few widely spaced teeth. Leaves become smaller as they ascend the stems. Basal leaves are broader and have winged stalks.

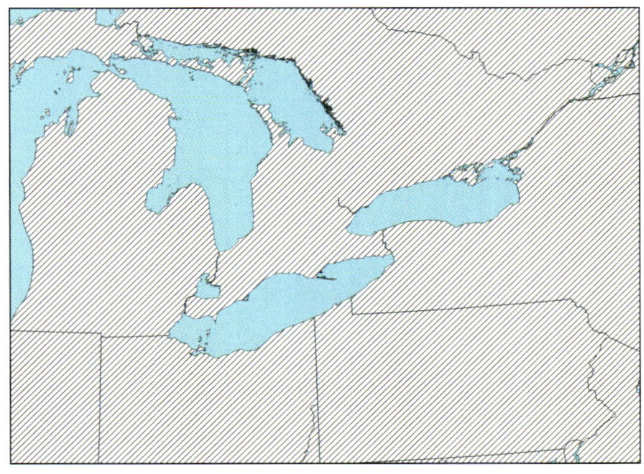

A defining feature of Panicled Aster is how the dead, withered leaves curl and persist on the stems. Branching clusters of stalked flowers are found

276

at the top of the plant and may contain 20 or more flowers per branch. Flowers also emerge from upper leaf axils (where the leaf meets the stem). Individual flowers measure 1 to 2 cm across and are made up of 16 to 50 narrow ray florets (petals) surrounding a yellow center that turns reddish with age. Petals point slightly downwards when the flower head is fully open. Bracts surrounding the base of the flowers are found in three to six layers and are lance shaped, smooth, and appressed (flattened on the stem) to slightly spreading, with the outer layers being shorter than the inner layers. Bracts are green but pale green at the base. Flowers give way to small, dry, brown seeds with tufts of white hairs that allow them to be carried off in the wind.

IN THE GARDEN Panicled Aster is a robust plant that boasts a profusion of elegant white flowers. It is easy to grow but spreads quickly via rhizomes to form colonies, so it may be better for large or less-formal gardens. The lower leaves wither away before the flowers emerge, so consider planting it with a companion that hides the scraggly base. The stems typically stand strong through the winter months.

SKILL LEVEL Beginner

LIFESPAN Perennial

EXPOSURE Full sun to part shade

SOIL TYPE Almost any soil, though they seem to do best in poorly drained areas where water temporarily pools then later dries out

MOISTURE Medium to moist

HEIGHT 30–150 cm

SPREAD 90 cm

BLOOM PERIOD Sep

COLOR White, pink, blue

FRAGRANT 

SHOWY FRUIT ✗

CUT FLOWER ✓

PESTS No serious insect or disease problems

Seed heads

NATURAL HABITAT Moist fields and other disturbed areas, edges of woodlands, and roadsides

WILDLIFE VALUE Nectar and pollen attract many native bees and butterflies; Ruffed Grouse (*Bonasa umbellus*) and Wild Turkey (*Meleagris gallopavo*) feed on the seeds and foliage, while American Tree Sparrows (*Spizelloides arborea*) and White-footed Mice (*Peromyscus leucopus*) feed on the seeds; it's also a favorite of White-tailed Deer (*Odocoileus virginianus*)

BUTTERFLY LARVA HOST PLANT FOR Silvery Checkerspot (*Chlosyne nycteis*), Tawny Crescent (*Phyciodes batesii*), Northern Crescent (*Phyciodes cocyta*), Pearl Crescent (*Phyciodes tharos*), Painted Lady (*Vanessa cardui*)

MOTH LARVA HOST PLANT FOR At least 40 species of moths, including members of the tiger moths, ribbed coccon-maker moths, case-bearer moths, twirler moths, geometer moths, leaf-blotch miner moths, slug caterpillar moths, owlet moths, clearwing moths, flower moths, trumpet leafminer moths, and tortrix moths

USDA HARDINESS ZONES 3–9

PROPAGATION No pretreatment of the seeds is needed for germination. They can be sown fresh in the fall or spring. Dividing the clumps can be done in the spring or fall.

ADDITIONAL INFO Panicled Aster is a good choice for wetland restoration. A similar species, the more aggressive Frost Aster (*Symphyotrichum pilosum*) can be differentiated by its hairier foliage and preference for drier habitats.

SAR STATUS N/A

Symphyotrichum lateriflorum
Calico Aster

QUICK GUIDE

PROPAGATION

 **PART SHADE**

 **MEDIUM**

Flowers

Whole plant

FAMILY *Asteraceae* (Aster Family)

ALTERNATE COMMON NAMES Goblet Aster, One-sided Aster, Side Flowering Aster, Small White Aster, Starved Aster, White Woodland Aster

PLANT DESCRIPTION Calico Aster features one to a few erect, green-to-red stems with soft, white hairs. Many long branches spread out from the main stems to give this plant a bushy appearance. Leaves are thin and lance shaped, measuring from 1.3 to 15 cm long and up to 4 cm wide. They are toothless (or with very shallow teeth), stalkless, and smooth to slightly rough, except for small hairs found underneath the leaf on the midvein. Basal leaves are broader and more spatula shaped with winged stalks. As the leaves ascend the stem, they become smaller and more lance shaped. Flowers are

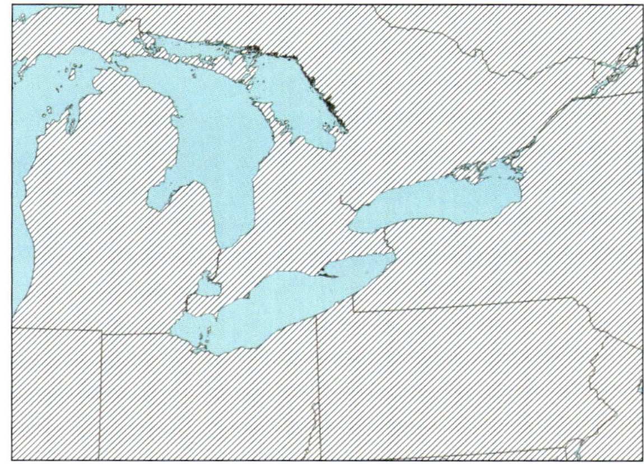

found at the top of the plant in branching clusters and also emerging from upper leaf axils (where the leaf meets the stem). Individual flowers are

278

borne on short stalks and measure 1.3 cm across
and have 8 to 16 unevenly distributed ray florets
(petals) surrounding a pale yellow disk that turns
a purplish-red color as the flower matures. Bracts
surround the base of each flower in three to four
layers. They are pressed closely together, have
sparse hairs, and are light green with dark green,
diamond-shaped tips. Flowers give way to small,
dry seeds with tufts of white hairs that allow them
to be carried by the wind.

IN THE GARDEN Calico Aster puts on a delightful display of small, but numerous, white flowers
that stand out well, especially in partly shaded
sites. As the flowers mature, their centers take on
a purplish-red color. The heavily branched habit
of this aster gives it a full, bushy appearance.
Calico Aster is not an aggressive spreader but may
self-seed into open spaces.

SKILL LEVEL Beginner

LIFESPAN Perennial

EXPOSURE Part shade (tolerates full sun if not too
dry)

SOIL TYPE Growth is best in rich, organic soil or a
moisture-retaining clay loam

MOISTURE Medium (tolerates periodic flooding)

HEIGHT 20–60 cm (occasionally to 120 cm)

SPREAD 60–90 cm

BLOOM PERIOD Aug, Sep, Oct

COLOR White, purple

FRAGRANT 

SHOWY FRUIT ⊗

CUT FLOWER ⊗

PESTS No serious insect or disease problems

NATURAL HABITAT Everywhere from moist
woods to dry savannas and areas with a history of
disturbance

WILDLIFE VALUE Calico Aster attracts a wide variety of insects, particularly in sunny areas, including
many native bees and butterflies; the short nectar
tubes are particularly attractive to mining bees

BUTTERFLY LARVA HOST PLANT FOR Silvery
Checkerspot (*Chlosyne nycteis*), Tawny Crescent

Basal leaves

Seed heads

(*Phyciodes batesii*), Northern Crescent (*Phyciodes
cocyta*), Pearl Crescent (*Phyciodes tharos*), Painted
Lady (*Vanessa cardui*)

MOTH LARVA HOST PLANT FOR At least 40
species of moths, including members of the tiger
moths, ribbed cocoon-maker moths, case-bearer
moths, twirler moths, geometer moths, leaf-blotch
miner moths, slug caterpillar moths, owlet moths,
clearwing moths, flower moths, trumpet leafminer
moths, and tortrix moths

USDA HARDINESS ZONES 3–9

PROPAGATION Seeds usually germinate in warm
locations without the need for cold stratification
and can be sown outdoors in spring or fall, or
started indoors. Seeds need light to germinate, so
do not cover with soil. If storing seeds, they need to
be stored in a dry, cold environment. Calico Asters
should be divided every three to four years — in
the fall after blooming has finished — to ensure the
vigor of the plant.

ADDITIONAL INFO Stems may be pinched back
in late spring to early summer if shorter plants are
desired. Calico Aster is sometimes confused with
White Heath Aster (*Symphyotrichum ericoides*), but
White Heath Aster is a more compact plant with
leaves that are shorter and narrower than those
of Calico Aster. In addition, White Heath Aster
is often found in open prairies whereas Calico
Aster is almost always found near woodlands and
semi-shaded wetland areas. Another similar species,
Frost Aster (*S. pilosum*), has larger flower heads
with more ray florets than those of Calico Aster.

SAR STATUS N/A

Symphyotrichum novae-angliae
New England Aster

FAMILY *Asteraceae* (Aster Family)

ALTERNATE COMMON NAMES First Flower, Hardy Aster, Michaelmas Daisy, Starwort

PLANT DESCRIPTION New England Aster features rigid central stems that are brownish red, are covered in short white hairs, and typically branch out towards the top of the plant. Leaves are borne on the stems in an alternate pattern and measure up to 10 cm long and up to 2.5 cm wide. They are oblanceolate, toothless, and covered in fine hairs. The leaves are stalkless with a pair of lobes at the base that extend back to clasp the stem — a key identifier. Branching clusters of stalked flowers are found at the top of the plant. Flower stalks also emerge from upper leaf axils (where the leaf meets the stem). Individual flowers are up to 4 cm across and consist of 40 to 100 narrow ray florets (petals) that surround a yellow center. The center turns reddish yellow with age. Petals can vary in color from

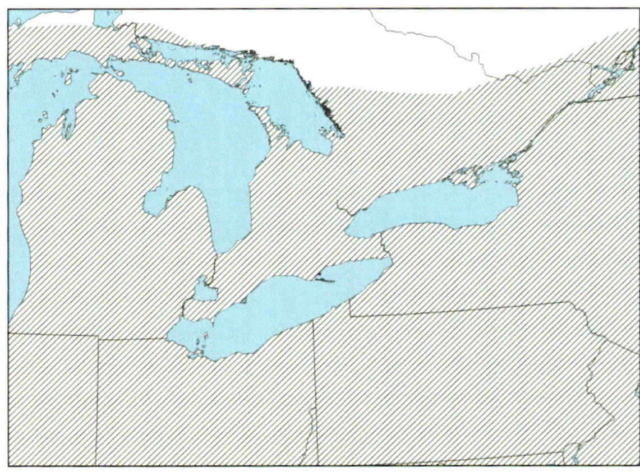

lavender to dark purple to pink. Bracts surrounding the base of each flower are found in three to five layers, narrow, widely spreading, and covered in fine hairs. They are green but may have a purple tinge. Flowers give way to dry, light brown seeds with tufts of white hairs that allow them to be carried off by the wind.

IN THE GARDEN New England Aster is a classic wildflower that puts on a dramatic floral display in fall. You can bring this display, and the butterflies that come with it, into your garden, too, as New England Aster is a reliable performer. This aster maintains a clumping habit but will freely self-seed into gaps in the garden.

SKILL LEVEL Beginner

LIFESPAN Perennial

EXPOSURE Full sun to part shade

SOIL TYPE Sandy loam to clay

MOISTURE Moist to medium

HEIGHT 30–120 cm (occasionally to 200 cm)

SPREAD 60–90 cm

BLOOM PERIOD Aug, Sep, Oct (until frost)

COLOR Pink, purple, blue

FRAGRANT ✕

SHOWY FRUIT ✕

CUT FLOWER ✓

PESTS No serious insect or disease problems, though there is some susceptibility to powdery mildew; aster wilt can also be an occasional problem, particularly if plants are grown in poorly drained clay soils

NATURAL HABITAT Moist, open wooded areas, meadows, mesic prairies, disturbed sites, and streambanks

WILDLIFE VALUE Bees and butterflies frequent this wildflower, and it is an important nectar source for Monarch Butterflies (*Danaus plexippus*)

BUTTERFLY LARVA HOST PLANT FOR Silvery Checkerspot (*Chlosyne nycteis*), Tawny Crescent (*Phyciodes batesii*), Northern Crescent (*Phyciodes cocyta*), Pearl Crescent (*Phyciodes tharos*), Painted Lady (*Vanessa cardui*)

MOTH LARVA HOST PLANT FOR At least 40 species of moths, including members of the tiger moths, ribbed cocoon-maker moths, case-bearer moths, twirler moths, geometer moths, leaf-blotch miner moths, slug caterpillar moths, owlet moths, clearwing moths, flower moths, trumpet leafminer moths, and tortrix moths

USDA HARDINESS ZONES 3–8

PROPAGATION Seeds do not need any pretreatment, but cold, moist stratification can speed germination. Plants can be propagated by stem cuttings taken in late spring and inserted into moist sand or rockwool (rockwool is an inert growing medium often used in hydroponics). Mature plants may also be divided in the spring by separating individual stems with their associated roots.

ADDITIONAL INFO New England Aster has a tendency to become root-bound and will benefit from dividing the plant every three to four years. Pinching back the stems a few times before mid-July will help to make the plant bushier and eliminate the need for staking. The lower leaves often die back by the time flowers emerge, so consider planting it with a companion that can hide the scraggly base.

SAR STATUS N/A

Symphyotrichum ontarionis
Ontario Aster

PART SHADE

MOIST　　MEDIUM

Flower

Whole plant

FAMILY *Asteraceae* (Aster Family)

ALTERNATE COMMON NAMES Bottomland Aster, Lake Ontario Aster

PLANT DESCRIPTION Ontario Aster features one to three erect stems (usually one) that are densely covered in short hairs. Stems typically branch out just below the middle. Leaves are borne in an alternate pattern and measure about 1.3 cm to 8 cm long and 2.5 cm wide. They are stalkless and have slightly toothed edges that become smoother as they ascend the stems. The upper leaf surface is green and rough textured, while the lower surface is light green and finely hairy. Basal leaves are spatula shaped with winged stalks and become more lance shaped as they ascend the stem. Upper stems are topped by widely spreading, elongated flower

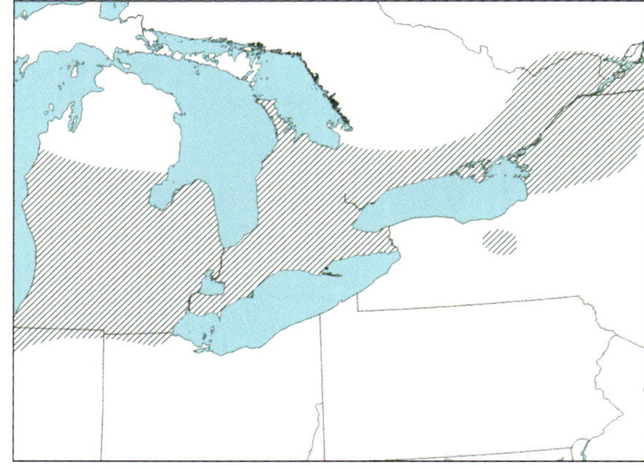

clusters made up of many stalked flowers, some of which arise from upper leaf axils (where the leaf meets the stem). Each flower measures 1 cm to

1.3 cm across and is characterized by 15 to 25 narrow ray florets (petals) surrounding a yellow center that turns reddish with age. Bracts surround the base of each plant in three to six layers. They are linear with pointed tips, evenly green from top to bottom and covered in short, sparse hairs. They are appressed (flattened) to slightly spreading. One to five leafy bracts are found below the flower. Flowers give way to small, dry seeds with tufts of white hairs that allow them to be carried by the wind.

IN THE GARDEN Ontario Aster features dainty, white flowers and a sturdy, upright form. It spreads via rhizomes and seeds to form colonies, so keep this in mind when choosing its location. The lower leaves wither away by flowering time so consider planting it with a companion that can hide its base.

SKILL LEVEL Beginner

LIFESPAN Perennial

EXPOSURE Part (dappled) shade

SOIL TYPE Fertile, loamy soil

MOISTURE Moist to medium

HEIGHT 20–120 cm

SPREAD 20–40 cm

BLOOM PERIOD Sep

COLOR White (occasionally light pink, blue or purple)

FRAGRANT 

SHOWY FRUIT 

CUT FLOWER ✓

PESTS No serious insect or disease problems

NATURAL HABITAT Moist meadows and woodlands

WILDLIFE VALUE Important for native bees, butterflies, and other insects

BUTTERFLY LARVA HOST PLANT FOR Unknown

MOTH LARVA HOST PLANT FOR At least 40 species of moths, including members of the tiger moths, ribbed cocoon-maker moths, case-bearer moths, twirler moths, geometer moths, leaf-blotch miner moths, slug caterpillar moths, owlet moths, clearwing moths, flower moths, trumpet leafminer moths, and tortrix moths

USDA HARDINESS ZONES 3–9

PROPAGATION No pretreatment is needed. Sow in the spring or fall and do not cover the seeds as they need light to germinate. Little is published about the cultivation of Ontario Aster, but given it is a colony-forming plant, it is likely safe to assume that plants may be divided in the spring or fall as with similar asters.

ADDITIONAL INFO This species is differentiated from Calico Aster (*Symphyotrichum lateriflorum*) by the hairy underside of its leaves, including the veins.

SAR STATUS NY – (not ranked)/R

Symphyotrichum oolentangiensis
Sky Blue Aster

QUICK GUIDE

PROPAGATION

FAMILY *Asteraceae* (Aster Family)

ALTERNATE COMMON NAMES Azure Aster, Blue Devils, Sky Blue American Aster

PLANT DESCRIPTION Sky Blue Aster features one to a few erect, light-green-to-red stems that are sparsely hairy to smooth. They are unbranched except for the upper part of the plant. Leaves are found in an alternate pattern, are smooth margined, and have a fine sandpaper-like feel to both surfaces. Basal and lower leaves are arrowhead to narrowly heart shaped, abruptly narrow at the base, and attached with winged stalks. They are up to 10 cm long and 4.5 cm wide. As the leaves ascend the stem they become more lance shaped with broadly winged stalks, becoming stalkless towards the top. Many-branched, open flower clusters are found at

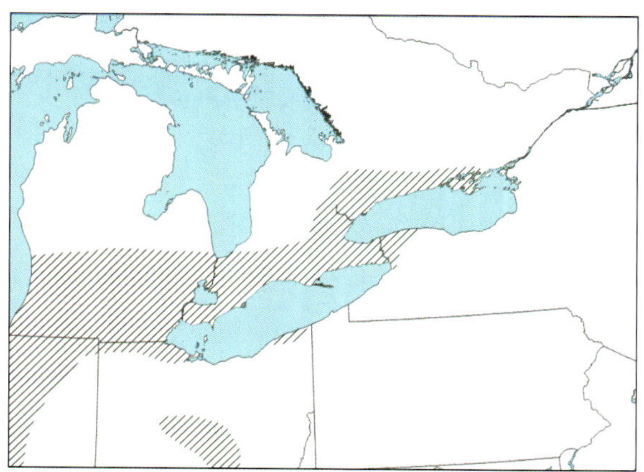

the top of the plant. They contain numerous individual flowers each measuring about 1.2 cm across. They are characterized by 10 to 25 ray florets (petals) surrounding a yellow center disk that turns reddish purple with age. Surrounding each flower are scale-like bracts found in three to five layers. They are appressed (flattened) together and light green, except for their dark green, diamond-shaped tips. They are hairless to minutely hairy. Flowers give way to small, dry seeds with tufts of white hairs that allow them to be carried off by the wind.

Note that Sky Blue Aster is similar to Blue Wood Aster (*Symphyotrichum cordifolium*), but it does not have the distinctive toothed, heart-shaped leaves. Also, it is usually found in sunnier conditions than what Blue Wood Aster will tolerate.

IN THE GARDEN The brilliant blue flowers of Sky Blue Aster are a welcoming sight for pollinators and gardeners alike. It is adaptable and drought tolerant, while its clumping habit makes it well suited to small or formal gardens. The lower leaves often wither away by flowering time, so consider planting it with a companion that can hide the scraggly base.

SKILL LEVEL Beginner

LIFESPAN Perennial

EXPOSURE Full sun

SOIL TYPE Most well-drained soils

MOISTURE Dry to moist

HEIGHT 20–80 cm (occasionally to 150 cm)

SPREAD 30–60 cm

BLOOM PERIOD Aug, Sep, Oct

COLOR Blue

FRAGRANT ❌

SHOWY FRUIT ❌

CUT FLOWER ✅

PESTS No serious insect or disease problems, though powdery mildew may occur

NATURAL HABITAT Dry sandy, loamy, or rocky soils, dry-to-moist (seasonally drying) prairies, and open woods

WILDLIFE VALUE Nectar and pollen attract several native bees, and many herbivorous mammals browse on the foliage

BUTTERFLY LARVA HOST PLANT FOR Silvery Checkerspot (*Chlosyne nycteis*), Tawny Crescent (*Phyciodes batesii*), Northern Crescent (*Phyciodes cocyta*), Pearl Crescent (*Phyciodes tharos*), Painted Lady (*Vanessa cardui*)

MOTH LARVA HOST PLANT FOR At least 40 species of moths, including members of the tiger moths, ribbed cocoon-maker moths, case-bearer moths, twirler moths, geometer moths, leaf-blotch miner moths, slug caterpillar moths, owlet moths, clearwing moths, flower moths, trumpet leafminer moths, and tortrix moths

USDA HARDINESS ZONES 3–8

PROPAGATION No pretreatment is needed. In fact, some sources suggest cold, moist stratification may actually be detrimental to germination for this species. Seeds need light to germinate, so surface sow seeds (or very lightly cover) outside in the fall or in a sunny location after the last spring frost. Plants may also be started from stem cuttings in late spring or by dividing mature clumps after any danger of frost in the spring.

ADDITIONAL INFO Pinch back one-third of the foliage twice before early summer to maintain a compact shape. The lower moisture requirements of Sky Blue Aster make it a good choice for containers and rock gardens.

SAR STATUS NY – S1/E

Symphyotrichum pilosum
Frost Aster

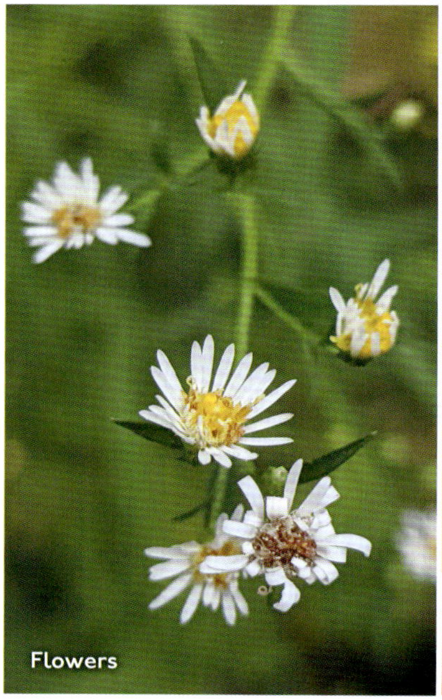

Flowers

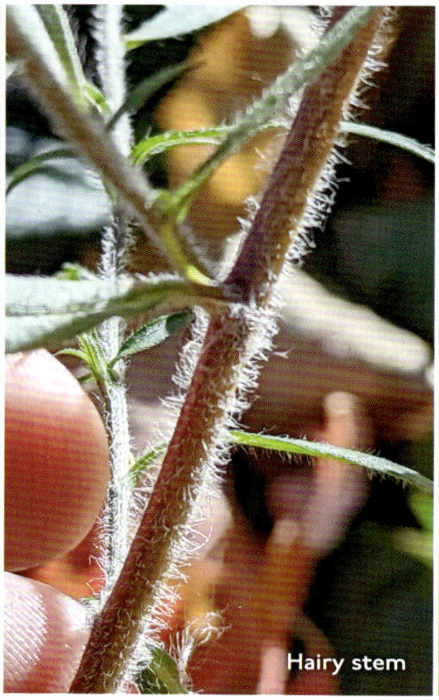

Hairy stem

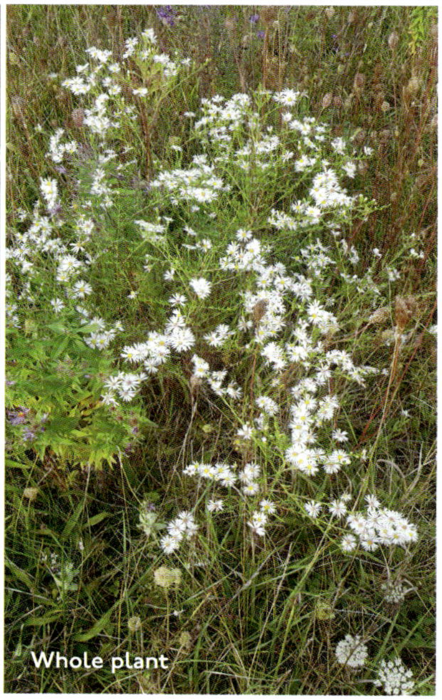

Whole plant

FAMILY *Asteraceae* (Aster Family)

ALTERNATE COMMON NAMES Awl Aster, Frostweed Aster, Hairy Oldfield Aster, Hairy White Oldfield Aster, Oldfield Aster, Pringle's Aster, White Oldfield Aster

PLANT DESCRIPTION Frost Aster features one or more erect, arching stems that are densely covered in spreading hairs. Lower stems may turn a reddish-brown color with age. Alternate leaves are lance-elliptic to lance-linear in shape (often widest just above the middle), toothless, and covered in long hairs. Leaves taper towards the base and are stalkless. They measure about 11 cm long and 2.5 cm wide. Clusters of smaller leaves emerge from the axils (where the leaf meets the stem). Basal leaves are spatula shaped with rounded tips and winged stalks.

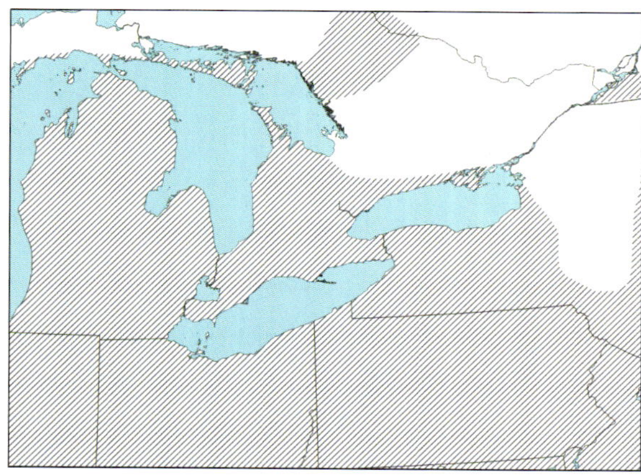

Upper stems terminate with branching clusters of stalked flowers. These clusters are usually widely spread out and have an arching appearance. Flower

clusters also arise from the upper leaf axils. Small, leafy bracts are common along the flower stalks. Flowers are usually found all on one side of the branch and each measure 1 to 2 cm across. Flowers are characterized by 15 to 35 ray florets (petals) surrounding a yellow disk that turns a reddish color with age. Bracts are found at the base of each flower in four to six layers. Each bract is appressed (flattened) to slightly spreading, light green with a long, dark green tip, and hairless to minutely hairy. Their edges may be rolled under, which gives them a narrow look. Flowers give way to small, linear seeds with tufts of white hairs that allow them to be carried off in the wind.

Leaves

Seed head

IN THE GARDEN Frost Aster puts on an elegant display of abundant, daisy-like flowers in fall. Dense hairs on the leaves and stems make it look as though it has been dusted by frost. It also blooms late into fall, which may explain how this aster got its common name. Mature plants can have a shrubby appearance, so it works well as a border plant. Note that the basal and lower stem leaves wither away by flowering time, so consider pairing it with a companion that can hide the base. Frost Aster may not be suitable for small gardens due to its ability to spread by seeds and rhizomes.

SKILL LEVEL Beginner

LIFESPAN Perennial

EXPOSURE Full sun

SOIL TYPE Most well-drained soils

MOISTURE Moist to dry

HEIGHT 20–120 cm

SPREAD 60–120 cm

BLOOM PERIOD Sep, Oct (to heavy frost)

COLOR White

FRAGRANT ❌

SHOWY FRUIT ❌

CUT FLOWER ✅

PESTS No serious insect or disease problems

NATURAL HABITAT Open, prairie-like habitats

WILDLIFE VALUE Attracts many kinds of bees, butterflies, and other insects; Ruffed Grouse (*Bonasa umbellus*) and Wild Turkeys (*Meleagris gallopavo*) eat the leaves and seeds while American Tree Sparrows (*Spizelloides arborea*) will eat the seeds during the winter; immature plants are a favorite of rabbits and deer

BUTTERFLY LARVA HOST PLANT FOR Silvery Checkerspot (*Chlosyne nycteis*), Tawny Crescent (*Phyciodes batesii*), Northern Crescent (*Phyciodes cocyta*), Pearl Crescent (*Phyciodes tharos*), Painted Lady (*Vanessa cardui*)

MOTH LARVA HOST PLANT FOR At least 40 species of moths, including members of the tiger moths, ribbed cocoon-maker moths, case-bearer moths, twirler moths, geometer moths, leaf-blotch miner moths, slug caterpillar moths, owlet moths, clearwing moths, flower moths, trumpet leafminer moths, and tortrix moths

USDA HARDINESS ZONES 3–9

PROPAGATION Sow seeds outside in the fall, or provide seeds with 60 days of cold, moist stratification. Frost Aster may also be propagated by stem cuttings taken in late spring.

ADDITIONAL INFO An aggressive spreader when compared to the common Panicled Aster (*Symphyotrichum lanceolatum*), Frost Aster has hairier foliage and prefers drier habitats. Another similar species, Calico Aster (*S. lateriflorum*) has smaller flower heads with fewer ray florets than those of the Frost Aster. Calico Aster also prefers shadier habitats, such as woodland borders and woodland openings. Yet another species, White Heath Aster (*S. ericoides*), typically found in prairies, also has smaller flower heads with fewer ray florets, and its leaves are smaller.

SAR STATUS N/A

Symphyotrichum puniceum
Purple-stemmed Aster

Flower

Whole plant

FAMILY *Asteraceae* (Aster Family)

ALTERNATE COMMON NAMES Glossy Leaved Aster, Red-stem Aster, Swamp Aster

PLANT DESCRIPTION Purple-stemmed Aster features erect, light-green-to-reddish-purple stems that are evenly covered with stiff hairs; however, they may be hairless on the lower stem. Leaves are found in an alternate pattern, measure 5 to 20 cm long and 8 mm to 3 cm wide, and become smaller as they ascend the stem. They are lance shaped with fine, widely spaced teeth and smooth except for the center vein, which is hairy under the leaf. Leaves are stalkless with a pair of lobes at the base that extend back to clasp the stem. Basal leaves are spatula shaped with winged stalks. The central stem terminates with an open cluster of flowers, which also arises from leaf axils

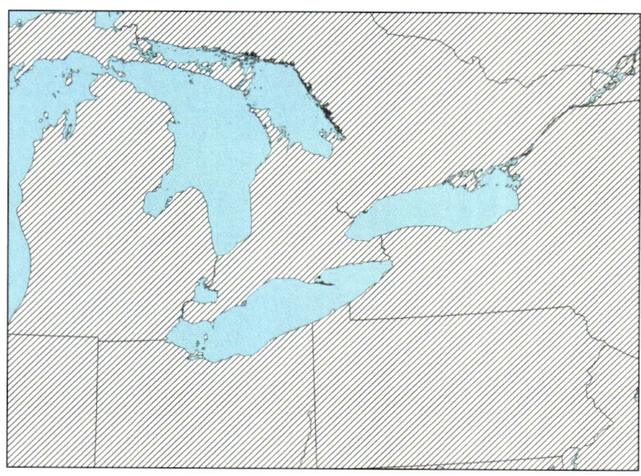

(where the leaf meets the stem). Individual flowers are 2 to 3 cm across and feature 30 to 50 narrow ray florets (petals) surrounding a yellow center disk that turns dull red with age. Surrounding each flower are four to six layers of narrow, hairless bracts. They are light green with dark green tips and spread outwards. Flowers give way to small, dry seeds with tufts of white hairs that allow them to be carried by the wind.

IN THE GARDEN Purple-stemmed Aster is a beautiful answer to wet or mucky soils. Its charming light-blue-to-purple flowers and reddish-purple stems add a lasting presence to the fall landscape. The stems are rigid enough to stand upright through the winter months. Note that lower leaves wither away by flowering time, especially if grown in drier soil than it prefers, so consider pairing it with a companion that can hide the base. The plant is clump forming but will self-seed freely.

SKILL LEVEL Beginner

LIFESPAN Perennial

EXPOSURE Full sun to part shade

SOIL TYPE The soil should contain some organic material to retain moisture, and it should be reasonably fertile and well drained

MOISTURE Wet to moist

HEIGHT 10–120 cm (occasionally up to 240 cm; the size of individual plants can be highly variable)

SPREAD 60–90 cm

BLOOM PERIOD Aug, Sep

COLOR Pale blue

FRAGRANT ✗

SHOWY FRUIT ✗

CUT FLOWER ✓

PESTS No serious insect or disease problems

NATURAL HABITAT Wet areas, such as soggy thickets, sedge meadows, and fens

WILDLIFE VALUE Nectar and pollen of the flower heads attract a wide variety of insects, such as wasps, native bees, and butterflies; Wild Turkeys (*Meleagris gallopavo*) eat the seeds and leaves occasionally; and White-tailed Deer (*Odocoileus virginianus*) and rabbits also enjoy the plant

BUTTERFLY LARVA HOST PLANT FOR Silvery Checkerspot (*Chlosyne nycteis*), Tawny Crescent (*Phyciodes batesii*), Northern Crescent (*Phyciodes*

Seed heads

Leaves

cocyta), Pearl Crescent (*Phyciodes tharos*), Painted Lady (*Vanessa cardui*)

MOTH LARVA HOST PLANT FOR At least 40 species of moths, including members of the tiger moths, ribbed cocoon-maker moths, case-bearer moths, twirler moths, geometer moths, leaf-blotch miner moths, slug caterpillar moths, owlet moths, clearwing moths, flower moths, trumpet leafminer moths, and tortrix moths

USDA HARDINESS ZONES 3–7

PROPAGATION Cold, moist stratify the seeds for 60 days if you are not sowing in the fall. Seeds require light to break dormancy, so do not cover. They are also slow to germinate — they can take a month or more to sprout.

ADDITIONAL INFO This plant tolerates clay soils and anaerobic conditions well. Drought tolerance is low, however. Purple-stemmed Aster is similar to New England Aster (*Symphyotrichum novae-angliae*), but the latter has more petals and is usually found in drier, upland sites.

SAR STATUS N//A

Symphyotrichum urophyllum
Arrowleaf Aster

Flowers

Leaf

Seed heads

FAMILY *Asteraceae* (Aster Family)

ALTERNATE COMMON NAMES Arrow-leaved Aster, White-arrow Aster, White Arrow Leaf Aster

PLANT DESCRIPTION Arrowleaf Aster features erect, light-green-to-yellowish stems that are smooth along the lower portion but have lines of hairs along the upper portion. Leaves are shallowly toothed with pointed tips, are found in an alternate pattern, measure 2.5 to 10 cm long and 1.2 to 5 cm wide, and become smaller and narrower as they ascend the stem. Lower leaves are oval–heart shaped on long winged stalks but become more lance-linear (narrow and pointed) and nearly stalkless towards the top. The upper surface is smooth to rough from small hairs, while the underside is hairy mainly along the central vein. Stems terminate with branching

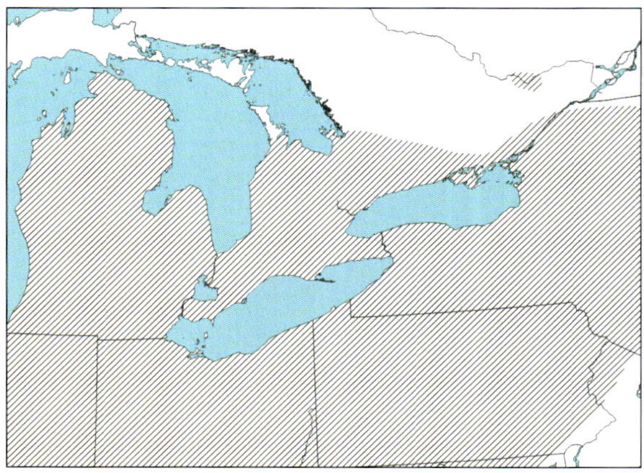

clusters of short-stalked flowers that also arise from upper leaf axils (the point where the leaf meets the stem). Clusters are tightly packed with flowers and

Whole plant

are taller than they are wide, which gives them a cylindrical appearance. Individual flowers are 1.2 cm across and feature eight to 15 widely spreading ray florets (petals) surrounding a yellow center that turns reddish with age. Bracts surround the base of each flower in four to six layers. Outer bracts are awl shaped while inner bracts are linear-lanceolate. They are sharply pointed, appressed (flattened) together or slightly spreading, and pale green with slender, dark-green-to-purplish tips. Flowers give way to small, dry seeds with tufts of white hairs that allow them to be carried by the wind.

IN THE GARDEN Arrowleaf Aster shows off with an extravagant display of white flowers throughout fall. It is easy to grow and tolerates a variety of growing conditions, but it produces more blooms in sunny sites. This aster has a clumping habit but readily self-seeds, so keep this in mind when choosing its location.

SKILL LEVEL Beginner

LIFESPAN Perennial

EXPOSURE Full sun to part shade

SOIL TYPE Sandy loam to clay loam

MOISTURE Dry to medium (should be watered during hot, dry spells of the summer)

HEIGHT 40–120 cm

SPREAD 45–60 cm

BLOOM PERIOD Aug, Sep, Oct

COLOR White

FRAGRANT ❌

SHOWY FRUIT ❌

CUT FLOWER ✅

PESTS No serious insect or disease problems

NATURAL HABITAT Most medium-to-dry locations, from woodlands to forest edges to prairies

WILDLIFE VALUE Bees, flies, wasps, butterflies, and skippers are attracted to this plant

BUTTERFLY LARVA HOST PLANT FOR Silvery Checkerspot (*Chlosyne nycteis*), Tawny Crescent (*Phyciodes batesii*), Northern Crescent (*Phyciodes cocyta*), Pearl Crescent (*Phyciodes tharos*), Painted Lady (*Vanessa cardui*)

MOTH LARVA HOST PLANT FOR At least 40 species of moths, including members of the tiger moths, ribbed cocoon-maker moths, case-bearer moths, twirler moths, geometer moths, leaf-blotch miner moths, slug caterpillar moths, owlet moths, clearwing moths, flower moths, trumpet leafminer moths, and tortrix moths

USDA HARDINESS ZONES 3–9

PROPAGATION Seeds need light to germinate, and you can either fall sow or cold, moist stratify for 60 days.

ADDITIONAL INFO This plant tends to get very tall and flop over if grown in rich soil under too much shade.

SAR STATUS N/A

Thalictrum dasycarpum
Purple Meadowrue

QUICK GUIDE

PROPAGATION

PART SHADE　　FULL SUN

WET　　MOIST

Female flowers　　Male flowers

FAMILY *Ranunculaceae* (Buttercup Family)

ALTERNATE COMMON NAMES Meadowrue, Tall Meadowrue

PLANT DESCRIPTION Purple Meadowrue's stems are smooth to sparingly hairy, purple, and unbranched below but become branched towards the top. Leaves are compounded three to five times into groups of three to five short-stalked leaflets and are found in an alternate arrangement along the stems. Each leaflet is oblong-ovate in shape, 5 cm long, and just under 4 cm wide and has two to three lobes (sometimes unlobed) with bluntly pointed tips. Leaflets are smooth above and finely hairy underneath and have toothless edges. You will notice that basal and lower stem leaves are stalked

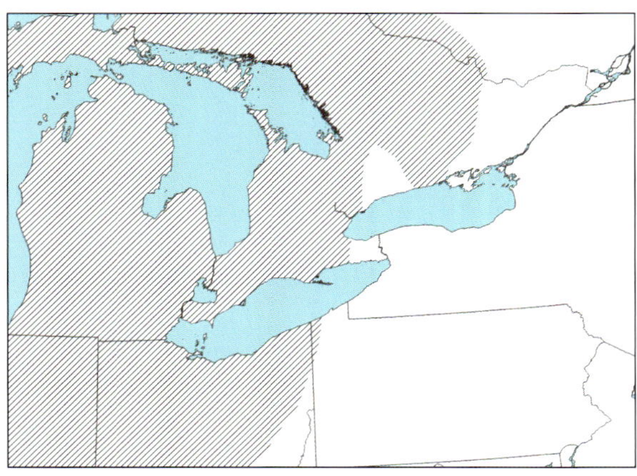

and are the largest of the leaves, but as the leaves ascend the stem they become stalkless and smaller.

Whole plant

Leaves

Seed heads

Stems are topped with loose, branching clusters of hanging flowers, with smaller clusters also emerging from upper leaf axils. Purple Meadowrue is a dioecious plant, meaning it has male and female flowers on separate plants. Male flowers are 8 mm wide and characterized by 15 slender, dangling, white stamens with yellow tips that turn darker with age. Female flowers are 8 mm wide and are characterized by 15 light green pistils. All flowers have four to five greenish-white sepals that fall off as soon as the flowers open. Female flowers mature into dry, oval seeds with four to six raised ridges on their sides.

IN THE GARDEN The delicate, hanging flowers of Purple Meadowrue stand tall and sway gracefully in the breeze. It makes for a unique structural plant in the garden, with interesting foliage, a clumping habit, and deep purple stems. As a bonus, herbivores tend to avoid eating this plant.

SKILL LEVEL Beginner

LIFESPAN Perennial

EXPOSURE Full sun to part (dappled) shade

SOIL TYPE Most soils, but prefers fertile, humus-rich, well-drained soils

MOISTURE Moist to wet

HEIGHT 150 cm

SPREAD 90–120 cm

BLOOM PERIOD May, Jun, Jul

COLOR Cream

FRAGRANT ❌

SHOWY FRUIT ❌

CUT FLOWER ✅

PESTS No serious insect or disease problems, though powdery mildew and rust may be occasional problems

NATURAL HABITAT Wet meadows, streambanks, and moist, wooded ravines

WILDLIFE VALUE Even though this is a wind-pollinated plant, small sweat bees, in particular, are attracted to the abundant pollen of male flowers

BUTTERFLY LARVA HOST PLANT FOR None

MOTH LARVA HOST PLANT FOR Canadian Owlet (*Calyptra canadensis*), Pink-patched Looper Moth (*Eosphoropteryx thyatyroides*), Meadowrue Borer Moth (*Papaipema unimoda*), Straight-lined Looper Moth (*Pseudeva purpurigera*)

USDA HARDINESS ZONES 2–8

PROPAGATION Cold, moist stratify seeds for 60 days before planting into cool soil for best germination results, or sow directly in the garden in late fall. Mature plants may be divided.

ADDITIONAL INFO Traditionally, Indigenous children would make flutes from the hollow stems of this plant.

SAR STATUS PA – SH/(not listed)

Thalictrum dioicum

Early Meadowrue

Female flowers

FAMILY *Ranunculaceae* (Buttercup Family)

ALTERNATE COMMON NAMES Feathered Columbine, Poor Man's Rhubarb, Quicksilver-weed, Shining Grass

PLANT DESCRIPTION Early Meadowrue has pale-green-to-purplish stems that are smooth and covered in a powdery bloom. Leaves are found in an alternate arrangement and measure up to 30 cm long. They are compounded one to four times into groups of three to five stalked leaflets. Each leaflet is just over 4 cm long and wide with three to 12 rounded, terminal lobes. Leaf surfaces are smooth. Stems are topped with loose clusters of flowers. Early Meadowrue is dioecious, meaning male and female flowers are found on separate plants. Male flowers are just over 1 cm long and have 10 or more

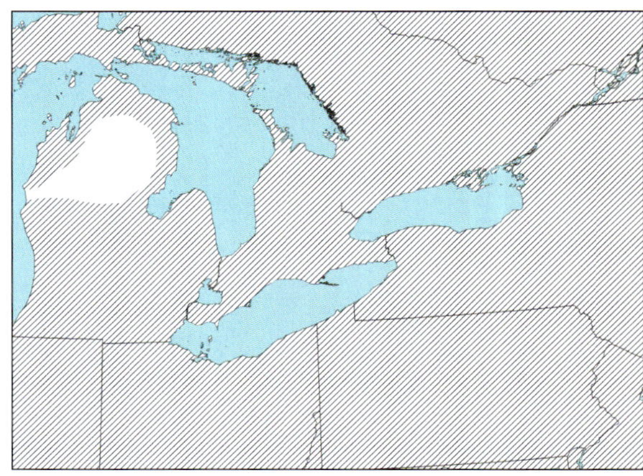

hanging stamens with greenish-yellow anthers that are longer than the stamens themselves. Female flowers are more erect and feature one to 16 whitish

Leaves

Whole plant

Seed heads

pistils with light purple stigmas. All flowers are backed by four to five greenish-purple sepals that fall away by the time the flowers open. Female flowers mature into dry, ribbed seeds.

IN THE GARDEN The airy blooms of Early Meadowrue rise above its delicate, fern-like foliage in early spring. These leaves last from spring to fall and are an excellent choice to fill in gaps where early spring woodland flowers have gone dormant for the season. This plant will self-seed non-aggressively to form colonies over time.

SKILL LEVEL Beginner

LIFESPAN Perennial

EXPOSURE Full shade to part shade

SOIL TYPE Rich loam or clay loam

MOISTURE Moist to medium

HEIGHT 80 cm

SPREAD 30–45 cm

BLOOM PERIOD May

COLOR Green

FRAGRANT ❌

SHOWY FRUIT ❌

CUT FLOWER ✅ (foliage)

PESTS No serious insect or disease problems, though powdery mildew and rust are occasional problems

NATURAL HABITAT Rich woods and ravines and alluvial terraces

WILDLIFE VALUE An important host plant for caterpillars of the Canadian Owlet (*Calyptra canadensis*), the Meadowrue Borer Moth (*Papaipema unimoda*), and the Straight-lined Looper Moth (*Pseudeva purpurigera*)

BUTTERFLY LARVA HOST PLANT FOR None

MOTH LARVA HOST PLANT FOR Canadian Owlet (*Calyptra canadensis*), Pink-patched Looper Moth (*Eosphoropteryx thyatyroides*), Meadowrue Borer Moth (*Papaipema unimoda*), Straight-lined Looper Moth (*Pseudeva purpurigera*)

USDA HARDINESS ZONES 2–8

PROPAGATION If starting indoors, seeds need 60 days of cold, moist stratification, but they have a short shelf life and are best planted outside in the fall. Seedlings will germinate in one year and flower in two or three years. Mature plants may be divided in the fall.

ADDITIONAL INFO Early Meadowrue is reported to be deer tolerant, but apparently not all deer have read the report. The flowers are wind pollinated.

SAR STATUS N/A

Thalictrum pubescens
Tall Meadowrue

FAMILY *Ranunculaceae* (Buttercup Family)

ALTERNATE COMMON NAMES King-of-the-meadow, Late Meadowrue, Meadow-weed, Muskrat-weed, Pubescent Meadowrue, Thalictrum

PLANT DESCRIPTION Tall Meadowrue features upright, hollow, green stems. Basal and lower stem leaves are borne on long stalks. Stem leaves are alternate and become stalkless towards the top. All leaves are compounded one to four times with round-to-ovate, three-lobed leaflets. Each leaflet measures up to 2.5 cm wide and is smooth. Flowers are found in clusters at the tops of the stems. Most Tall Meadowrue plants are dioecious (male and female flowers are found on separate plants); however, sometimes both will occur on the same plant. Male flowers have seven to 30, white-to-purplish,

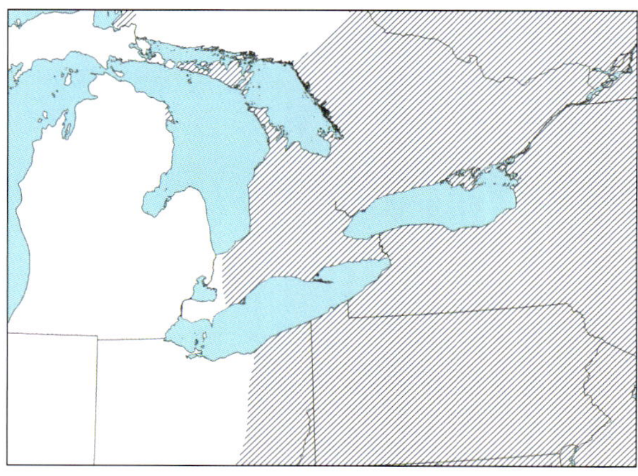

thread-like stamens topped with small, pale yellow anthers. Female flowers produce one to 16 pistils and also have stamens, but with sterile pollen. All

Leaves

Whole plant

Seed heads

flowers have four to six whitish-purple sepals that fall off as the flowers open. Female flowers mature into dry, ovoid, ribbed seeds.

IN THE GARDEN Tall Meadowrue makes an excellent structural plant for shady gardens with its tall, airy blooms rising above fine-textured foliage. Although they don't produce petals, the thread-like stamens on male flowers are very showy and sway delightfully in the breeze.

SKILL LEVEL Beginner

LIFESPAN Perennial

EXPOSURE Full shade to part shade (will tolerate full sun if soil is sufficiently moist)

SOIL TYPE Average soil

MOISTURE Medium to wet

HEIGHT 200 cm (occasionally to 270 cm)

SPREAD 60–90 cm

BLOOM PERIOD Jun, Jul, Aug

COLOR White

FRAGRANT ❌

SHOWY FRUIT ❌

CUT FLOWER ✅ (foliage)

PESTS No serious insect or disease problems, though powdery mildew, smut (a type of fungus), and rust may occasionally appear

NATURAL HABITAT Rich woods, swamps, wet meadows, and streambanks

WILDLIFE VALUE Although Tall Meadowrue is wind pollinated, the male flowers are an important source of pollen for many insects

BUTTERFLY LARVA HOST PLANT FOR None

MOTH LARVA HOST PLANT FOR Canadian Owlet (*Calyptra canadensis*), Pink-patched Looper Moth (*Eosphoropteryx thyatyroides*), Straight-lined Looper Moth (*Pseudeva purpurigera*)

USDA HARDINESS ZONES 3–8

PROPAGATION If starting indoors, seeds need 60 days of cold, moist stratification, but they have a short shelf life and are best planted outside as soon as they ripen. Seeds may not germinate until the second year. New plants may be started by separating offsets while the plant is dormant in spring or fall.

ADDITIONAL INFO Tall Meadowrue (*Thalictrum pubescens*) and Purple Meadowrue (*T. dasycarpum*) are easily confused — and to add to the confusion, they have both been called Tall Meadowrue. One way to differentiate them is you can measure the anthers in the flowers: in *T. pubescens* they are less than 1.5 mm long, whereas in *T. dasycarpum* they are more than 1.5 mm long. The bloom periods are also different (albeit overlapping), with *T. pubescens* flowering later (June to August) than *T. dasycarpum* (May to July). And though there is a fair amount of variability in the leaf shape, I have observed that *T. pubescens* leaves are often more pointed at the tips and *T. dasycarpum*'s are more rounded.

SAR STATUS IN – S3/T; MI – (not ranked)/SC

Tiarella stolonifera
Foamflower

Flower stalk

Flowers with immature seed capsules

FAMILY *Saxifragaceae* (Saxifrage Family)

ALTERNATE COMMON NAMES Allegheny Foamflower, Coolwort, False Miterwort, Heartleaf Foamflower, Heart-leaved False Miterwort

PLANT DESCRIPTION Foamflower has basal leaves born on hairy stalks. Leaves are up to 10 cm wide, heart shaped at the base, and sharply toothed and have five to seven shallow lobes. Leaf veins are usually hairy and sometimes have a burgundy color along them. Flowering stems rise above the leaves and terminate with a spike of five to 70 stalked, star-shaped flowers that each measure about 0.6 cm wide. Stems are hairy and leafless but may have a single, small leaf. Each flower features five white, ovate-to-lanceolate petals, five blunt, white sepals, and slender, protruding stamens. Flowers mature into small capsules that split open into two parts to release many small, shiny, black seeds.

IN THE GARDEN Foamflower is valued by gardeners as a low-growing, semi-evergreen groundcover. It spreads non-aggressively by runners to form dense patches over time and sends up feathery plumes of white flowers in the spring.

SKILL LEVEL Intermediate

LIFESPAN Perennial

EXPOSURE Full shade to part shade

SOIL TYPE Well-drained, humus-rich soils

MOISTURE Moist (plant should not dry out; however, moisture-saturated soils, particularly in the winter, can be fatal)

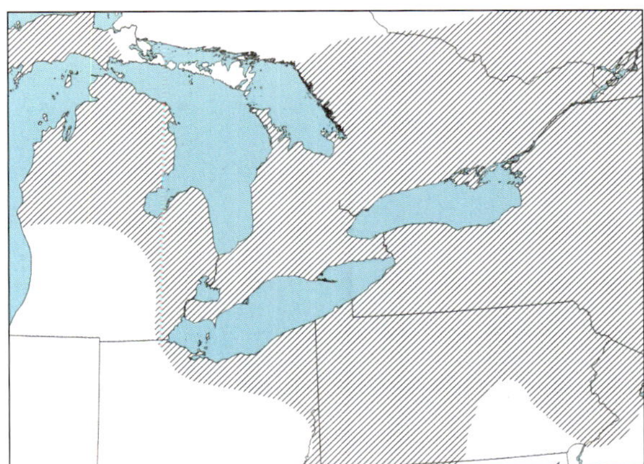

HEIGHT 15–40 cm

SPREAD 30–60 cm

BLOOM PERIOD May

COLOR White

FRAGRANT ✗

SHOWY FRUIT ✗

CUT FLOWER ✓ (foliage)

PESTS No serious insect or disease problems

NATURAL HABITAT Cool, moist deciduous woods and streambanks

WILDLIFE VALUE An important source of pollen and nectar for early spring pollinators, and later in the year Ruffed Grouse (*Bonasa umbellus*) and White-footed Mice (*Peromyscus leucopus*) eat the seeds

BUTTERFLY LARVA HOST PLANT FOR None

MOTH LARVA HOST PLANT FOR None

USDA HARDINESS ZONES 4–9

PROPAGATION Seed viability drops if the seeds dry out, so if not sowing immediately when harvested, they should be stored in a resealable plastic bag with dampened sand or sphagnum moss and sown as soon as possible. Fresh-sown seeds take about a month to germinate. Propagation is easiest by dividing the runners or crowns in the fall or spring and ensuring that each cutting has at least one leaf.

ADDITIONAL INFO *Tiarella stolonifera* is semi-evergreen in cold climates, with the leaves turning a deep purple during the winter months. These leaves should be left on the plant over winter so it can start photosynthesizing early, otherwise flowering will be delayed. This plant's scientific name was recently changed from *Tiarella cordifolia*. Many books, websites, and native plant nurseries may still be using the old name.

SAR STATUS N/A

Tradescantia ohiensis
Ohio Spiderwort

Flowers

Whole plant

FAMILY *Commelinaceae* (Spiderwort Family)

ALTERNATE COMMON NAMES Bluejacket, Common Spiderwort

PLANT DESCRIPTION Ohio Spiderwort has smooth, unbranched stems that are sometimes covered with a powdery bloom. Leaves are grass like, blue green, and found in an alternate pattern along the stem. You will notice that the base of each leaf wraps around the stem. Leaves measure up to 38 cm long and 2 cm wide and are flat with grooved midveins. Flowers are found in small clusters at the top of the stem and are borne on hairless flower stalks. Each flower is under 4 cm wide and characterized by three egg-shaped, violet-blue petals surrounding six yellow-tipped stamens with hairs at their bases. Each flower is backed by two conspicuous leafy bracts. Flowers mature into three-parted seed capsules that contain small seeds with unevenly textured surfaces.

Ohio Spiderwort is very similar to Virginia Spiderwort (*Tradescantia virginiana*); however, the latter has conspicuous hairs on its flower stalks and a shorter stature.

IN THE GARDEN Ohio Spiderwort is a vigorous and adaptable plant that is admired for its grass-like foliage and unique, blue-violet flowers. Only a few flowers bloom at a time, thus extending its blooms over a long period.

SKILL LEVEL Beginner to intermediate

LIFESPAN Perennial

EXPOSURE Full sun to part shade

SOIL TYPE Most well-drained soils

MOISTURE Moist to medium

HEIGHT 100 cm

SPREAD 50–60 cm

BLOOM PERIOD May (mid), Jun, Jul

COLOR Blue

FRAGRANT ✗

SHOWY FRUIT ✗

CUT FLOWER ✗

PESTS No serious insect or disease problems, though young shoots are susceptible to snail damage

NATURAL HABITAT Moist meadows, prairies, and thickets and especially in disturbed soils such as along roadsides and railroad rights-of-way

WILDLIFE VALUE Attracts long-tongued bees, especially bumblebees, and deer and rabbits eat the foliage

BUTTERFLY LARVA HOST PLANT FOR None

MOTH LARVA HOST PLANT FOR None

USDA HARDINESS ZONES 4–9

PROPAGATION Cold, moist stratification may or may not enhance germination when starting seeds indoors, but seeds readily germinate when sown outdoors in the fall in cool soil. Plants may also be started from stem cuttings at any time through the season and from division of mature plants in early spring or after flowering.

ADDITIONAL INFO Foliage tends to sprawl in an unattractive manner by midsummer, so it is recommended to cut it back to 15 to 30 cm in midsummer to encourage new growth and a possible fall bloom. The intriguing, blue-violet flowers tend to open in morning but often shrivel to a fluid jelly in the heat of the day.

SAR STATUS ON – S2/(not listed)

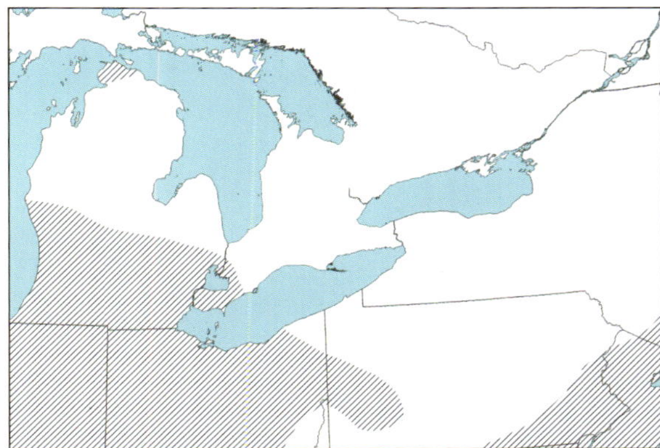

Range map for *Tradescantia ohiensis*

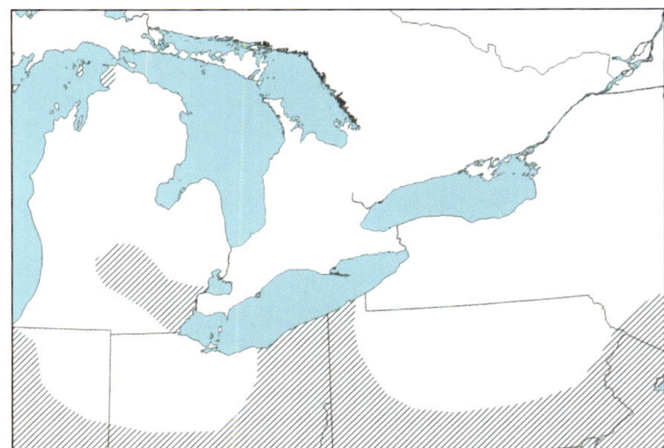

Range map for *Tradescantia virginiana*

Trillium erectum

Red Trillium

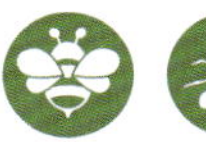

Flower

Whole plant

FAMILY *Liliaceae* (Lily Family)

ALTERNATE COMMON NAMES American True-love, Bethroot, Birthroot, Birthwort, Bumblebee Root, Ill-scented Wake-robin, Indian Shamrock, Purple Trillium, Purple Wakerobin, Red Wakerobin, Stinking Benjamin, Stinking Willie, Threeleaf Nightshade, Wet Dog Trillium, Wet Dog Wakerobin

PLANT DESCRIPTION Red Trillium has a smooth, greenish-red stem with three widely spreading, stalkless leaves found in a whorl at the top of it. Each leaf is up to 20 cm long and wide, is ovate-diamond in shape, and has smooth edges. A single flower is borne on an erect-to-leaning flower stalk at the top of the stem, above the leaves. This flower measures about 5 cm wide and has three maroon, lance-shaped petals surrounding six stamens topped with yellowish anthers. Three light green, pointed sepals are found in between the petals. Flowers mature to a six-sided, dark maroon fruit containing several seeds.

IN THE GARDEN Red Trillium is a sought-after addition to woodland gardens for its crimson-to-maroon flowers that rise above bold leaves. It is a spring ephemeral, meaning it blooms early in the spring, before the trees above it have leafed out, and goes dormant by mid to late summer. You can expect a healthy plant to slowly form a large clump over time.

SKILL LEVEL Beginner to intermediate

LIFESPAN Perennial (spring ephemeral)

Seeds with elaiosomes attached, with a dime for scale

Leaves

Fruit

EXPOSURE Part (dappled) shade (this species must be grown in an area where the plant will receive ample sun in early spring)

SOIL TYPE Humus-rich soils

MOISTURE Moist to medium

HEIGHT 20–60 cm

SPREAD 30 cm

BLOOM PERIOD Apr, May, (Jun)

COLOR Red (maroon), may also be white, pink, yellow, or green

FRAGRANT ✓ (but it's not pleasant)

SHOWY FRUIT ✗

CUT FLOWER ✗ (picking flowers may result in the death of the plant or, if the plant survives, render the plant unable to flower for potentially another seven years)

PESTS No serious pests, but may be affected by smut (a type of fungus), rust, fungal spots, slugs, and/or snails

NATURAL HABITAT Rich deciduous woods

WILDLIFE VALUE Flies are attracted to the rotting-meat color and smell, and ants carry home the seeds to consume the nutrient-rich elaiosomes that are attached (thereby dispersing the seeds to new locations)

BUTTERFLY LARVA HOST PLANT FOR None

MOTH LARVA HOST PLANT FOR None

USDA HARDINESS ZONES 3–8

PROPAGATION Seeds must remain moist and have a very short period of viability. They should be planted as soon as harvested. They will break dormancy after about three months of winter temperatures and put down a root. Typically, they will not send up a leaf till after the second winter, after which they can take anywhere from three to eight more years to produce a flower. Red Trilliums may be propagated by division when plants are dormant in late summer or early fall, though this process can be traumatic for the parent plant, and it may not bloom the following year.

ADDITIONAL INFO Red Trillium has no nectar and is pollinated by carrion flies and beetles, which are attracted to the odor of rotting meat that is given off by the petals of the flowers. (Note that the rotting-meat smell is only noticeable if you stick your nose in the flower.)

SAR STATUS NY – (not ranked)/EV

Trillium grandiflorum
Large White Trillium

Flowers

FAMILY *Liliaceae* (Lily Family)

ALTERNATE COMMON NAMES Great White Trillium, Large-flowered Trillium, Large-flowered Wakerobin, Snow Trillium, White Trillium, White Wakerobin

PLANT DESCRIPTION Large White Trillium has a central stem that is pale green and smooth. Three broadly ovate leaves are found in a whorl at the very top of the stem. They are up to 15 cm long and 12 cm wide and have smooth edges and prominent veins. Each plant produces a single, large flower borne on a 7.5 cm long stalk that is either upright or slightly leaning. The flower measures up to 10 cm wide and is characterized by three white petals surrounding six stamens, each with a yellow tip. Petals are elliptical in shape with a pointed tip. Three green sepals are found in between the petals. Note that the white petals usually fade to pink with age. Flowers mature into a three-sided, ridged, green berry that contains several seeds.

IN THE GARDEN Large White Trillium is an iconic spring wildflower that carpets woodlands with broad green leaves and elegant white blooms, before the trees above them have even leafed out. It makes an excellent garden plant but requires patience as it is slow to mature and spread. Being a spring ephemeral, it will go dormant by midsummer. This gives you the opportunity to co-plant it with a later-blooming plant, thus increasing the wildlife value and beauty you can pack into your yard.

SKILL LEVEL Beginner

LIFESPAN Perennial (spring ephemeral)

EXPOSURE Full shade to part shade

SOIL TYPE Humus-rich, sandy loam

MOISTURE Moist to medium

HEIGHT 30 cm

SPREAD 20–30 cm

BLOOM PERIOD May

COLOR White (turning pink as they age)

FRAGRANT ✗

SHOWY FRUIT ✗

CUT FLOWER ✗ (will result in the death of the plant)

PESTS No serious insect or disease problems, but watch for slugs and snails; potential disease problems include leaf spot, smut (a type of fungus), and rust

NATURAL HABITAT Rich, mixed woods, thickets, and swamps

WILDLIFE VALUE While the flowers are rarely visited by insects, the seeds of this and other trilliums are carried off by ants who consume the nutrient-rich elaiosomes and discard the seeds, thereby propagating plants away from the parent; Large White Trillium, in particular, is regularly eaten by White-tailed Deer (*Odocoileus virginianus*)

BUTTERFLY LARVA HOST PLANT FOR None

MOTH LARVA HOST PLANT FOR Black-patched Clepsis (*Clepsis melaleucana*), American Angle Shades (*Euplexia benesimilis*)

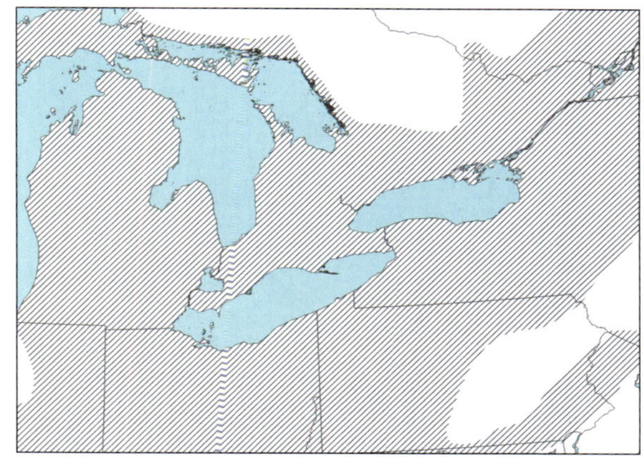

USDA HARDINESS ZONES 3–8

PROPAGATION As with other trilliums, the seeds should be kept moist or planted immediately (sow 2 mm deep) and not be allowed to dry out as seeds have a very short viability. Store for short periods only by packing the whole berry in moist sphagnum sealed in a refrigerated container. Seeds often require two winters to germinate, and even those that germinate in the first year may not send up a shoot till the following year. Plants may be divided with great care after they have gone dormant in the fall, but doing so may result in the parent not blooming in the subsequent year.

ADDITIONAL INFO Trilliums are slow growers requiring a minimum of seven years from seed to flowering. Applying leaf mulch in fall will help to maintain soil moisture, add nutrients, and protect the plant. *Trillium grandiflorum* is the floral emblem of Ontario and the state wildflower for Ohio.

SAR STATUS NY – (not ranked)/EV

305

Uvularia grandiflora
Large-flowered Bellwort

Flowers

FAMILY *Liliaceae* (Lily Family)

ALTERNATE COMMON NAMES Bellwort, Fairy Bells, Large Merrybells, Merrybells

PLANT DESCRIPTION Large-flowered Bellwort features a central stem that may be branched into two or three droopy side stems. Stems are smooth with a powdery bloom and are light green to reddish. Leaves are found in an alternate pattern, with the base of each leaf completely surrounding the stem, and measure about 15 cm long and 5 cm wide. Each leaf is oblong-ovate to elliptic in shape with smooth edges that may be rolled under. You will notice that upper leaf surfaces are smooth and green, while lower surfaces are light green and finely hairy. Droopy, yellow flowers hang down on slender flower stalks. Each flower is about 5 cm long and

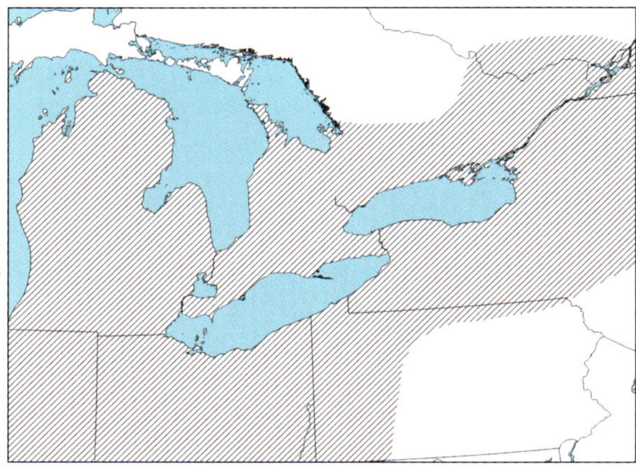

features six droopy, twisted petals. Flowers mature into three-sectioned, pyramidal seed capsules that contain several seeds.

IN THE GARDEN Large-flowered Bellwort lends a whimsical look to the woodland garden, especially in the springtime when its droopy, bell-shaped flowers are in bloom. It has an attractive clumping habit and keeps its leaves much later in the season than other spring wildflowers, making it a great foliage plant.

SKILL LEVEL Beginner

LIFESPAN Perennial

EXPOSURE Part shade to full shade

SOIL TYPE Humus-rich soils

MOISTURE Moist

HEIGHT 60 cm

SPREAD 30–45 cm

BLOOM PERIOD May

COLOR Yellow

FRAGRANT ✖

SHOWY FRUIT ✖

CUT FLOWER ✖

PESTS No serious insect or disease problems, though young growth is susceptible to slugs

NATURAL HABITAT Rich deciduous woods and thickets

WILDLIFE VALUE Pollen and nectar attract native bees, and seeds are collected by ants who cart them back to their nests to eat the nutrient-rich elaiosomes then discard the seeds; a favored plant of White-tailed Deer (*Odocoileus virginianus*)

BUTTERFLY LARVAL HOST PLANT FOR None

MOTH LARVA HOST PLANT FOR None

USDA HARDINESS ZONES 4–8

PROPAGATION Keep seeds moist and do not let them dry out. If not planting immediately, store in a resealable plastic bag in dampened sphagnum moss. *Uvularia* flowers and sets seeds early, therefore its seeds have evolved to need a period of warmth followed by a period of cold, moist stratification. Mature clumps may be divided in spring or fall, though mature specimens of this species have a thick, clustered rhizome that can be difficult to separate.

ADDITIONAL INFO The genus name, *Uvularia*, comes from the appearance of the dangling flowers that resemble the uvula — that dangling piece of soft tissue at the back of your mouth.

SAR STATUS N/A

Verbena hastata
Blue Vervain

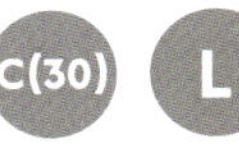  

Flowers

Flower cluster

FAMILY *Verbenaceae* (Verbena Family)

ALTERNATE COMMON NAMES Blue Verbena, False Vervain, Simpler's Joy, Swamp Verbena, Wild Hyssop

PLANT DESCRIPTION Blue Vervain stems are greenish red, four sided, and covered in short, appressed (closely pressed together) hairs. The stems branch occasionally at the top. Leaves are borne on short leaf stalks and are found in an alternate arrangement, measuring about 18 cm long and 2.5 cm wide. They are lanceolate in shape with a pointed tip, coarsely toothed edges, and prominent veins. Violet-blue flowers are densely packed into spiked clusters at the top of the stem. Each flower measures about 0.6 cm wide and features five spreading petals that are fused at the base to form a tube. Each flower matures into a four-chambered nutlet that contains four small, oblong seeds.

Blue Vervain is similar to Hoary Vervain (*Verbena stricta*), but the latter is found in drier habitats with larger flowers and stalkless leaves.

IN THE GARDEN Blue Vervain puts on a majestic display of candelabra-shaped flower clusters in midsummer, filling the garden with accents of violet-blue. It maintains a clumping habit and makes a great structural plant. The rigid stems and seed heads stand tall through the winter to provide excellent seasonal interest.

SKILL LEVEL Beginner

LIFESPAN Perennial (may be short lived)

EXPOSURE Full sun to part shade

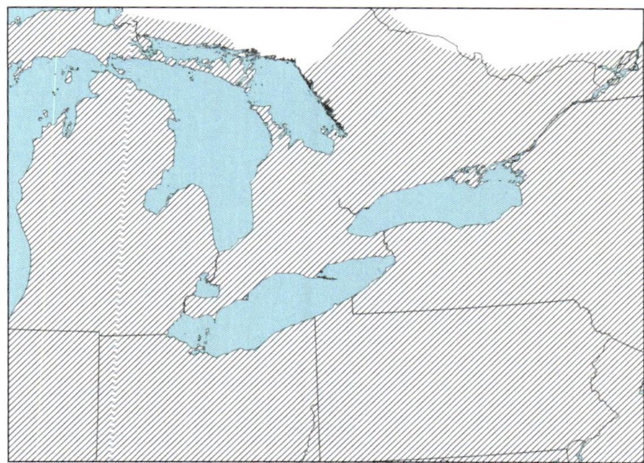

SOIL TYPE Sand to clay, but prefers soil consisting of fertile loam or wet muck

MOISTURE Wet to medium (this plant tolerates standing water, if it is temporary)

HEIGHT 150 cm

SPREAD 60–75 cm

BLOOM PERIOD Jul, Aug

COLOR Blue

FRAGRANT ✗

SHOWY FRUIT ✗

CUT FLOWER ✓

PESTS No serious insect or disease problems

NATURAL HABITAT Moist meadows, marshes, damp thickets, shores, wet woods, and wet wastelands

WILDLIFE VALUE The flowers attract many kinds of long-tongued and short-tongued bees, butterflies, and wasps; it is a preferred nectar plant for the Least Skipper (*Ancyloxypha numitor*), Broad-winged Skipper (*Poanes viator*), and Peck's Skipper (*Polites peckius*); various songbirds occasionally eat the seeds, including Northern Cardinals (*Cardinalis cardinalis*), Swamp Sparrows (*Melospiza georgiana*), Field Sparrows (*Spizella pusilla*), Song Sparrows (*Melospiza melodia*), and Slate-colored Juncos (*Junco hyemalis*), and small mammals are known to eat the new shoots

BUTTERFLY LARVA HOST PLANT FOR Common Buckeye (*Junonia coenia*)

MOTH LARVA HOST PLANT FOR Verbena Moth (*Crambodes talidiformis*), Verbena Bud Moth (*Endothenia hebesana*), Turtlehead Borer Moth (*Papaipema nepheleptena*), Sparganothis Leafroller Moth (*Sparganothis sulfureana*)

USDA HARDINESS ZONES 3–8

PROPAGATION Seeds need at least 30 days of cold, moist stratification (some sources say up to 90 days) and light to germinate, so do not cover the seeds with soil. Fall sowing works well — this plant readily self-seeds in my garden, coming up in unlikely and often inhospitable places. Stem cuttings may be started in the summer and work well.

ADDITIONAL INFO This short-lived perennial is a great choice for landscaping in wet areas, though in some jurisdictions it is considered invasive because of its prolific self-seeding.

SAR STATUS N/A

Verbena stricta
Hoary Vervain

FAMILY *Verbenaceae* (Verbena Family)

ALTERNATE COMMON NAMES Hoary Verbena, Tall Vervain, Woolly Verbena

PLANT DESCRIPTION Hoary Vervain features upright stems that are light green to reddish and densely covered in white hairs. Stalkless leaves are found in an opposite arrangement along the stem, measure about 10 cm long and 7.5 cm wide, and are oval to egg shaped. They have pointed tips and coarsely toothed edges and are densely covered in hairs. Small, 1.2 cm wide flowers are densely packed into spiked clusters at the top of the plant. Each flower is characterized by five petals that are fused at the base to form a short tube. Note how the lower lobe has a notched tip. Each flower matures into a four-chambered nutlet that contains four small, oblong seeds.

Hoary Vervain might be confused with Blue Vervain (*Verbena hastata*), but the latter plant has smaller flowers and stalked leaves that are longer and proportionately much narrower, and it prefers moist habitats.

IN THE GARDEN Even in the driest of sites, Hoary Vervain still manages to put on a charming display of long-lasting, purple flower spikes. It maintains an upright, clumping form, which makes it a good choice for a structural plant. In addition, the rigid stems and attractive seed heads persist through the winter months to maintain visual appeal and wildlife value in the garden. Herbivores avoid eating this plant.

SKILL LEVEL Beginner

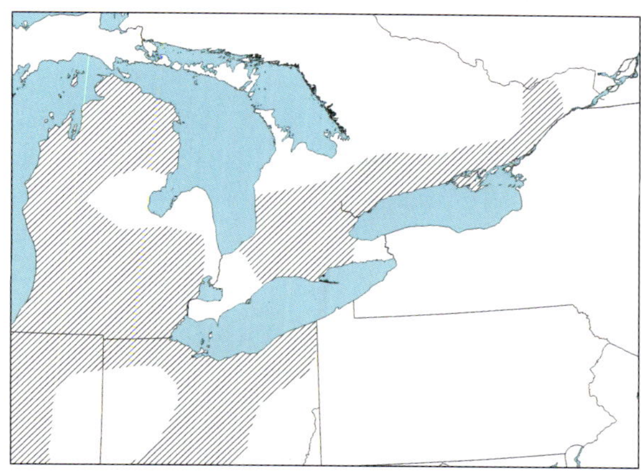

Leaves

Seed heads

LIFESPAN Perennial (may be short lived in rich soils, where it may behave as an annual or biennial)

EXPOSURE Full sun

SOIL TYPE Most well-drained soils

MOISTURE Medium to dry (extremely drought tolerant once established)

HEIGHT 60–100 cm

SPREAD 45–60 cm

BLOOM PERIOD Jun, Jul, Aug

COLOR Purple

FRAGRANT ✖

SHOWY FRUIT ✖

CUT FLOWER ✔

PESTS No serious insect or disease problems

NATURAL HABITAT Dry fields, prairies, and sandy hills, and degraded prairies

WILDLIFE VALUE Many kinds of bees, butterflies, and wasps are attracted to the flowers, and various songbirds occasionally eat the seeds

BUTTERFLY LARVA HOST PLANT FOR Common Buckeye (*Junonia coenia*)

MOTH LARVA HOST PLANT FOR Fine-lined Sallow (*Catabena lineolata*), Verbena Moth (*Crambodes talidiformis*), Verbena Bud Moth (*Endothenia hebesana*), Sparganothis Leafroller Moth (*Sparganothis sulfureana*)

USDA HARDINESS ZONES 3–8

PROPAGATION The small seeds require 30 to 60 days of cold, moist stratification to break dormancy, and they need light to germinate so simply press them lightly into the soil without covering or scatter the seeds in the fall. Plants usually bloom in the second year after seeding.

ADDITIONAL INFO For Hoary Vervain, the drier the better, as even mesic soils tend to result in shorter lifespans. It does not handle competition well.

SAR STATUS N/A

Vernonia gigantea
Tall Ironweed

	PART SHADE	FULL SUN
	MOIST	MEDIUM

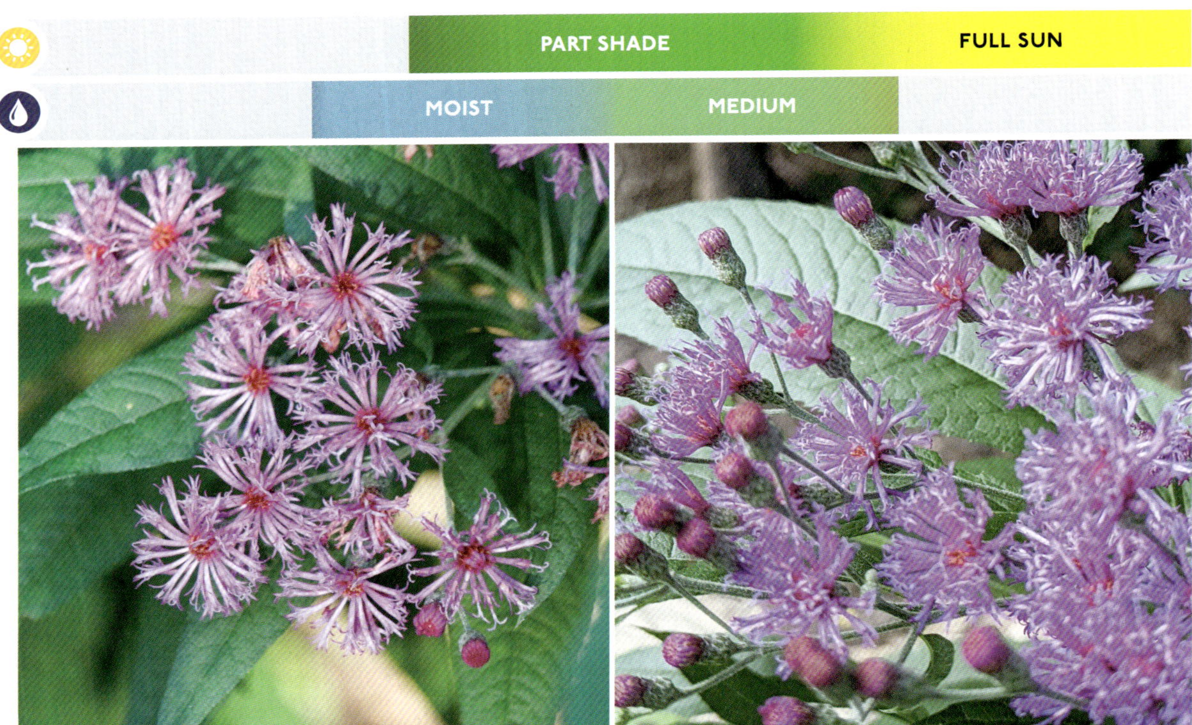

Flowers

Flowers

FAMILY *Asteraceae* (Aster Family)

ALTERNATE COMMON NAMES Giant Ironweed

PLANT DESCRIPTION Tall Ironweed has light green stems that are smooth to finely hairy. They are unbranched, except for near the flower cluster. Leaves are found in an alternate pattern, measure up to 25 cm long and 6 cm wide, are lanceolate to lanceolate-ovate in shape, and are borne on short-to-no leaf stalk. Leaf edges are irregularly serrated, upper leaf surfaces are smooth, and lower leaf surfaces are smooth to finely hairy. Stems terminate in loose, flat-headed clusters of 2.5 cm wide, purple flowers. Each flower has 10 to 20 disk flowers, no ray florets (petals), five recurved lobes, and a protruding style. Flowers mature into small, bullet-shaped seeds with tufts of hairs that allow them to be distributed by the wind.

IN THE GARDEN Tall Ironweed lives up to its name, towering over the garden to show off clusters of vivid purple flowers that attract butterflies from far and wide. The strong stems are topped with fluffy, rust-colored seed heads and persist through-out the winter months. Its height and clumping form make it useful as a structural plant.

SKILL LEVEL Beginner

LIFESPAN Perennial

EXPOSURE Full sun to part shade

SOIL TYPE Soil containing loam, clay loam, silt loam, or sandy loam

MOISTURE Moist to medium

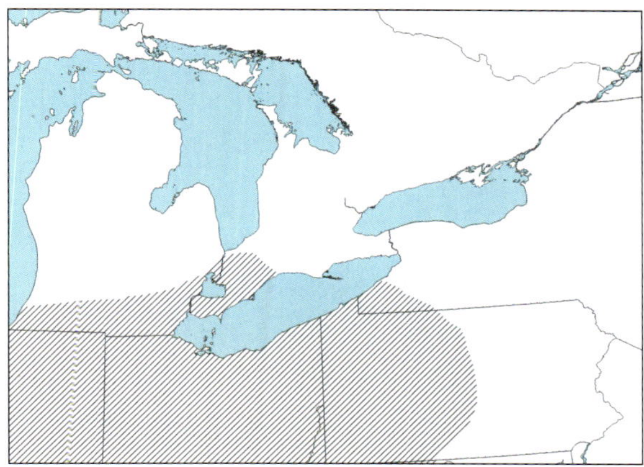

HEIGHT 90–210 cm (to 270 cm or more in ideal conditions)

SPREAD 90–150 cm

BLOOM PERIOD Aug, Sep

COLOR Purple

FRAGRANT ❌

SHOWY FRUIT ❌

CUT FLOWER ✅

PESTS No serious insect of disease problems

NATURAL HABITAT Damp areas in fields or along forest edges and streams

WILDLIFE VALUE Nectar-rich flowers attract a variety of native bees and butterflies, and some bees also collect pollen for their larvae; because of the bitter foliage it is seldom eaten by mammalian herbivores

BUTTERFLY LARVA HOST PLANT FOR Crossline Skipper (*Polites origenes*), Painted Lady (*Vanessa cardui*), American Lady (*Vanessa virginiensis*)

MOTH LARVA HOST PLANT FOR Eupatorium Borer Moth (*Carmenta bassiformis*), Parthenice Tiger Moth (*Grammia parthenice*), Ironweed Borer Moth (*Papaipema cerussata*), Ironweed Borer Moth (*Papaipema limpida*), Red Groundling (*Perigea xanthioides*), Ruby Tiger Moth (*Phragmatobia fuliginosa*), Ironweed Root Moth (*Polygrammodes flavidalis*), Ironweed Moth (*Polygrammodes langdonalis*)

USDA HARDINESS ZONES 5–8

PROPAGATION Seeds need 60 days of cold, moist stratification to germinate, or you can direct sow in the garden in the fall. William Cullina advises that five to seven node cuttings root well but don't always overwinter well.

ADDITIONAL INFO The stems and leaves are less pubescent (smoother) than those of the similar-looking Missouri Ironweed (*Vernonia missurica*).

SAR STATUS ON – S1/(not listed); NY – S1/E

Vernonia missurica
Missouri Ironweed

FULL SUN

MOIST **MEDIUM**

Flowers

Hairy stem

FAMILY *Asteraceae* (Aster Family)

ALTERNATE COMMON NAMES None

PLANT DESCRIPTION Missouri Ironweed has reddish-green stems that are densely covered in fine white hairs. It is unbranched except for near the flowers. Leaves are alternate in arrangement, short stalked to stalkless, and lanceolate to narrowly ovate and measure up to 18 cm long and 5 cm wide. They have serrated edges and are densely hairy on their lower surfaces. Magenta-to-purple flowers are found in flat-topped clusters at the top of the stem. They measure about 2 cm wide and feature 30 to 60 disk flowers (petals) and no ray florets. Flowers mature into small, bullet-shaped seeds with tufts of hairs that allow them to be distributed by the wind.

Missouri Ironweed can be distinguished from other ironweeds by its hairy stems, hairy leaf undersides, and greater number of disk flowers.

IN THE GARDEN Missouri Ironweed puts on a remarkable late-summer display of magenta flowers that are as appealing to butterflies as they are to gardeners. This is a good choice for a structural plant as it has an upright, clumping habit. The strong stems are topped with showy, rust-colored seed heads and stand tall through the winter months.

SKILL LEVEL Beginner

LIFESPAN Perennial

EXPOSURE Full sun

SOIL TYPE Clay loam to sandy loam

MOISTURE Medium to moist

HEIGHT 90–150 cm

SPREAD 90–120 cm

BLOOM PERIOD Jul, Aug, (Sep)

COLOR Purple

FRAGRANT ❌

SHOWY FRUIT ❌

CUT FLOWER ✅

PESTS No serious insect or disease problems, though powdery mildew may occasionally affect plants in the fall

NATURAL HABITAT Prairies, streambanks, bottomland forests, and disturbed areas

WILDLIFE VALUE The flowers attract primarily long-tongued bees, butterflies, and skippers, but herbivores tend to avoid this plant because its leaves are quite bitter

BUTTERFLY LARVA HOST PLANT FOR Crossline Skipper (*Polites origenes*), Painted Lady (*Vanessa cardui*), American Lady (*Vanessa virginiensis*)

MOTH LARVA HOST PLANT FOR Parthenice Tiger Moth (*Grammia parthenice*), Ironweed Borer Moth (*Papaipema cerussata*), Red Groundling (*Perigea xanthioides*), Ruby Tiger Moth (*Phragmatobia fuliginosa*), Ironweed Root Moth (*Polygrammodes*

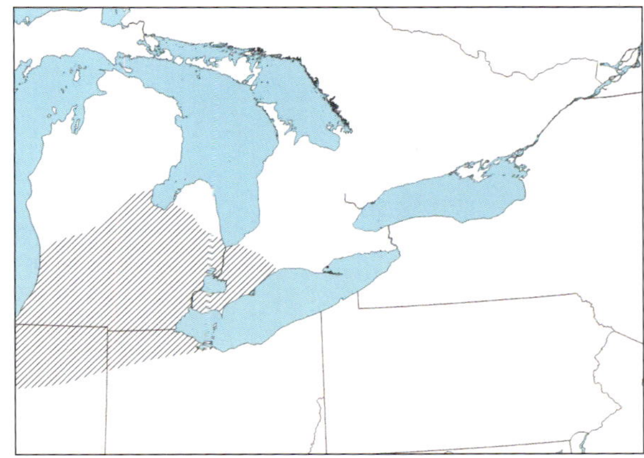

flavidalis), Ironweed Moth (*Polygrammodes langdonalis*)

USDA HARDINESS ZONES 4–9

PROPAGATION Seeds germinate after about 60 days of cold, moist stratification, or plants may be started from stem cuttings. William Cullina advises that five to seven node cuttings root well but don't always overwinter well.

ADDITIONAL INFO Missouri Ironweed tolerates periodic flooding, and it is moderately drought tolerant when established. In overgrazed pastures, ironweed is an "increaser" because it is one of the last plants to be eaten.

SAR STATUS ON – S3/(not listed); OH – S2/E

Veronicastrum virginicum
Culver's Root

Flowers

Flower clusters

Leaf formation along stem

FAMILY *Plantaginaceae* (Plantain family)

ALTERNATE COMMON NAMES Beaumont Root, Blackroot, Bourman's Root, Bowman's Root, Culver's Physic, Oxadaddy, Physic Root, Quitch, Tall Speedwell, Tall Veronica, Virginia Culver's-root

PLANT DESCRIPTION Culver's Root has smooth (sometimes finely hairy), green stems that are unbranched, except near the flowers. Along the stem are widely spaced groups of three to seven whorled leaves borne on short-to-no leaf stalks. Each leaf measures up to 20 cm long and just under 4 cm wide, is ovate to narrowly ovate in shape, and tapers to a point at both ends. Leaves have well-defined veins and are hairy underneath. Tubular flowers are densely packed into spiked clusters at the top of the stem. Immature plants will have only one terminal

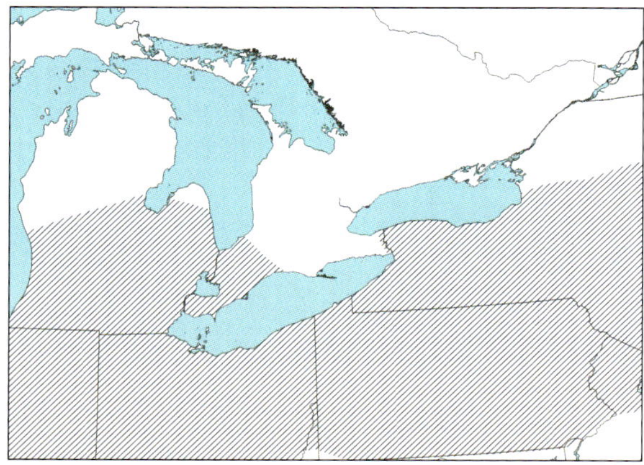

spike, while more mature plants will have many smaller, additional spikes emerging from the stem below the terminal spike. Each flower is about 0.6 cm

Leaf

Whole plant

Seed head

long and four parted and features two protruding stamens with yellow tips. The flowers mature into capsules that contain many tiny, brown seeds.

IN THE GARDEN The slender, candelabra-like flower heads of Culver's Root lend an elegant look to the landscape, while the whorled leaves add interesting texture. It is quite versatile in the garden and makes for an excellent structural plant with its upright, clumping habit. The rigid stems persist through the winter months to extend seasonal interest and, as a bonus, herbivores rarely bother with this plant.

SKILL LEVEL Beginner

LIFESPAN Perennial

EXPOSURE Full sun to part shade

SOIL TYPE Rich, loamy, well-drained soil

MOISTURE Moist to medium

HEIGHT 150–180 cm

SPREAD 60–120 cm

BLOOM PERIOD Jun, Jul, Aug

COLOR White to pale purple/blue

FRAGRANT ❌

SHOWY FRUIT ❌

CUT FLOWER ✅

PESTS No serious insect or disease problems

NATURAL HABITAT Open woods, thickets, and moist meadows and prairies

WILDLIFE VALUE Most common visitors to the flowers are long-tongued and short-tongued bees, which collect pollen or drink nectar; you may also notice butterflies, moths, and syrphid flies visiting

BUTTERFLY LARVA HOST PLANT FOR None

MOTH LARVA HOST PLANT FOR Culver's Root Borer Moth (*Papaipema sciata*)

USDA HARDINESS ZONES 3–8

PROPAGATION No pretreatment of seeds is necessary, though some sources suggest 30 days of cold, moist stratification improves the germination rate. The seeds need light to break dormancy, so surface sow and do not cover with soil. The seeds are also hydrophilic and should not be allowed to dry out, therefore if not sowing immediately they should be stored in slightly dampened sphagnum moss in a resealable plastic bag. However, Culver's Root is most easily propagated by root divisions in late fall or early spring — each piece of root must have a bud. You can also start plants from two-to-three-node stem cuttings taken in late spring, which root easily.

ADDITIONAL INFO Deadhead spent flowers to extend bloom period and cut back plants after flowering to stimulate new growth and a possible late-summer or fall bloom.

SAR STATUS ON – S2/(not listed); NY – S2/T

Viola blanda
Sweet White Violet

Flowers

FAMILY *Violaceae* (Violet Family)

ALTERNATE COMMON NAMES Large-leaved White Violet, Smooth White Violet, Woodland White Violet

PLANT DESCRIPTION Sweet White Violet features heart-shaped basal leaves borne on reddish stalks that emerge from nodes along its underground rhizomes. Each leaf measures about 6 cm wide with a deeply cleft base and rounded-to-pointed tips. They may have sparse white hairs on the upper surface. A single flower is borne on top of a bare, reddish stem. This stem emerges from the underground rhizomes and usually rises slightly higher than the leaves. Individual flowers are just over 1 cm wide and feature five white petals. Upper petals curve backwards, while the lower petals are more

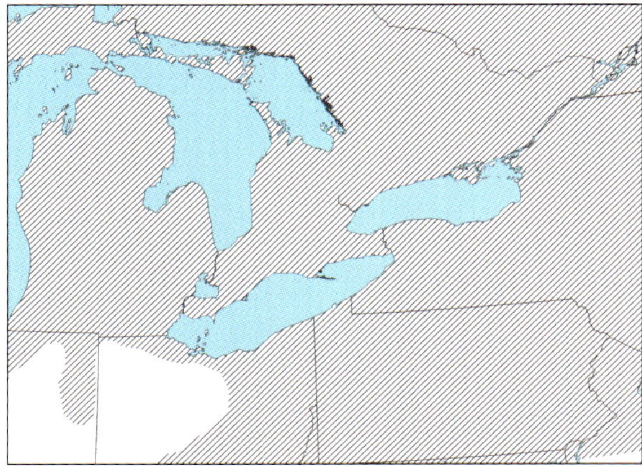

protruding with a few hairs at their bases. Lower petals have brownish-purple veins along them. Each flower matures into an oval seed capsule that splits

Leaf

Whole plant

Seed capsules

open to release many small, brown seeds.

This species is similar to Northern White Violet (*Viola pallens*), but Northern White Violet doesn't have red stems and grows in wetter habitats.

IN THE GARDEN Sweet White Violet spreads by rhizomes to carpet the ground with its heart-shaped leaves. The plant is resistant to herbivores.

SKILL LEVEL Beginner

LIFESPAN Perennial

EXPOSURE Part shade

SOIL TYPE Well-drained, humus-rich, moisture-retentive soils

MOISTURE Moist

HEIGHT 7–15 cm

SPREAD 20–30 cm

BLOOM PERIOD Apr, May

COLOR White

FRAGRANT ✓

SHOWY FRUIT ✗

CUT FLOWER ✗

PESTS No serious insect or disease problems

NATURAL HABITAT Dry-to-very-moist woods, thickets, and clearings in deciduous and mixed forests

WILDLIFE VALUE An important source of nectar and pollen for a variety of bees, including bumblebees, mason bees, and sweat bees, and the nectar attracts butterflies, moths, and flies, among other insects

BUTTERFLY LARVA HOST PLANT FOR Fritillary butterflies (*Bolloria* spp. and *Speyeria* spp.)

MOTH LARVA HOST PLANT FOR Nais Tiger Moth (*Apantesis nais*), Omniverous Leafroller Moth (*Archips purpurana*), Grateful Midget (*Elaphria grata*), The Beggar (*Eubaphe mendica*), Giant Leopard Moth (*Hypercompe scribonia*), Striped Garden Caterpillar (*Trichordestra legitima*), Smith's Dart (*Xestia smithii*)

USDA HARDINESS ZONES 2–7

PROPAGATION Seeds need 60 days of cold, moist stratification as well as light to germinate. Plants may also be divided in spring or fall.

ADDITIONAL INFO The lovely flowers of Sweet White Violet emerge in spring and are quite fragrant for their size, so it is worth getting up close to smell them.

SAR STATUS IN – S3/WL

Viola canadensis
Canada Violet

QUICK GUIDE

PROPAGATION

| PART SHADE |
| MOIST | MEDIUM |

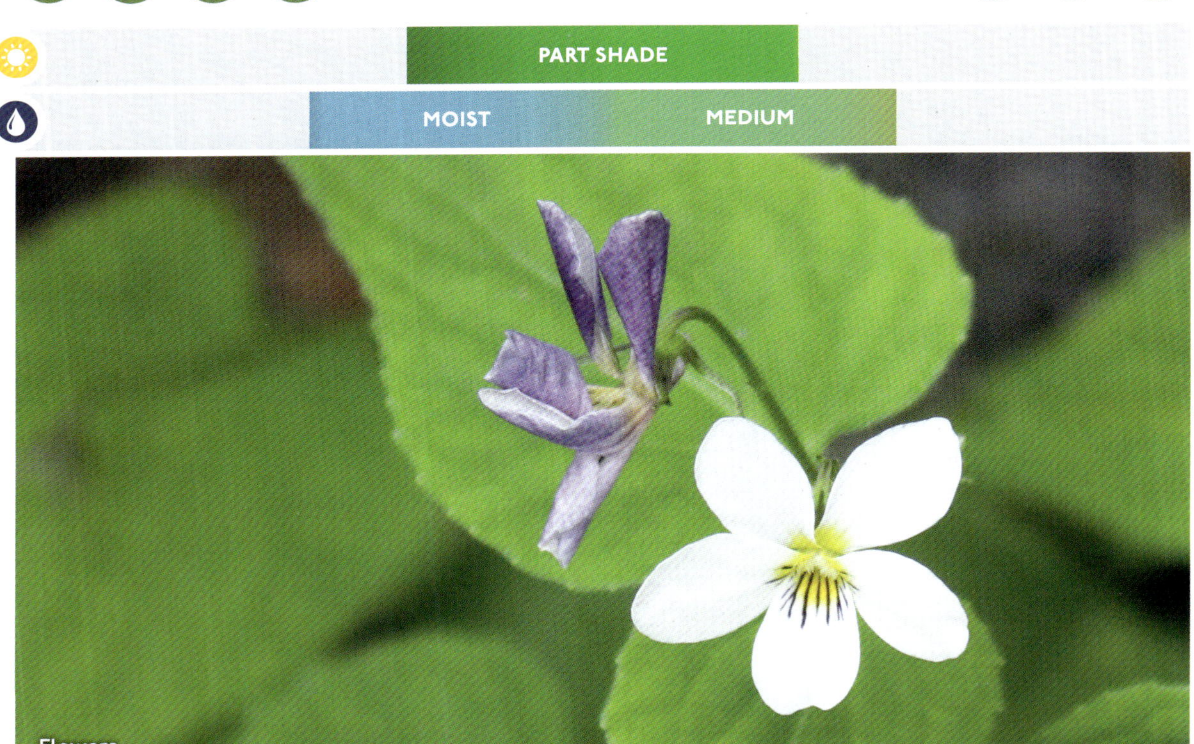

Flowers

FAMILY *Violaceae* (Violet Family)

ALTERNATE COMMON NAMES Canada White Violet, Canadian Violet, Tall White Violet

PLANT DESCRIPTION Canada Violet features angular, finely hairy stems that are green with purple tinges. You will notice both basal leaves and alternate stem leaves. Both types of leaves are up to 10 cm long, 7.5 cm wide and heart shaped with pointed tips and have shallowly toothed edges. Stem leaves differ only slightly as their shape is more elongated. A single flower sits atop a leafless flower stalk that arises from upper leaf axils. This stalk sticks out only slightly above the leaves. Individual flowers are about 2.5 cm wide and feature five white petals with distinctive yellow bases. The lower petals have dark purple lines, and the side

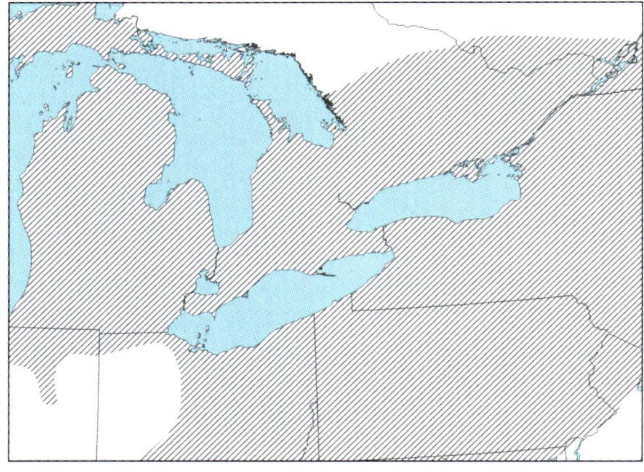

petals have hairs. Each flower matures into an oval seed capsule, just over 1 cm long, that is covered in fine hairs. When ripe, it turns brown and splits

Leaves

Whole plant

Seed capsules

Whole plant

open into three sections to release small, brown seeds.

IN THE GARDEN Canada Violet can be used as a robust groundcover for shady sites, where it will spread readily by self-seeding. The dainty white flowers bloom over a long period and contrast nicely with its dark green leaves.

SKILL LEVEL Beginner to intermediate

LIFESPAN Perennial

EXPOSURE Part (dappled) shade

SOIL TYPE Rich soil containing loam and decaying organic matter

MOISTURE Moist to medium

HEIGHT 15–40 cm

SPREAD 30 cm (can spread by rhizomes to 60 cm)

BLOOM PERIOD Apr, May, Jun, Jul

COLOR White

FRAGRANT ✅ (very mild)

SHOWY FRUIT ❌

CUT FLOWER ❌

PESTS No serious insect and disease problems, and deer resistant

NATURAL HABITAT Moist, open deciduous forests

WILDLIFE VALUE Attracts primarily bees and butterflies that seek nectar

BUTTERFLY LARVA HOST PLANT FOR Fritillary butterflies (*Bolloria* spp. and *Speyeria* spp.)

MOTH LARVA HOST PLANT FOR Nais Tiger Moth (*Apantesis nais*), Omniverous Leafroller Moth (*Archips purpurana*), Grateful Midget (*Elaphria grata*), The Beggar (*Eubaphe mendica*), Giant Leopard Moth (*Hypercompe scribonia*), Striped Garden Caterpillar (*Trichordestra legitima*), Smith's Dart (*Xestia smithii*)

USDA HARDINESS ZONES 2–8

PROPAGATION Direct sow in the fall or cold, moist stratify for 60 days. Seeds require light to germinate. Offshoots of plants may be separated and planted. The plant will readily self-seed in suitable conditions.

ADDITIONAL INFO Canada Violet differs from other native white violets in that it has yellow at the bases of the petals and a purple tint to the back of the petals.

SAR STATUS N/A

Viola pubescens
Downy Yellow Violet

FAMILY *Violaceae* (Violet Family)

ALTERNATE COMMON NAMES Common Yellow Violet, Downy Violet, Hairy Yellow Violet, Yellow Forest Violet, Yellow Violet

PLANT DESCRIPTION Downy Yellow Violet has both basal and stem leaves. Basal leaves are broadly heart shaped, measure up to 10 cm long and 8 cm wide, and are borne on 7.5 cm long hairy leaf stalks that are erect to leaning. Leaves are coarsely toothed or with scalloped edges and smooth above and finely hairy below, especially along the veins. Stem leaves are found in an alternate arrangement along the smooth-to-hairy stems. They are similar to the basal leaves but are slightly more elongated and smaller. There are one to two stems per root-stock. Individual flowers sit atop flower stalks that

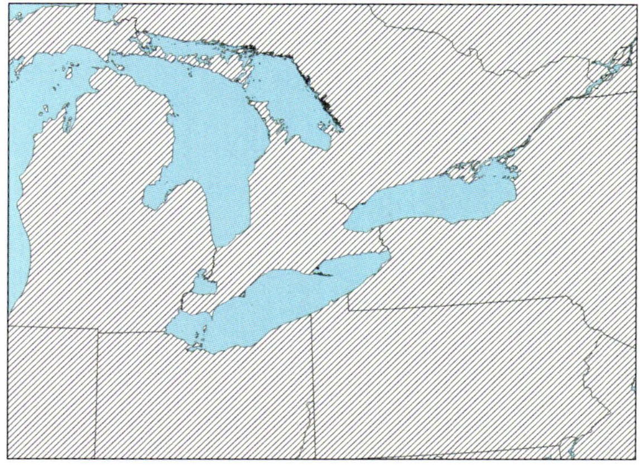

arise from leaf axils. Each flower is 2 cm wide and features five yellow petals. The lower petal has many brownish-purple lines along it, while the other petals

Open seed capsule

Whole plant

have only a few to none. Note how the side petals have hairs near their bases. Each flower matures into a hanging, egg-shaped seed capsule. When ripe, the seed capsule turns brown, becomes upright, and splits into three sections to eject its seeds outward.

IN THE GARDEN Downy Yellow Violet speckles the woodland garden with enchanting yellow flowers in early spring. It makes for a fine woodland groundcover with its broad, heart-shaped leaves and spreads non-aggressively by seed to form colonies. The leaves usually persist throughout the summer and are not often bothered by herbivores.

SKILL LEVEL Beginner

LIFESPAN Perennial

EXPOSURE Full shade to part shade

SOIL TYPE Relatively loose soil containing loam or sandy loam with some decaying organic matter (e.g., fallen leaves)

MOISTURE Moist to dry

HEIGHT 10–45 cm

SPREAD 15–30 cm

BLOOM PERIOD Apr, May, Jun

COLOR Yellow

FRAGRANT ✓ (very mild)

SHOWY FRUIT ✗

CUT FLOWER ✗

PESTS No serious insect or disease problems

NATURAL HABITAT Rich, sandy woodlands and areas along woodland paths

WILDLIFE VALUE Provides nectar and pollen for many small native bees, and ants take the seeds back to their nests to consume the nutrient-rich elaiosomes, thus spreading the plants to new areas

BUTTERFLY LARVA HOST PLANT FOR Fritillary butterflies (*Boiloria* spp. and *Speyeria* spp.)

MOTH LARVA HOST PLANT FOR Nais Tiger Moth (*Apantesis nais*), Omniverous Leafroller Moth (*Archips purpurana*), Grateful Midget (*Elaphria grata*), The Beggar (*Eubaphe mendica*), Giant Leopard Moth (*Hypercompe scribonia*), Striped Garden Caterpillar (*Trichordestra legitima*), Smith's Dart (*Xestia smithii*)

USDA HARDINESS ZONES 3–7

PROPAGATION Surface sow as seeds need light to break dormancy, and seeds require 60 days of cold, moist stratification. Seeds are hydrophilic and should not be allowed to dry out. They are best planted outdoors in fall.

ADDITIONAL INFO This violet doesn't tolerate mowing. It is better able to grow in areas that have pine needles as ground litter than many other plants. Some authorities treat *Viola pubescens* as two separate species, dividing it into *V. pubescens* (Downy Yellow Violet) and *V. eriocarpa* (Smooth Yellow Violet), while others treat Smooth Yellow Violet as simply a subspecies. There are subtle differences between them: *V. eriocarpa* tends to have multiple basal leaves, it is more apt to lay down on the ground, and the foliage is generally less hairy than *V. pubescens*. But, given that their range and habitat is very similar, either one would make a great addition to a shade garden.

SAR STATUS N/A

Viola rostrata
Long-spurred Violet

Flower

The flower's long spur

FAMILY *Violaceae* (Violet Family)

ALTERNATE COMMON NAMES Longspur Violet

PLANT DESCRIPTION Long-spurred Violet
has both basal and stem leaves. Basal leaves are
broadly heart shaped, up to 4 cm wide, and borne
on leaf stalks that reach up to 9 cm long. They have
serrated edges and finely hairy surfaces but may be
smooth. Stem leaves are similar to the basal leaves
but have shorter leaf stalks and gradually taper at
the tips. The stems themselves are smooth and erect
to leaning. Flower stalks emerge from leaf axils with
a single, 1 cm wide, five-petaled flower atop each
stalk. The petals are beardless (lack hairs) and are
pale purple with the lower three petals having dark
lines. Note the distinct spur that sticks out behind
the flower, which is as long as the petals or longer,

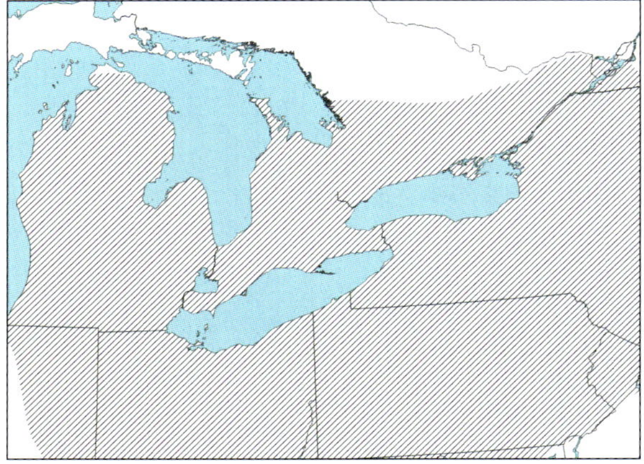

usually about 1.2 cm. Flowers mature into smooth,
egg-shaped seed capsules that split open to release
light brown seeds.

IN THE GARDEN In early spring, the pastel-purple flowers of Long-spurred Violet rise above its dark green leaves. Like the tail of a shooting star, a long nectar spur trails behind each flower, adding to the unique charm of this violet. It has a clumping habit and spreads by self-seeding.

SKILL LEVEL Beginner to intermediate

LIFESPAN Perennial

EXPOSURE Part shade to full shade

SOIL TYPE Acidic-to-neutral, rich, well-drained loam, sand, gravel, or rock

MOISTURE Moist to medium

HEIGHT 15–20 cm

SPREAD 10–20 cm

BLOOM PERIOD Apr, May, Jun

COLOR Pale purple

FRAGRANT ❌

SHOWY FRUIT ❌

CUT FLOWER ❌

PESTS No serious insect or disease problems

NATURAL HABITAT Rich hardwood forests, in limey soil

WILDLIFE VALUE Nectar source for bees and butterflies

BUTTERFLY LARVA HOST PLANT FOR Fritillary butterflies (*Bolloria* spp. and *Speyeria* spp.)

MOTH LARVA HOST PLANT FOR Nais Tiger Moth (*Apantesis nais*), Omniverous Leafroller Moth (*Archips purpurana*), Grateful Midget (*Elaphria grata*), The Beggar (*Eubaphe mendica*), Giant Leopard Moth (*Hypercompe scribonia*), Striped Garden Caterpillar (*Trichordestra legitima*), Smith's Dart (*Xestia smithii*)

USDA HARDINESS ZONES 3–7

PROPAGATION Surface sow as seeds need light to break dormancy, and seeds require 60 days of cold, moist stratification. Seeds are hydrophilic and should not be allowed to dry out. They are best planted outdoors in fall. This violet spreads by rhizomes, and the plantlets can be divided from the parent in early spring or late fall.

ADDITIONAL INFO *Viola rostrata* has the longest spur of any North American violet species.

SAR STATUS N/A

Zizia aurea
Golden Alexander

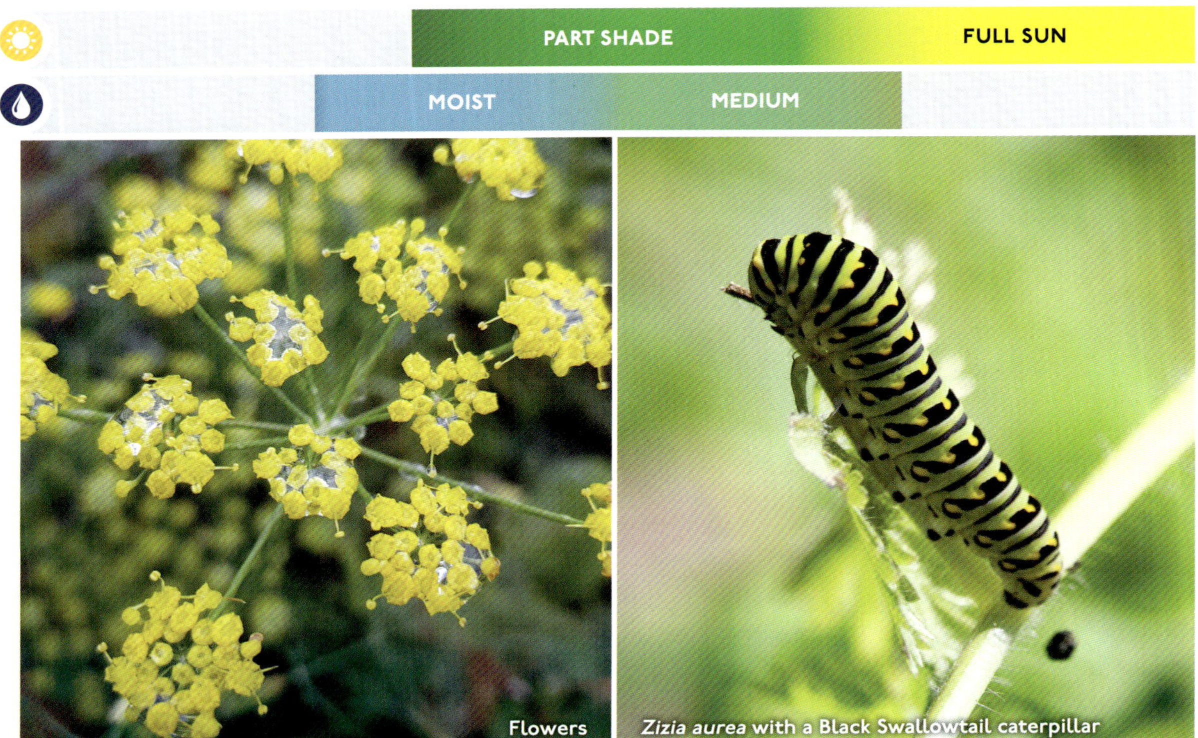

Flowers

Zizia aurea with a Black Swallowtail caterpillar

FAMILY *Apiaceae* (Carrot Family)

ALTERNATE COMMON NAMES Golden Zizia, Golden Meadow Parsnip

PLANT DESCRIPTION Golden Alexander features light green stems that are smooth and slightly shiny. Leaves are alternate in arrangement and compounded two to three times into groups of three leaflets. Each leaflet is about 7.5 cm long, 5 cm wide (the end leaflet is the largest), toothed, and variable in shape — but usually it's lanceolate, oblong, or ovate with pointed tips. Lower leaves have long stalks, but they become shorter as they climb the stem. Flower heads are found at the tops of the stems, measure about 7.5 cm wide, and are flat to slightly rounded in shape. These flower heads are made up of 10 to 18 groups of up to 21 small, yellow flowers. Each flower

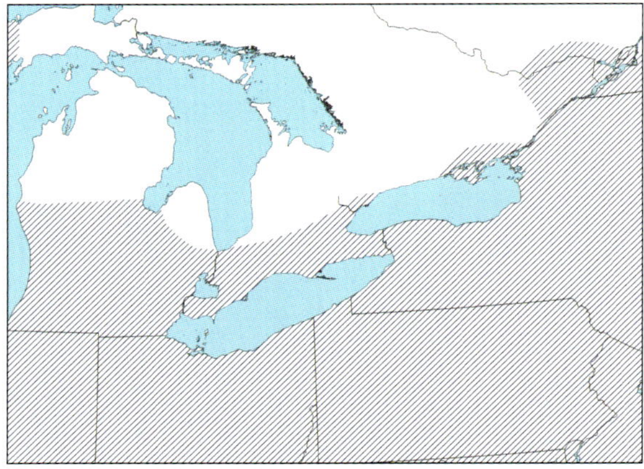

is about 3 mm wide and has five incurved petals and five yellow stamens. Flowers mature into dark brown, oblong seeds, about 3 mm long.

IN THE GARDEN Golden Alexander blooms in profusion to fill the late-spring garden with a vibrant yellow. It is easy to grow and adaptable, and it maintains a lush mound of leaves even after the flowers fade. In the fall, the leaves often take on attractive shades of maroon. Note that individual plants maintain a clumping habit but will freely self-seed.

SKILL LEVEL Beginner

LIFESPAN Short-lived perennial

EXPOSURE Full sun to part shade

SOIL TYPE Most well-drained soils

MOISTURE Moist to medium

HEIGHT 45–90 cm

SPREAD 30–60 cm

BLOOM PERIOD May, Jun

COLOR Yellow

FRAGRANT ✔ (mild)

SHOWY FRUIT ✘

CUT FLOWER ✔

PESTS No serious insect or disease problems

NATURAL HABITAT Moist prairies, thickets, fens, and moist open woodlands

WILDLIFE VALUE The flowers are attractive to many kinds of insects seeking pollen or nectar, especially short-tongued bees

BUTTERFLY LARVA HOST PLANT FOR Black Swallowtail (*Papilio polyxenes*)

MOTH LARVA HOST PLANT FOR Rigid Sunflower Borer Moth (*Papaipema rigida*)

USDA HARDINESS ZONES 3–8

PROPAGATION The seeds can be difficult to germinate and should be cold, moist stratified from 60 to 120 days (some sources suggest even longer), or you can plant the seeds in fall. Seeds germinate best in cool soil. Plants can also be divided.

ADDITIONAL INFO Golden Alexander tolerates clay soils.

SAR STATUS N/A

Leaves

Seed heads

Whole plant

Appendix A: Landscape Use

Scientific Name	Common Name	Bog	Border/Edging	Boulevard Garden	Butterfly Garden	Container	Cut Flower	Foundation Planting	Green Roof	Groundcover	Kitchen Garden	Moon Garden	Naturalizing	Prairie / Meadow	Rain Garden	Rock Garden	Woodland	Wood's Edge	Foreground	Midground	Background	Specimen	Chelsea Chop*
Achillea millefolium	Common Yarrow			X	X	X			X	X			X	X		X				X			X
Actaea pachypoda	White Baneberry																X			X			
Actaea racemosa	Black Cohosh				X	X									X			X		X			
Actaea rubra	Red Baneberry																X			X			
Agastache foeniculum	Anise Hyssop			X	X	X	X	X			X		X	X						X			
Agastache nepetoides	Yellow Giant Hyssop				X		X						X	X				X			X	X	
Agastache scrophulariifolia	Purple Giant Hyssop										X		X	X				X			X		
Ageratina altissima	White Snakeroot			X		X														X			
Allium cernuum	Nodding Wild Onion		X	X	X	X			X		X		X	X		X			X				
Allium tricoccum	Wild Leek					X											X	X	X				
Anaphalis margaritacea	Pearly Everlasting		X	X	X	X	X	X	X	X			X	X		X				X			
Anemonastrum canadense	Canada Anemone			X		X				X			X	X				X		X			
Anemone cylindrica	Thimbleweed						X						X	X		X		X		X			
Anemone quinquefolia	Wood Anemone												X				X	X	X				
Anemone virginiana	Tall Thimbleweed												X				X	X		X			
Antennaria neglecta	Field Pussytoes			X	X	X			X	X			X			X			X				
Antennaria parlinii	Parlin's Pussytoes			X	X	X			X	X			X			X	X	X	X				
Aquilegia canadensis	Wild Columbine		X	X	X	X	X	X	X				X			X		X		X			
Aralia nudicaulis	Wild Sarsparilla												X				X	X		X			
Aralia racemosa	Spikenard								X				X				X	X			X	X	
Arisaema triphyllum	Jack-in-the-pulpit					X														X			
Asarum canadense	Wild Ginger			X	X	X				X			X				X	X	X				
Asclepias exaltata	Poke Milkweed		X	X	X	X							X				X	X			X		
Asclepias incarnata	Swamp Milkweed	X		X	X	X							X		X					X			X
Asclepias purpurascens	Purple Milkweed			X	X	X				X			X	X				X		X			
Asclepias syriaca	Common Milkweed		X	X	X	X							X	X	X			X			X		X
Asclepias tuberosa	Butterfly Milkweed		X	X	X	X	X	X					X	X		X				X			
Asclepias verticillata	Whorled Milkweed			X	X	X			X				X	X		X				X			
Asclepias viridiflora	Green Milkweed				X	X							X			X				X			
Caltha palustris	Marsh Marigold	X											X						X				
Campanula rotundifolia	Harebell			X		X			X				X			X		X	X				X
Campanulastrum americanum	Tall Bellflower	X											X				X	X			X		
Capnoides sempervirens	Pale Corydalis		X	X		X							X			X				X		X	
Cardamine concatenata	Cut-leaved Toothwort				X								X				X		X				
Cardamine diphylla	Two-leaved Toothwort				X								X				X		X				
Caulophyllum giganteum	Giant Blue Cohosh					X											X	X		X			
Caulophyllum thalictroides	Blue Cohosh					X											X	X		X			
Chelone glabra	White Turtlehead	X		X	X	X	X					X	X		X			X		X			
Claytonia caroliniana	Carolina Spring Beauty												X			X	X	X	X				

* The "Chelsea chop" (so named because it is usually done in late spring/early summer, around the time of the Chelsea Flower Show in Britain) is the process of cutting back plants to make them produce more blossoms and to reduce their overall height in the garden. Late May to early July is often the best time to do this, depending on the bloom period.

SCIENTIFIC NAME	COMMON NAME	Bog	Border/Edging	Boulevard Garden	Butterfly Garden	Container	Cut Flower	Foundation Planting	Green Roof	Groundcover	Kitchen Garden	Moon Garden	Naturalizing	Prairie / Meadow	Rain Garden	Rock Garden	Woodland	Wood's Edge	Foreground	Midground	Background	Specimen	Chelsea Chop*
Claytonia virginica	Virginia Spring Beauty															X	X	X	X				
Coreopsis lanceolata	Lanceleaf Coreopsis		X	X		X	X	X		X			X	X	X	X		X	X				X
Coreopsis tripteris	Tall Tickseed		X			X	X	X					X	X	X			X			X	X	X
Dicentra canadensis	Squirrel Corn																X		X				
Dicentra cucullaria	Dutchman's Breeches																X		X				
Doellingeria umbellata	Flat-topped White Aster		X		X		X	X				X	X	X	X						X		
Echinacea purpurea	Purple Coneflower		X	X	X	X	X	X	X				X	X							X		X
Erythronium americanum	Trout Lily																X	X	X				
Eupatorium perfoliatum	Boneset	X		X		X	X	X				X	X		X			X			X		X
Eurybia macrophylla	Large Leaf Aster			X					X											X			X
Euthamia graminifolia	Grass-leaved Goldenrod				X	X	X						X	X							X		X
Eutrochium maculatum	Spotted Joe Pye Weed	X				X	X						X		X			X			X		X
Eutrochium purpureum	Sweet Joe Pye Weed					X	X						X	X	X		X	X			X		X
Fragaria vesca	Woodland Strawberry			X	X	X			X	X	X		X			X	X	X	X				
Fragaria virginiana	Wild Strawberry		X	X	X	X			X	X	X		X	X	X	X			X				
Gentiana andrewsii	Bottle Gentian	X				X	X								X		X	X		X			
Geranium maculatum	Wild Geranium		X	X		X		X	X	X			X	X	X	X		X	X				
Geum fragarioides	Barren Strawberry			X						X						X		X	X				
Geum rivale	Water Avens	X							X				X		X				X				
Geum triflorum	Prairie Smoke		X	X		X	X	X		X			X	X		X			X				
Helenium autumnale	Sneezeweed			X		X	X						X		X			X			X		X
Helianthus divaricatus	Woodland Sunflower			X	X	X	X			X			X	X			X	X			X		X
Helianthus giganteus	Tall Sunflower					X	X						X	X	X						X		X
Helianthus tuberosus	Jerusalem Artichoke				X	X	X				X							X				X	X
Heliopsis helianthoides	False Sunflower				X	X	X	X					X	X	X			X			X		X
Hepatica acutiloba	Sharp-lobed Hepatica		X	X		X							X			X	X	X	X				
Hepatica americana	Round-lobed Hepatica		X			X							X			X	X	X	X				
Houstonia longifolia	Long-leaved Bluet					X			X							X			X				
Hydrophyllum virginianum	Virginia Waterleaf							X					X		X		X	X		X			
Impatiens capensis	Spotted Jewelweed	X				X							X		X		X	X			X		
Impatiens pallida	Pale Touch-me-not	X				X							X		X		X	X			X		
Iris versicolor	Blue Flag Iris	X				X	X						X					X		X			
Liatris cylindracea	Ontario Blazing Star		X	X	X	X			X					X		X			X				
Liatris spicata	Dense Blazing Star			X	X	X	X	X					X					X		X		X	
Lilium canadense	Canada Lily			X	X	X							X	X				X			X	X	
Lilium michiganense	Michigan Lily			X	X	X							X	X	X			X			X	X	
Lilium philadelphicum	Wood Lily			X	X	X							X	X	X		X	X		X			
Lobelia cardinalis	Cardinal Flower	X			X	X	X						X		X			X			X	X	
Lobelia siphilitica	Blue Lobelia	X		X		X							X					X			X		

* The "Chelsea chop" (so named because it is usually done in late spring/early summer, around the time of the Chelsea Flower Show in Britain) is the process of cutting back plants to make them produce more blossoms and to reduce their overall height in the garden. Late May to early July is often the best time to do this, depending on the bloom period.

Scientific Name	Common Name	Bog	Border/Edging	Boulevard Garden	Butterfly Garden	Container	Cut Flower	Foundation Planting	Green Roof	Groundcover	Kitchen Garden	Moon Garden	Naturalizing	Prairie / Meadow	Rain Garden	Rock Garden	Woodland	Wood's Edge	Foreground	Midground	Background	Specimen	Chelsea Chop*
Lupinus perennis	Wild Lupine				X				X				X	X						X			
Maianthemum canadense	Canada Mayflower		X			X				X			X				X	X	X				
Maianthemum racemosum	False Solomon's Seal		X	X		X		X		X			X		X		X	X		X			
Maianthemum stellatum	Starry False Solomon's Seal			X		X		X		X			X	X	X		X	X	X				
Mertensia virginica	Virginia Bluebells												X		X		X	X		X			
Mimulus ringens	Monkey Flower	X											X							X			
Monarda didyma	Bee Balm			X	X	X	X	X	X				X	X	X						X		X
Monarda fistulosa	Wild Bergamot			X	X	X	X	X	X		X		X	X	X						X		X
Monarda punctata	Spotted Bee Balm			X	X	X			X				X							X			X
Oenothera biennis	Evening Primrose			X		X					X		X	X	X			X			X		X
Oenothera pilosella	Prairie Sundrops			X		X		X			X		X	X	X	X		X		X			
Packera aurea	Golden Ragwort	X					X			X			X					X		X			
Packera paupercula	Balsam Groundsel					X							X	X	X	X			X				
Pedicularis lanceolata	Swamp Betony	X				X							X					X		X			
Penstemon digitalis	Foxglove Beardtongue			X		X	X					X	X	X	X	X	X	X		X			X
Penstemon hirsutus	Hairy Beardtongue			X	X	X	X	X					X	X	X	X	X	X		X			X
Phlox divaricata	Wild Blue Phlox		X	X		X		X	X	X			X		X	X		X	X				X
Physostegia virginiana	Obedient Plant			X	X	X	X						X	X				X			X		X
Podophyllum peltatum	Mayapple												X				X	X	X				
Polygonatum biflorum	Solomon's Seal			X		X							X		X			X		X			
Pycnanthemum tenuifolium	Slender Mountain Mint		X	X	X	X	X	X			X		X	X	X	X		X		X			
Pycnanthemum virginianum	Virginia Mountain Mint	X		X	X	X					X		X		X			X		X			
Ranunculus fascicularis	Early Buttercup		X										X			X	X	X	X				
Ratibida pinnata	Gray-headed Coneflower		X	X	X	X	X	X	X				X	X	X	X					X		X
Rudbeckia fulgida	Orange Coneflower		X	X	X	X	X	X					X	X				X		X			X
Rudbeckia hirta	Black-eyed Susan			X	X	X	X	X					X	X				X		X			X
Rudbeckia laciniata	Green-headed Coneflower					X	X	X					X	X			X	X			X		X
Sanguinaria canadensis	Bloodroot					X				X							X	X	X				
Silphium laciniatum	Compass Plant			X		X	X						X	X							X	X	
Silphium perfoliatum	Cup Plant					X							X	X							X		
Sisyrinchium angustifolium	Blue-eyed Grass		X			X							X	X	X				X				
Sisyrinchium montanum	Mountain Blue-eyed Grass		X	X		X							X	X	X				X				
Solidago bicolor	Silverrod					X	X	X			X		X			X	X	X		X			X
Solidago caesia	Blue-stemmed Goldenrod					X	X	X					X		X	X		X		X			X
Solidago flexicaulis	Zigzag Golenrod			X		X	X						X		X	X		X		X			X
Solidago hispida	Hairy Goldenrod					X	X	X					X	X		X		X		X			X
Solidago juncea	Early Goldenrod					X	X						X		X			X			X		X
Solidago nemoralis	Gray Goldenrod			X		X		X					X	X	X			X		X			X

* The "Chelsea chop" (so named because it is usually done in late spring/early summer, around the time of the Chelsea Flower Show in Britain) is the process of cutting back plants to make them produce more blossoms and to reduce their overall height in the garden. Late May to early July is often the best time to do this, depending on the bloom period.

Scientific Name	Common Name	Bog	Border/Edging	Boulevard Garden	Butterfly Garden	Container	Cut Flower	Foundation Planting	Green Roof	Groundcover	Kitchen Garden	Moon Garden	Naturalizing	Prairie / Meadow	Rain Garden	Rock Garden	Woodland	Wood's Edge	Foreground	Midground	Background	Specimen	Chelsea Chop*
Solidago ohioensis	Ohio Goldenrod	X				X									X					X			X
Solidago ptarmicoides	Upland White Goldenrod		X	X		X	X		X			X	X	X		X				X			X
Solidago riddellii	Riddell's Goldenrod	X				X							X	X	X					X			X
Solidago rigida	Stiff Goldenrod			X		X	X	X	X					X		X					X		X
Solidago rugosa	Rough-stemmed Goldenrod					X	X						X	X	X			X			X		X
Solidago speciosa	Showy Goldenrod			X		X			X				X	X		X				X			X
Solidago uliginosa	Bog Goldenrod	X				X	X								X						X		X
Symphyotrichum cordifolium	Blue Wood Aster				X	X															X		X
Symphyotrichum ericoides	White Heath Aster		X	X	X	X	X		X				X			X				X			X
Symphyotrichum laeve	Smooth Aster		X	X	X	X	X	X	X				X	X		X				X			X
Symphyotrichum lanceolatum	Panicled Aster			X	X	X							X	X	X						X		X
Symphyotrichum lateriflorum	Calico Aster			X	X	X	X		X				X		X			X		X			X
Symphyotrichum novae-angliae	New England Aster			X	X	X	X	X					X	X	X						X		X
Symphyotrichum ontarionis	Ontario Aster					X															X		X
Symphyotrichum oolentangiense	Sky Blue Aster		X	X	X	X	X	X	X				X	X		X				X			X
Symphyotrichum pilosum	Frost Aster					X							X	X							X		X
Symphyotrichum puniceum	Purple-stemmed Aster	X				X	X	X					X								X		X
Symphyotrichum urophyllum	Arrowleaf Aster			X	X	X							X	X	X		X	X			X		X
Thalictrum dasycarpum	Purple Meadowrue												X		X			X			X		
Thalictrum dioicum	Early Meadowrue						X						X		X		X	X		X			
Thalictrum pubescens	Tall Meadowrue												X		X		X	X			X		
Tiarella stolonifera	Foamflower			X		X				X					X		X	X	X				
Tradescantia ohiensis	Ohio Spiderwort			X		X	X	X					X	X				X		X			
Tradescantia virginiana	Virginia Spiderwort			X		X							X	X	X			X		X			
Trillium erectum	Red Trillium																X	X			X		
Trillium grandiflorum	Large White Trillium											X					X	X	X				
Uvularia grandiflora	Large-flowered Bellwort		X			X							X		X	X	X	X	X	X		X	
Verbena hastata	Blue Vervain	X		X		X	X						X	X	X						X	X	X
Verbena stricta	Hoary Vervain			X	X	X	X	X					X	X		X				X			
Vernonia gigantea	Tall Ironweed						X							X	X						X	X	X
Vernonia missurica	Missouri Ironweed				X	X	X						X	X	X						X	X	X
Veronicastrum virginicum	Culver's Root					X	X	X			X		X	X	X			X			X	X	X
Viola blanda	Sweet White Violet		X	X	X	X				X					X	X	X	X	X				
Viola canadensis	Canada Violet		X	X	X	X				X			X		X		X	X	X				
Viola pubescens	Downy Yellow Violet		X	X	X	X		X					X				X	X	X				
Viola rostrata	Long-spurred Violet		X	X	X												X	X	X				
Zizia aurea	Golden Alexander			X	X	X		X					X	X	X			X		X			

* The "Chelsea chop" (so named because it is usually done in late spring/early summer, around the time of the Chelsea Flower Show in Britain) is the process of cutting back plants to make them produce more blossoms and to reduce their overall height in the garden. Late May to early July is often the best time to do this, depending on the bloom period.

SCIENTIFIC NAME	COMMON NAME	Gravelly	Sandy	Loam	Clay	Humus rich	Acidic	Neutral	Alkaline
Achillea millefolium	Common Yarrow		X	X	X				
Actaea pachypoda	White Baneberry		X	X		X	X		
Actaea racemosa	Black Cohosh			X		X			
Actaea rubra	Red Baneberry		X	X	X	X	X		
Agastache foeniculum	Anise Hyssop		X				X		
Agastache nepetoides	Yellow Giant Hyssop		X	X	X				
Agastache scrophulariifolia	Purple Giant Hyssop		X	X	X				
Ageratina altissima	White Snakeroot		X						
Allium cernuum	Nodding Wild Onion		X	X	X	X	(X)	X	X
Allium tricoccum	Wild Leek		X	X		X			
Anaphalis margaritacea	Pearly Everlasting	X	X						
Anemonastrum canadense	Canada Anemone		X	X	(X)				
Anemone cylindrica	Thimbleweed	X	X	X	(X)				
Anemone quinquefolia	Wood Anemone		X	X					
Anemone virginiana	Tall Thimbleweed			X			X	X	
Antennaria neglecta	Field Pussytoes	X	X				X		
Antennaria parlinii	Parlin's Pussytoes	X	X				X		
Aquilegia canadensis	Wild Columbine	X	X						
Aralia nudicaulis	Wild Sarsaparilla		X	X					
Aralia racemosa	Spikenard	X	X	X	X				
Arisaema triphyllum	Jack-in-the-pulpit		X			X			
Asarum canadense	Wild Ginger		X	X	X	X	X		
Asclepias exaltata	Poke Milkweed		X	X					
Asclepias incarnata	Swamp Milkweed			X	X		X	X	
Asclepias purpurascens	Purple Milkweed		X	X					
Asclepias syriaca	Common Milkweed	X	X	X	(X)				
Asclepias tuberosa	Butterfly Milkweed	X	X						
Asclepias verticillata	Whorled Milkweed		X	X	(X)				
Asclepias viridiflora	Green Milkweed	X	X						
Caltha palustris	Marsh Marigold			X		X			
Campanula rotundifolia	Harebell		X	X					
Campanulastrum americanum	Tall Bellflower		X	X	X			X	
Capnoides sempervirens	Pale Corydalis	X	X						
Cardamine concatenata	Cut-leaved Toothwort		X	X		X			
Cardamine diphylla	Two-leaved Toothwort		X	X		X			
Caulophyllum giganteum	Giant Blue Cohosh		X	X	(X)	X			
Caulophyllum thalictroides	Blue Cohosh		X	X		X			
Chelone glabra	White Turtlehead		X	X		X			
Claytonia caroliniana	Carolina Spring Beauty		X	X		X			
Claytonia virginica	Virginia Spring Beauty		X	X	(X)	X			
Coreopsis lanceolata	Lanceleaf Coreopsis	X	X	X	X				
Coreopsis tripteris	Tall Tickseed	X	X	X					
Dicentra canadensis	Squirrel Corn		X	X		X	X	X	
Dicentra cucullaria	Dutchman's Breeches		X	X	X	X	X	X	
Doellingeria umbellata	Flat-topped White Aster		X	X					
Echinacea purpurea	Purple Coneflower	X	X	X					
Erythronium americanum	Trout Lily		X	X	X	X			
Eupatorium perfoliatum	Boneset	X	X	X					
Eurybia macrophylla	Large Leaf Aster		X	X					
Euthamia graminifolia	Grass-leaved Goldenrod	X	X	X	(X)				
Eutrochium maculatum	Spotted Joe Pye Weed	X	X	X	X				

(X) indicates that plants will grow in the soil type but may not thrive.

SCIENTIFIC NAME	COMMON NAME	Gravelly	Sandy	Loam	Clay	Humus rich	Acidic	Neutral	Alkaline
Eutrochium purpureum	Sweet Joe Pye Weed	X	X	X		X			
Fragaria vesca	Woodland Strawberry		X	X	X				
Fragaria virginiana	Wild Strawberry		X	X	X				
Gentiana andrewsii	Bottle Gentian		X	X		X	X	X	
Geranium maculatum	Wild Geranium	X	X	X	X	X			
Geum fragarioides	Barren Strawberry		X	X	X	X	X	X	
Geum rivale	Water Avens	X	X	X					
Geum triflorum	Prairie Smoke	X	X						
Helenium autumnale	Sneezeweed			X	X				
Helianthus divaricatus	Woodland Sunflower	X	X	X					
Helianthus giganteus	Tall Sunflower		X						
Helianthus tuberosus	Jerusalem Artichoke		X	X					
Heliopsis helianthoides	False Sunflower	X	X	X					
Hepatica acutiloba	Sharp-lobed Hepatica		X	X				X	
Hepatica americana	Round-lobed Hepatica		X	X			X	X	
Houstonia longifolia	Long-leaved Bluet	X	X						
Hydrophyllum virginianum	Virginia Waterleaf			X					
Impatiens capensis	Spotted Jewelweed		(X)	X	X				
Impatiens pallida	Pale Touch-me-not		X	X	(X)				
Iris versicolor	Blue Flag Iris	(X)	(X)	X	X				
Liatris cylindracea	Ontario Blazing Star		X	X					
Liatris spicata	Dense Blazing Star		X	X					
Lilium canadense	Canada Lily		X	X					
Lilium michiganense	Michigan Lily		X	X					
Lilium philadelphicum	Wood Lily		X	X		X			
Lobelia cardinalis	Cardinal Flower		X	X	X	X			
Lobelia siphilitica	Blue Lobelia			X					
Lupinus perennis	Wild Lupine		X						
Maianthemum canadense	Canada Mayflower	X	X				X		
Maianthemum racemosum	False Solomon's Seal	(X)	(X)	X		X			
Maianthemum stellatum	Starry False Solomon's Seal		X	X	(X)				
Mertensia virginica	Virginia Bluebells		X	X					
Mimulus ringens	Monkey Flower			X					
Monarda didyma	Bee Balm		X	X	X				
Monarda fistulosa	Wild Bergamot		X	X	X		X	X	X
Monarda punctata	Spotted Bee Balm		X						
Oenothera biennis	Evening Primrose	X	X						
Oenothera pilosella	Prairie Sundrops		X	X	X				
Packera aurea	Golden Ragwort			X					
Packera paupercula	Balsam Groundsel	X	X	X					
Pedicularis lanceolata	Swamp Betony		X	X	(X)				
Penstemon digitalis	Foxglove Beardtongue		X	X	(X)				
Penstemon hirsutus	Hairy Beardtongue		X						
Phlox divaricata	Wild Blue Phlox		X	X	(X)	X			
Physostegia virginiana	Obedient Plant		X	X	X				
Podophyllum peltatum	Mayapple		X	X		X			
Polygonatum biflorum	Solomon's Seal			X		X			
Pycnanthemum tenuifolium	Slender Mountain Mint		X	X	X				
Pycnanthemum virginianum	Virginia Mountain Mint		X	X	X				
Ranunculus fascicularis	Early Buttercup	X	X	(X)					
Ratibida pinnata	Gray-headed Coneflower		(X)	X	(X)				

(X) indicates that plants will grow in the soil type but may not thrive.

Scientific Name	Common Name	Gravelly	Sandy	Loam	Clay	Humus rich	Acidic	Neutral	Alkaline
Rudbeckia fulgida	Orange Coneflower		X	X	X				
Rudbeckia hirta	Black-eyed Susan		X	X	X				
Rudbeckia laciniata	Green-headed Coneflower		X	X	X				
Sanguinaria canadensis	Bloodroot		X	X		X			
Silphium laciniatum	Compass Plant	X	X	X					
Silphium perfoliatum	Cup Plant		(X)	X	(X)				
Sisyrinchium angustifolium	Blue-eyed Grass		X						
Sisyrinchium montanum	Mountain Blue-eyed Grass		X						
Solidago bicolor	Silverrod		X	X	X				
Solidago caesia	Blue-stemmed Goldenrod	X	X	X					
Solidago flexicaulis	Zigzag Golenrod		X	X	X				
Solidago hispida	Hairy Goldenrod		X	X	X				
Solidago juncea	Early Goldenrod		X	X	X				
Solidago nemoralis	Gray Goldenrod	X	X	X	X				
Solidago ohioensis	Ohio Goldenrod		X	X					
Solidago ptarmicoides	Upland White Goldenrod		X	X				X	
Solidago riddellii	Riddell's Goldenrod	X	X						
Solidago rigida	Stiff Goldenrod	X	X	X	(X)				
Solidago rugosa	Rough-stemmed Goldenrod		X	X					
Solidago speciosa	Showy Goldenrod		X	X					
Solidago uliginosa	Bog Goldenrod		X	(X)					X
Symphyotrichum cordifolium	Blue Wood Aster	X		X	X				
Symphyotrichum ericoides	White Heath Aster		X	X					
Symphyotrichum laeve	Smooth Aster		X	X					
Symphyotrichum lanceolatum	Panicled Aster		X	X	X				
Symphyotrichum lateriflorum	Calico Aster			X	X	X			
Symphyotrichum novae-angliae	New England Aster		(X)	X	X				
Symphyotrichum ontarionis	Ontario Aster			X					
Symphyotrichum oolentangiense	Sky Blue Aster		X	X					
Symphyotrichum pilosum	Frost Aster		X	X					
Symphyotrichum puniceum	Purple-stemmed Aster		X	X					
Symphyotrichum urophyllum	Arrowleaf Aster		(X)	X	(X)				
Thalictrum dasycarpum	Purple Meadowrue		X	X	(X)	X			
Thalictrum dioicum	Early Meadowrue			X	(X)	X			
Thalictrum pubescens	Tall Meadowrue			X					
Tiarella stolonifera	Foamflower		X	X		X			
Tradescantia ohiensis	Ohio Spiderwort		X	X					
Tradescantia virginiana	Virginia Spiderwort		X	X					
Trillium erectum	Red Trillium		X	X		X			
Trillium grandiflorum	Large White Trillium		X	X		X			
Uvularia grandiflora	Large-flowered Bellwort		X	X	X	X			
Verbena hastata	Blue Vervain		(X)	X	(X)				
Verbena stricta	Hoary Vervain		X	X					
Vernonia gigantea	Tall Ironweed		(X)	X	(X)				
Vernonia missurica	Missouri Ironweed		(X)	X	(X)				
Veronicastrum virginicum	Culver's Root		X	X					
Viola blanda	Sweet White Violet			X		X			
Viola canadensis	Canada Violet			X		X			
Viola pubescens	Downy Yellow Violet		X	X		X			
Viola rostrata	Long-spurred Violet	X	X	X		X	X		
Zizia aurea	Golden Alexander		X	X					

(X) indicates that plants will grow in the soil type but may not thrive.

Appendix C: **Propagation**

SCIENTIFIC NAME	COMMON NAME	SEED TREATMENT							OTHER PROPAGATION			
GRAY/BOOTH PROPAGATION CODE:		NT	C	L	WC	CWC	S	M	D	LYR	CR	CS
Achillea millefolium	Common Yarrow	X		X					X			
Actaea pachypoda	White Baneberry					X		X	(X)			
Actaea racemosa	Black Cohosh					X		X	X			
Actaea rubra	Red Baneberry					X		X	X			
Agastache foeniculum	Anise Hyssop		(60)	X					X			
Agastache nepetoides	Yellow Giant Hyssop		60	X					X			
Agastache scrophulariifolia	Purple Giant Hyssop		60	X					X			
Ageratina altissima	White Snakeroot		60	X					X			
Allium cernuum	Nodding Wild Onion		60	?					X			
Allium tricoccum	Wild Leek		60		X			X	X			
Anaphalis margaritacea	Pearly Everlasting	X		X					X			X
Anemonastrum canadense	Canada Anemone					X		X	X		X	
Anemone cylindrica	Thimbleweed	X		X					X			
Anemone quinquefolia	Wood Anemone					X			X			
Anemone virginiana	Tall Thimbleweed			X					X			
Antennaria neglecta	Field Pussytoes	X	(60)	X				X	X			
Antennaria parlinii	Parlin's Pussytoes	X		X					X			X
Aquilegia canadensis	Wild Columbine		60	X					(X)			
Aralia nudicaulis	Wild Sarsparilla		90		(X)		X	X	X			
Aralia racemosa	Spikenard		(90)				X	X	X		X	
Arisaema triphyllum	Jack-in-the-pulpit					X		X	X			
Asarum canadense	Wild Ginger				X			X	X			X
Asclepias exaltata	Poke Milkweed		60–90									
Asclepias incarnata	Swamp Milkweed		30–90						X			
Asclepias purpurascens	Purple Milkweed		30						X			
Asclepias syriaca	Common Milkweed		30						X			
Asclepias tuberosa	Butterfly Milkweed		30								X	
Asclepias verticillata	Whorled Milkweed		30						X			
Asclepias viridiflora	Green Milkweed		30				X					
Caltha palustris	Marsh Marigold	X						X	X			
Campanula rotundifolia	Harebell		30	X					X	X		X
Campanulastrum americanum	Tall Bellflower	X		X								
Capnoides sempervirens	Pale Corydalis		30									
Cardamine concatenata	Cut-leaved Toothwort		60					X	X			
Cardamine diphylla	Two-leaved Toothwort	X						X	X			
Caulophyllum giganteum	Giant Blue Cohosh					X		X	X			
Caulophyllum thalictroides	Blue Cohosh					X		X	X			
Chelone glabra	White Turtlehead		120	X				X	X			X
Claytonia caroliniana	Carolina Spring Beauty				X			X	X			
Claytonia virginica	Virginia Spring Beauty				X			X	X			
Coreopsis lanceolata	Lanceleaf Coreopsis		30	X					X			
Coreopsis tripteris	Tall Tickseed		60	X					X			
Dicentra canadensis	Squirrel Corn	X						X	X			
Dicentra cucullaria	Dutchman's Breeches				X			X	X			
Doellingeria umbellata	Flat-topped White Aster		60	X					X			
Echinacea purpurea	Purple Coneflower		30						X			
Erythronium americanum	Trout Lily				X			X	X			
Eupatorium perfoliatum	Boneset		30	X					X			X
Eurybia macrophylla	Large Leaf Aster	X	(60)	X					X			X
Euthamia graminifolia	Grass-leaved Goldenrod	(X)	(60)	X					X			
Eutrochium maculatum	Spotted Joe Pye Weed		30	X					X		X	X

( ) indicates that there is conflicting information in the literature about these characteristics.

SCIENTIFIC NAME	COMMON NAME	NT	C	L	WC	CWC	S	M	D	LYR	CR	CS
GRAY/BOOTH PROPAGATION CODE:												
Eutrochium purpureum	Sweet Joe Pye Weed		30	X					X			X
Fragaria vesca	Woodland Strawberry		60						X			X
Fragaria virginiana	Wild Strawberry		60						X			X
Gentiana andrewsii	Bottle Gentian		60	X					X			
Geranium maculatum	Wild Geranium		60					X	X			
Geum fragarioides	Barren Strawberry	X		X					X			
Geum rivale	Water Avens	X							X			
Geum triflorum	Prairie Smoke	X	(60)						X		X	
Helenium autumnale	Sneezeweed	X		X					X			X
Helianthus divaricatus	Woodland Sunflower		30						X			
Helianthus giganteus	Tall Sunflower		30						X			X
Helianthus tuberosus	Jerusalem Artichoke								X			
Heliopsis helianthoides	False Sunflower		30–60						X			X
Hepatica acutiloba	Sharp-lobed Hepatica		30					M	X			
Hepatica americana	Round-lobed Hepatica		30					M	X			
Houstonia longifolia	Long-leaved Bluet	X		X					X			
Hydrophyllum virginianum	Virginia Waterleaf		60					M	X			
Impatiens capensis	Spotted Jewelweed			X		X		M				
Impatiens pallida	Pale Touch-me-not			X		X		M				
Iris versicolor	Blue Flag Iris		120					M	X			
Liatris cylindracea	Ontario Blazing Star		60						X			
Liatris spicata	Dense Blazing Star		60				S		X			
Lilium canadense	Canada Lily				X				X			
Lilium michiganense	Michigan Lily				X				X			
Lilium philadelphicum	Wood Lily		60						X			
Lobelia cardinalis	Cardinal Flower		60	X					X	X		X
Lobelia siphilitica	Blue Lobelia		60	X					X			X
Lupinus perennis	Wild Lupine		10				S		X			
Maianthemum canadense	Canada Mayflower		90					M	X			
Maianthemum racemosum	False Solomon's Seal			X		X		M	X			
Maianthemum stellatum	Starry False Solomon's Seal					X		M	X			
Mertensia virginica	Virginia Bluebells		60					M	X		X	
Mimulus ringens	Monkey Flower		60	X					X			X
Monarda didyma	Bee Balm	X							X			X
Monarda fistulosa	Wild Bergamot	X		X					X			X
Monarda punctata	Spotted Bee Balm	X		X								X
Oenothera biennis	Evening Primrose	X		X								X
Oenothera pilosella	Prairie Sundrops	X									X	
Packera aurea	Golden Ragwort	X		X					X			
Packera paupercula	Balsam Groundsel		60									
Pedicularis lanceolata	Swamp Betony		30	X				M				
Penstemon digitalis	Foxglove Beardtongue		30	X					X	X		X
Penstemon hirsutus	Hairy Beardtongue		60	X					X			
Phlox divaricata	Wild Blue Phlox		60						X	X		X
Physostegia virginiana	Obedient Plant		60						X			X
Podophyllum peltatum	Mayapple		90					M	X			
Polygonatum biflorum	Solomon's Seal					X		M	X		X	
Pycnanthemum tenuifolium	Slender Mountain Mint	X		X					X			X
Pycnanthemum virginianum	Virginia Mountain Mint	X		X					X			X
Ranunculus fascicularis	Early Buttercup		60					M				
Ratibida pinnata	Gray-headed Coneflower		30	X					X			

( ) indicates that there is conflicting information in the literature about these characteristics.

SCIENTIFIC NAME	COMMON NAME	SEED TREATMENT							OTHER PROPAGATION			
GRAY/BOOTH PROPAGATION CODE:		NT	C	L	WC	CWC	S	M	D	LYR	CR	CS
Rudbeckia fulgida	Orange Coneflower		60						X			
Rudbeckia hirta	Black-eyed Susan		30	X								
Rudbeckia laciniata	Green-headed Coneflower		30						X			
Sanguinaria canadensis	Bloodroot					X		X	X			
Silphium laciniatum	Compass Plant		60									
Silphium perfoliatum	Cup Plant		60									
Sisyrinchium angustifolium	Blue-eyed Grass		60	X					X			
Sisyrinchium montanum	Mountain Blue-eyed Grass		60	X					X			
Solidago bicolor	Silverrod		60	X								
Solidago caesia	Blue-stemmed Goldenrod		90	X					X			X
Solidago flexicaulis	Zigzag Golenrod		60	X					X			
Solidago hispida	Hairy Goldenrod		60	X					X			
Solidago juncea	Early Goldenrod		60	X					X			
Solidago nemoralis	Gray Goldenrod		60	X					X			X
Solidago ohioensis	Ohio Goldenrod		30	X								
Solidago ptarmicoides	Upland White Goldenrod	X		X								
Solidago riddellii	Riddell's Goldenrod		60	X					X			
Solidago rigida	Stiff Goldenrod	(X)	(60)	X					X			
Solidago rugosa	Rough-stemmed Goldenrod	X		X					X			X
Solidago speciosa	Showy Goldenrod		60	X					X			
Solidago uliginosa	Bog Goldenrod	X		X					X			
Symphyotrichum cordifolium	Blue Wood Aster		60	X					X			X
Symphyotrichum ericoides	White Heath Aster	X		X					X			
Symphyotrichum laeve	Smooth Aster	X		X							X	
Symphyotrichum lanceolatum	Panicled Aster	X		X					X			
Symphyotrichum lateriflorum	Calico Aster	X		X					X			
Symphyotrichum novae-angliae	New England Aster	X		X					X			X
Symphyotrichum ontarionis	Ontario Aster	X		X					X			
Symphyotrichum oolentangiense	Sky Blue Aster	X		X					X			X
Symphyotrichum pilosum	Frost Aster	X		X								X
Symphyotrichum puniceum	Purple-stemmed Aster		60	X								
Symphyotrichum urophyllum	Arrowleaf Aster		60	X								
Thalictrum dasycarpum	Purple Meadowrue		60						X			
Thalictrum dioicum	Early Meadowrue		60						X			
Thalictrum pubescens	Tall Meadowrue		60						X			
Tiarella stolonifera	Foamflower	X						X	X			
Tradescantia ohiensis	Ohio Spiderwort		(120)						X			X
Tradescantia virginiana	Virginia Spiderwort		(120)						X			X
Trillium erectum	Red Trillium					X		X				
Trillium grandiflorum	Large White Trillium					X		X				
Uvularia grandiflora	Large-flowered Bellwort				X			X	X			
Verbena hastata	Blue Vervain		30	X					X			X
Verbena stricta	Hoary Vervain		60	X								
Vernonia gigantea	Tall Ironweed		60									
Vernonia missurica	Missouri Ironweed		60									X
Veronicastrum virginicum	Culver's Root	(X)	(30)	X					X			X
Viola blanda	Sweet White Violet		60	X				X	X			
Viola canadensis	Canada Violet		60	X				X	X			
Viola pubescens	Downy Yellow Violet		60	X				X				
Viola rostrata	Long-spurred Violet		60	X				X	X			
Zizia aurea	Golden Alexander		60–120						X			

( ) indicates that there is conflicting information in the literature about these characteristics.

Appendix D: **Seed Collection and Prep**

SCIENTIFIC NAME	COMMON NAME	FRUIT COLLECTION	OPTIMAL SEED COLLECTION MONTH(S)	CURE 2 TO 7 DAYS	SEED APPEARANCE	SEED PROCESSING AND CLEANING	SCREEN AND WINNOW (SEPARATE SEED FROM CHAFF)
Achillea millefolium	Common Yarrow	Semi-dry, mostly brown seed heads	Late summer/fall	Room temp	Small, gray, flattened, elongated triangular	Shake dry heads in a hard container or crush	Large screen/winnow
Actaea pachypoda	White Baneberry	Ripe white berries	Aug/Sep	Humid shade	Medium, dark brown half-moon wedge with rough surface	Macerate and float off pulp and skins **	Sound seed sinks
Actaea racemosa	Black Cohosh	Dry, brown capsules	Sep/Oct	Humid shade	Medium, dark brown half-moon wedge with rough surface	Shake/gently beat dry seed heads in a bag or hard container	Large screen
Actaea rubra	Red Baneberry	Ripe red berries	Aug/Sep	Humid shade	Medium, dark brown, half-moon wedge	Macerate and float off pulp and skins **	Sound seed sinks
Agastache foeniculum	Anise Hyssop	Semi-dry, mostly brown seed heads	Sep, Oct	Room temp	Small, oval, brown	Shake dried seed heads in a hard container or crush	Winnow
Agastache nepetoides	Yellow Giant Hyssop	Semi-dry, mostly brown seed heads	Late Sep, Oct	Room temp	Small, oval, brown	Shake dried seed heads in a hard container or crush	Winnow
Agastache scrophulariifolia	Purple Giant Hyssop	Semi-dry, mostly brown seed heads	Late Sep, Oct	Room temp	Small, oval, brown	Shake dried seed heads in a hard container or crush	Winnow
Ageratina altissima	White Snakeroot	Dry fluff	Late Sep, Oct	Room temp	Small, black, elongated with linear ridges	Thresh on fine screen to de-fluff	Winnow
Allium cernuum	Nodding Wild Onion	Dry capsules with exposed black seeds	Late Sep, Oct	Warm, dry	Small, black, stout half-moon wedge with wrinkly surface	Crush dry heads	Screen and winnow
Allium tricoccum	Wild Leek	Dry capsules with exposed black seeds	Aug/Sep	Room temp	Small, black, shiny, irregular round with smooth-to-wrinkly surface	Crush dry heads	Screen and winnow
Anaphalis margaritacea	Pearly Everlasting	Dry seed heads as they begin to fluff up	Late Sep/early Oct	Room temp	Tiny, dark brown, elongated	Shake in a paper bag to release seeds; can be sown with fluff attached	Winnow
Anemonastrum canadense	Canada Anemone	Dry, brown seed clusters	Aug	Room temp	Large, brown, flat, roughly triangular with broad wings and beaked tip	Crush heads	Hand
Anemone cylindrica	Thimbleweed	Brown seed heads (ripe seeds will detach easily)	Late Aug to Oct	Room temp	Small, brown, flat oval with cottony fluff	Rub over fine screen to de-fluff; difficult without mechanization	Fine screen, winnow
Anemone quinquefolia	Wood Anemone	Greenish-yellow seed heads	May/Jun	Humid shade	Small, greenish-brown, oval with beaked tip	Crush heads	Hand
Anemone virginiana	Tall Thimbleweed	Brown seed heads (ripe seeds will detach easily)	Sep, Oct	Room temp	Small, brown with cottony fluff	Rub over fine screen to de-fluff; difficult without mechanization	Fine screen, winnow
Antennaria neglecta	Field Pussytoes	Dry seed heads as they begin to fluff up	Late May, early Jun	Room temp	Very tiny, brown with tufts of white hair	Rub over fine screen to de-fluff; can be sown with fluff attached	Fine screen, winnow
Antennaria parlinii	Parlin's Pussytoes	Dry seed heads as they begin to fluff up	Late May, early Jun	Room temp	Very tiny, brown with tufts of white hair	Rub over fine screen to de-fluff; can be sown with fluff attached	Fine screen, winnow
Aquilegia canadensis	Wild Columbine	Dry, brown capsules	Jul	Room temp	Small, round, black, shiny	Shake/gently beat dry seed heads in a bag or hard container	Hand
Aralia nudicaulis	Wild Sarsparilla	Purple/bluish-black berries	Aug, Sep	Humid shade	Small, half-moon wedge, gray, rough surface	Macerate and float off pulp and skins	Viable seed sinks
Aralia racemosa	Spikenard	Purple-red, glossy berries	Aug, Sep	Humid shade	Small, oval, light brown to tan	Macerate and float off pulp and skins	Viable seed sinks
Arisaema triphyllum	Jack-in-the-pulpit	Fleshy, red berries	Sep	Humid shade	Medium, tan/light brown, globular	Macerate and float off pulp and skins **	Viable seed sinks
Asarum canadense	Wild Ginger	Light brown pods (should split apart easily)	Jun	Humid shade	Large, obovoid, caramel/dark brown with amber elaiosome	Extract by hand *	Hand
Asclepias exaltata	Poke Milkweed	Greenish-yellow pods (should split easily and contain brown seed)	Oct	Room temp	Large, brown, flat, winged disk	Vacuum fluff through a small screen	Small screen
Asclepias incarnata	Swamp Milkweed	Greenish-yellow pods (should split easily and contain brown seed)	Late Sep, Oct	Room temp	Large, brown, flat, winged disk	Vacuum fluff through a small screen	Small screen
Asclepias purpurascens	Purple Milkweed	Greenish-yellow pods (should split easily and contain brown seed)	Late Sep, Oct	Room temp	Large, brown, flat, winged disk	Vacuum fluff through a small screen	Small screen
Asclepias syriaca	Common Milkweed	Greenish-yellow pods (should split easily and contain brown seed)	Late Sep, Oct	Room temp	Large, brown, flat, winged disk	Vacuum fluff through a small screen	Small screen
Asclepias tuberosa	Butterfly Milkweed	Greenish-yellow pods (should split easily and contain brown seed)	Late Sep, Oct	Room temp	Large, brown, flat, winged disk	Vacuum fluff through a small screen	Small screen
Asclepias verticillata	Whorled Milkweed	Greenish-yellow pods (should split easily and contain brown seed)	Late Sep, Oct	Room temp	Large, brown, flat, winged disk	Vacuum fluff through a small screen	Small screen
Asclepias viridiflora	Green Milkweed	Greenish-yellow pods (should split easily and contain brown seed)	Late Sep, Oct	Room temp	Large, brown, flat, winged disk	Vacuum fluff through a small screen	Small screen
Caltha palustris	Marsh Marigold	Greenish-yellow pods (should split easily)	Jun	Room temp	Small, elongated ovoid, brown	Allow pods to dry and open, shake seeds loose in paper bag	Large screen/hand
Campanula rotundifolia	Harebell	Dry, mostly brown pods	Aug, Sep	Room temp	Tiny, brown, oblong	Rub seed heads between finger or thresh on fine screen	Winnow
Campanulastrum americanum	Tall Bellflower	Dry, mostly brown pods	Late Sep, Oct	Room temp	Tiny, brown, ovoid	Rub seed heads between finger or thresh on fine screen	Winnow
Capnoides sempervirens	Pale Corydalis	Semi-dry, yellowish-brown pods	Jul to Sep	Room temp	Small, round, black, shiny	Shake in a hard container or crush	Small screen/winnow

* Recommended that you cover ripening seed heads in a muslin or organza bag to facilitate seed collection
** Recommended that you use gloves when processing and cleaning the seeds

SCIENTIFIC NAME	COMMON NAME	FRUIT COLLECTION	OPTIMAL SEED COLLECTION MONTH(S)	CURE 2 TO 7 DAYS	SEED APPEARANCE	SEED PROCESSING AND CLEANING	SCREEN AND WINNOW (SEPARATE SEED FROM CHAFF)
Cardamine concatenata	Cut-leaved Toothwort	Semi-dry, yellowing pods	Jun	Room temp	Small, brown/light brown, flat ovoid	Dry mostly brown pods in paper bag until they eject seeds *	Hand
Cardamine diphylla	Two-leaved Toothwort	Semi-dry, yellowing pods	Jun	Room temp	Small, brown/light brown, flat ovoid	Dry mostly brown pods in paper bag until they eject seeds *	Hand
Caulophyllum giganteum	Giant Blue Cohosh	Ripe blue fruit	Late Aug to Sep	Humid shade	Large, brown, globular	Macerate and float off pulp and skins	Hand
Caulophyllum thalictroides	Blue Cohosh	Ripe blue fruit	Sep	Humid shade	Large, brown, globular	Macerate and float off pulp and skins	Hand
Chelone glabra	White Turtlehead	Mostly brown capsules	Sep	Room temp	Tiny, brown, flat, oval with surrounding lighter colored wing	Allow pods to dry and open, shake seeds loose in paper bag	Hand
Claytonia caroliniana	Carolina Spring Beauty	Mostly brown capsules	Late May, early Jun	Humid shade	Small, black, round and shiny	Dry capsules in a paper bag to contain catapulting seeds *	Hand
Claytonia virginica	Virginia Spring Beauty	Mostly brown capsules	Late May, early Jun	Humid shade	Small, black, round and shiny	Dry capsules in a paper bag to contain catapulting seeds *	Hand
Coreopsis lanceolata	Lanceleaf Coreopsis	Mostly brown seed heads	Late Aug, Sep	Room temp	Medium, black, concave-oval with surrounding wings	Thresh over large screen	Winnow
Coreopsis tripteris	Tall Tickseed	Mostly brown seed heads	Late Sep, Oct	Room temp	Medium, thin, concave-oblong with narrow wings	Thresh over large screen	Winnow
Dicentra canadensis	Squirrel Corn	Green capsule (should split easily)	Late May to early Jun	Humid shade	Small, black, shiny, kidney-shaped with white elaiosome	Split open capsules and remove seeds by hand *	Hand
Dicentra cucullaria	Dutchman's Breeches	Green capsule (should split easily)	Late May to early Jun	Humid shade	Small, black, shiny, kidney-shaped with white elaiosome	Split open capsules and remove seeds by hand *	Hand
Doellingeria umbellata	Flat-topped White Aster	Dry fluff	Oct	Room temp	Small, brown, elongated obovoid	Thresh over 1/2" to 1/4" screen to de-fluff	Winnow
Echinacea purpurea	Purple Coneflower	Dry, mostly brown seed heads	Late Sep to Oct	Room temp	Medium, light brown, elongated triangular	Shake dry seed heads in hard container or crush	Large screen/hand
Erythronium americanum	Trout Lily	Yellowish-green pods	Late May, early Jun	Humid shade	Small, brown, half-moon with white elaiosome	Split capsules open and remove seeds by hand *	Hand
Eupatorium perfoliatum	Boneset	Dry fluff	Oct	Room temp	Small, black or brown, narrow oblong	Thresh over 0.64–1.25 cm (0.25–0.5 in) screen to de-fluff	Winnow
Eurybia macrophylla	Large Leaf Aster	Dry fluff	Oct	Room temp	Small, triangular, beige (+/-smooth)	Thresh over 0.64–1.25 cm (0.25–0.5 in) screen to de-fluff	Winnow
Euthamia graminifolia	Grass-Leaved Goldenrod	Dry fluff	Oct	Room temp	Tiny, ovoid, brown to light brown	Thresh over 0.64–1.25 cm (0.25–0.5 in) screen to de-fluff	Winnow
Eutrochium maculatum	Spotted Joe Pye Weed	Dry fluff	Late Sep, Oct	Room temp	Small, narrow and linear, dark brown/black	Thresh over 0.64–1.25 cm (0.25–0.5 in) screen to de-fluff	Winnow
Eutrochium purpureum	Sweet Joe Pye Weed	Dry fluff	Late Sep, Oct	Room temp	Small, narrow and linear, dark brown/black	Thresh over 0.64–1.25 cm (0.25–0.5 in) screen to de-fluff	Winnow
Fragaria vesca	Woodland Strawberry	Ripe, red, fragrant berries	Jul	Room temp	Tiny, round, beige	Macerate and float off pulp and skins	Viable seed sinks
Fragaria virginiana	Wild Strawberry	Ripe, red, fragrant berries	Jul	Room temp	Tiny, round, beige	Macerate and float off pulp and skins	Viable seed sinks
Gentiana andrewsii	Bottle Gentian	Mostly brown capsules	Oct	Room temp	Tiny, brown, flat, oval with surrounding lighter colored wing	Shake/gently beat dry seed heads in a bag or hard container	Large screen/hand
Geranium maculatum	Wild Geranium	Browning pods start	Jun	Room temp	Small, dark brown, oval	Dry capsules in a paper bag to contain catapulting seeds *	Hand
Geum fragarioides	Barren Strawberry	Dry fruit (achene) once styles fall off	Jun	Room temp	Small, flattened, roughly oval, finely hairy	Crush dry heads	Hand
Geum rivale	Water Avens	Dry, upright, fluffy seed heads	Late Jul, Aug	Room temp	Small, brown, oval, hairy with long tail	Rub to de-fluff	Winnow
Geum triflorum	Prairie Smoke	Dry, upright, fluffy seed heads	Late May, Jun	Cool, dry	Small, brown, oval, hairy with long tail	Rub to de-fluff	Winnow
Helenium autumnale	Sneezeweed	Dry, mostly brown seed heads	Sep, Oct	Room temp	Small, light brown, triangular, finely hairy	Thresh over small screen	Winnow
Helianthus divaricatus	Woodland Sunflower	Semi-dry, mostly brown seed heads	Sep, early Oct	Room temp	Large, flat triangular, brown	Crush dry heads	Large screen
Helianthus giganteus	Tall Sunflower	Semi-dry, mostly brown seed heads	Sep, early Oct	Room temp	Large, flat triangular, light brown to black	Crush dry heads	Large screen
Helianthus tuberosus	Jerusalem Artichoke	Semi-dry, mostly brown seed heads	Sep, early Oct	Room temp	Large, flat triangular, light brown to black	Crush dry heads	Large screen
Heliopsis helianthoides	False Sunflower	Semi-dry, mostly brown seed heads	Sep, early Oct	Room temp	Large, triangular, dark brown to black	Crush dry heads	Large screen
Hepatica acutiloba	Sharp-lobed Hepatica	Dry fruit (achene) when stalks droop	Late May, Jun	Humid shade	Medium, greenish-brown, ovoid with attached elaiosome	Ripe fruit should detach from plant upon light touch *	Hand
Hepatica americana	Round-lobed Hepatica	Dry fruit (achene) when stalks droop	Late May, Jun	Humid shade	Medium, greenish-brown, ovoid with attached elaiosome	Ripe fruit should detach from plant upon light touch *	Hand
Houstonia longifolia	Long-leaved Bluet	Dry capsules	Sep	Room temp	Small, black, irregular-round-to-ellipsoid with wrinkled surface	Shake/gently beat dry seed capsules in a bag or hard container	Small screen/winnow
Hydrophyllum virginianum	Virginia Waterleaf	Plump, greenish-yellow capsules	Late Jun, early Jul	Room temp	Large, brown/light brown, irregular-rounded, wrinkled surface	Shake/gently beat dry seed capsules in a bag or hard container *	Large screen/hand
Impatiens capensis	Spotted Jewelweed	Plump green pods	Late Aug, Sep, early Oct	Humid shade	Medium, green or brown, ovoid with ridges	Carefully cover pod with bag then lightly touch to trigger their release *	Hand
Impatiens pallida	Pale Touch-me-not	Plump green pods	Late Aug, Sep, early Oct	Humid shade	Medium, green or brown, ovoid with ridges	Carefully cover pod with bag then lightly touch to trigger their release *	Hand

* Recommended that you cover ripening seed heads in a muslin or organza bag to facilitate seed collection

** Recommended that you use gloves when processing and cleaning the seeds

SCIENTIFIC NAME	COMMON NAME	FRUIT COLLECTION	OPTIMAL SEED COLLECTION MONTH(S)	CURE 2 TO 7 DAYS	SEED APPEARANCE	SEED PROCESSING AND CLEANING	SCREEN AND WINNOW (SEPARATE SEED FROM CHAFF)
Iris versicolor	Blue Flag Iris	Dry, mostly brown capsules	Late Aug, Sep	Room temp	Large, brown, roughly half-moon (variable)	Shake/gently beat dry capsules in a bag or hard container	Hand
Liatris cylindracea	Ontario Blazing Star	Dry fluff	Mid-Sep, Oct	Room temp	Large, triangular/narrow, rough, gray	Thresh over 0.32–0.64 cm (0.125–0.25 in) screen	Winnow
Liatris spicata	Dense Blazing Star	Dry fluff	Mid-Sep, Oct	Room temp	Large, triangular/narrow, rough, gray	Thresh over 0.32–0.64 cm (0.125–0.25 in) screen	Winnow
Lilium canadense	Canada Lily	Dry, mostly brown capsules	Sep, early Oct	Room temp	Large, brown, triangular, flat with surrounding papery wing	Shake/gently beat dry capsules in a bag or hard container	Large screen/hand
Lilium michiganense	Michigan Lily	Dry, mostly brown capsules	Sep, Oct	Room temp	Large, brown, triangular, flat with surrounding papery wing	Shake/gently beat dry capsules in a bag or hard container	Large screen/hand
Lilium philadelphicum	Wood Lily	Dry, mostly brown capsules	Late Aug, Sep	Room temp	Large, brown, triangular, flat with surrounding papery wing	Shake/gently beat dry capsules in a bag or hard container	Large screen/hand
Lobelia cardinalis	Cardinal Flower	Dry fluff	Sep, early Oct	Room temp	Very tiny, brown, oval with wrinkled surface	Shake/gently beat dry seed heads in a bag or hard container	Large screen/hand
Lobelia siphilitica	Blue Lobelia	Dry fluff	Sep, early Oct	Room temp	Very tiny, brown, oval with wrinkled surface	Shake/gently beat dry seed heads in a bag or hard container	Large screen/hand
Lupinus perennis	Wild Lupine	Semi-dry, browning pods	Late Jun, Jul	Warm, dry	Large, white or light brown or gray, roughly bean-shaped	Allow pods to dry in a paper bag until they eject seeds *	Large screen
Maianthemum canadense	Canada Mayflower	Red/mostly red berries	Aug/Sep	Cool, dry	Small, irregular round, light brown	Macerate and float off pulp and skins	Sound seed sinks
Maianthemum racemosum	False Solomon's Seal	Red/mostly red berries	Aug/Sep	Cool, dry	Small, irregular round, light brown	Macerate and float off pulp and skins	Sound seed sinks
Maianthemum stellatum	Starry False Solomon's Seal	Red/mostly red berries	Aug/Sep	Cool, dry	Small, irregular round, dark-light brown	Macerate and float off pulp and skins	Sound seed sinks
Mertensia virginica	Virginia Bluebells	Whole stems with browning nutlets	Jun	Room temp	Large, rounded-triangular, greenish-white-to-brown, rough surface	Allow stems to dry in a paper bag then shake to release seeds *	Hand/large screen
Mimulus ringens	Monkey Flower	Mostly brown capsules	Aug, Sep	Room temp	Very tiny, oval, light brown/tan	Shake/gently beat dry seed heads in a bag or hard container	Large screen/hand
Monarda didyma	Bee Balm	Semi-dry, mostly brown seed heads	Sep	Room temp	Small, oval, dark brown	Crush heads	Winnow
Monarda fistulosa	Wild Bergamot	Semi-dry, mostly brown seed heads	Late Aug, Sep	Room temp	Small, oval, dark brown	Crush heads	Winnow
Monarda punctata	Spotted Bee Balm	Semi-dry, mostly brown seed heads	Sep	Room temp	Small, oval, dark brown	Crush heads	Winnow
Oenothera biennis	Evening Primrose	Mostly brown capsules as the tip starts to split open	Sep, Oct	Room temp	Small, brown, irregular/angled	Shake/gently beat dry seed heads in a bag or hard container	Large screen/hand
Oenothera pilosella	Prairie Sundrops	Mostly brown capsules as the tip starts to split open	Aug, Sep	Room temp	Tiny, light brown, oval	Shake/gently beat dry seed heads in a bag or hard container	Large screen/hand
Packera aurea	Golden Ragwort	Dry fluff (yellow-green seed heads ok too)	Late Jun, Jul	Room temp	Small, narrow and linear, brown	Thresh over 0.32–0.64 cm (0.125–0.25 in) screen	Winnow
Packera paupercula	Balsam Groundsel	Dry fluff (yellow-green seed heads ok too)	Late Jun, Jul	Room temp	Small, narrow and linear, brown	Thresh over 0.32–0.64 cm (0.125–0.25 in) screen	Winnow
Pedicularis lanceolata	Swamp Betony	Whole stem with browning pods	Oct	Room temp	Tiny, brown, irregular oval	Shake/gently beat dry seed heads in a bag or hard container	Large screen/hand
Penstemon digitalis	Foxglove Beardtongue	Dry, brown capsules	Sep	Room temp	Small, square, brown / beige	Crush heads	Winnow/small screen
Penstemon hirsutus	Hairy Beardtongue	Dry, brown capsules	Sep	Room temp	Small, square, brown / beige	Crush heads	Winnow/small screen
Phlox divaricata	Wild Blue Phlox	Green-beige capsules	Jul	Room temp	Large, oval, brown/dark brown, wrinkled surface	Allow capsules to dry in a paper bag until they release seeds *	Hand
Physostegia virginiana	Obedient Plant	Green flower stalks	Oct	Room temp	Medium, reddish-brown-to-dark brown, three-sided	Shake/gently beat dry seed heads in a bag or hard container	Large screen/hand
Podophyllum peltatum	Mayapple	Fragrant, yellow fruit	Aug	Humid shade	Large, flattened oval, dark brown	Macerate and float off pulp and skins *	Hand
Polygonatum biflorum	Solomon's Seal	Ripe blue fruit	Sep	Humid shade	Large, light brown, irregular oval,	Macerate and float off pulp and skins	Sound seed sinks
Pycnanthemum tenuifolium	Slender Mountain Mint	Dry, brown seed heads	Oct	Room temp	Tiny, oval, almost black	Crush seed heads or shake in a hard container	Large screen/winnow
Pycnanthemum virginianum	Virginia Mountain Mint	Dry, brown seed heads	Oct	Room temp	Tiny, oval, almost black	Crush seed heads or shake in a hard container	Large screen/winnow
Ranunculus fascicularis	Early Buttercup	Semi-dry, browning seed heads	Jun	Room temp	Medium, brown, more or less round and flat with a beaked tip	Crush heads	Hand
Ratibida pinnata	Gray-headed Coneflower	Dry seed heads	Sep, early Oct	Room temp	Small, triangular, black	Crush heads	Winnow
Rudbeckia fulgida	Orange Coneflower	Semi-dry seed heads	Sep	Room temp	Small, linear black	Crush heads	Fine screen/winnow
Rudbeckia hirta	Black-eyed Susan	Semi-dry seed heads	Sep, Oct	Room temp	Small, linear black	Crush heads	Fine screen/winnow
Rudbeckia laciniata	Green-headed Coneflower	Semi-dry seed heads	Sep	Room temp	Small, linear black	Crush heads	Winnow
Sanguinaria canadensis	Bloodroot	Plump, green pods (must split easily)	Jun	Humid shade	Small, shiny, reddish-brown, oval with attached elaiosome	Split pods open and remove seeds by hand *	Hand
Silphium laciniatum	Compass Plant	Semi-dry seed head	Oct	Room temp	Large, dark brown, flat, rounded-triangular with surrounding papery wing	Crush dry heads	Winnow
Silphium perfoliatum	Cup Plant	Semi-dry seed head	Late Sep, Oct	Room temp	Large, dark brown, flat, rounded-triangular with surrounding papery wing	Crush dry heads	Winnow
Sisyrinchium angustifolium	Blue-eyed Grass	Semi-dry capsules	Jul	Room temp	Small, black, round	Crush dry heads	Winnow
Sisyrinchium montanum	Mountain Blue-eyed Grass	Semi-dry capsules	Aug	Room temp	Small, black, round	Crush dry heads	Winnow

* Recommended that you cover ripening seed heads in a muslin or organza bag to facilitate seed collection
** Recommended that you use gloves when processing and cleaning the seeds

SCIENTIFIC NAME	COMMON NAME	FRUIT COLLECTION	OPTIMAL SEED COLLECTION MONTH(S)	CURE 2 TO 7 DAYS	SEED APPEARANCE	SEED PROCESSING AND CLEANING	SCREEN AND WINNOW (SEPARATE SEED FROM CHAFF)
Solidago bicolor	Silverrod	Dry fluff	Oct	Room temp	Tiny, triangular, gray (+/-rough)	Rub to de-fluff	Winnow
Solidago caesia	Blue-stemmed Goldenrod	Dry fluff	Oct	Room temp	Tiny, triangular, gray (+/-rough)	Rub to de-fluff	Winnow
Solidago flexicaulis	Zigzag Golenrod	Dry fluff	Oct, early Nov	Room temp	Tiny, triangular, gray (+/-rough)	Rub to de-fluff	Winnow
Solidago hispida	Hairy Goldenrod	Dry fluff	Oct	Room temp	Tiny, triangular, gray (+/-rough)	Rub to de-fluff	Winnow
Solidago juncea	Early Goldenrod	Dry fluff	Oct	Room temp	Tiny, triangular, gray (+/-rough)	Rub to de-fluff	Winnow
Solidago nemoralis	Gray Goldenrod	Dry fluff	Oct, early Nov	Room temp	Tiny, triangular, gray (+/-rough)	Rub to de-fluff	Winnow
Solidago ohioensis	Ohio Goldenrod	Dry fluff	Oct	Room temp	Tiny, triangular, gray (+/-rough)	Rub to de-fluff	Winnow
solidago ptarmicoides	Upland White Goldenrod	Dry fluff	Mid-Sep, Oct	Room temp	Tiny, triangular, gray (+/-rough)	Rub to de-fluff	Winnow
Solidago riddellii	Riddell's Goldenrod	Dry fluff	Oct	Room temp	Tiny, triangular, gray (+/-rough)	Rub to de-fluff	Winnow
Solidago rigida	Stiff Goldenrod	Dry fluff	Oct	Room temp	Tiny, triangular, gray (+/-rough)	Rub to de-fluff	Winnow
Solidago rugosa	Rough-stemmed Goldenrod	Dry fluff	Oct	Room temp	Tiny, triangular, gray (+/-rough)	Rub to de-fluff	Winnow
Solidago speciosa	Showy Goldenrod	Dry fluff	Oct, early Nov	Room temp	Tiny, triangular, gray (+/-rough)	Rub to de-fluff	Winnow
Solidago uliginosa	Bog Goldenrod	Dry fluff	Oct	Room temp	Tiny, triangular, gray (+/-rough)	Rub to de-fluff	Winnow
Symphyotrichum cordifolium	Blue Wood Aster	Dry fluff	Oct, early Nov	Room temp	Small, triangular, gray (+/-smooth)	Rub to de-fluff	Winnow
Symphyotrichum ericoides	White Heath Aster	Dry fluff	Late Oct, early Nov	Room temp	Small, triangular, gray (+/-smooth)	Rub to de-fluff	Winnow
Symphyotrichum laeve	Smooth Aster	Dry fluff	Late Oct, early Nov	Room temp	Small, triangular, gray (+/-smooth)	Rub to de-fluff	Winnow
Symphyotrichum lanceolatum	Panicled Aster	Dry fluff	Oct	Room temp	Small, triangular, gray (+/-smooth)	Rub to de-fluff	Winnow
Symphyotrichum lateriflorum	Calico Aster	Dry fluff	Late Oct, early Nov	Room temp	Small, triangular, gray (+/-smooth)	Rub to de-fluff	Winnow
Symphyotrichum novae-angliae	New England Aster	Dry fluff	Late Oct, Nov	Room temp	Small, triangular, gray (+/-smooth)	Rub to de-fluff	Winnow
Symphyotrichum ontarionis	Ontario Aster	Dry fluff	Oct	Room temp	Small, triangular, gray (+/-smooth)	Rub to de-fluff	Winnow
Symphyotrichum oolentangiense	Sky Blue Aster	Dry fluff	Late Oct, early Nov	Room temp	Small, triangular, gray (+/-smooth)	Rub to de-fluff	Winnow
Symphyotrichum pilosum	Frost Aster	Dry fluff	Oct	Room temp	Small, triangular, gray (+/-smooth)	Rub to de-fluff	Winnow
Symphyotrichum puniceum	Purple-stemmed Aster	Dry fluff	Oct	Room temp	Small, triangular, gray (+/-smooth)	Rub to de-fluff	Winnow
Symphyotrichum urophyllum	Arrowleaf Aster	Dry fluff	Oct	Room temp	Small, triangular, gray (+/-smooth)	Rub to de-fluff	Winnow
Thalictrum dasycarpum	Purple Meadowrue	Dry fruit (achene)	Early Aug to mid-Sep	Room temp	Small, dark brown, ovoid with linear ridges, pointed at both ends	Ripe/dry fruit easily seperates from plant	Hand
Thalictrum dioicum	Early Meadowrue	Dry fruit (achene)	Jun	Room temp	Small, dark brown, ovoid with linear ridges, pointed at both ends	Ripe/dry fruit easily seperates from plant	Hand
Thalictrum pubescens	Tall Meadowrue	Dry fruit (achene)	Late Aug, Sep	Room temp	Small, dark brown, ovoid with linear ridges, pointed at both ends	Ripe/dry fruit easily seperates from plant	Hand
Tiarella stolonifera	Foamflower	Dry, mostly brown capsules	Late May, Jun	Humid shade	Small, black, oval, shiny	Shake/gently beat dry seed heads in a bag or hard container	Large screen/hand
Tradescantia ohiensis	Ohio Spiderwort	Semi-dry capsules	Mid-Jul, Aug	Room temp	Small, gray-to-black, oval, sunken center with ribbed sides	Thresh over 0.64–1.25 cm (0.25–0.5 in) screen to de-fluff **	Winnow
Tradescantia virginiana	Virginia Spiderwort	Semi-dry capsules	Mid-Jul, Aug	Room temp	Small, gray-to-black, oval, sunken center with ribbed sides	Thresh over 0.64–1.25 cm (0.25–0.5 in) screen to de-fluff **	Winnow
Trillium erectum	Red Trillium	Plump, red capsules	Late Jul, Aug	Humid shade	Small, reddish-brown, oval with attached elaiosome	Macerate and float off pulp and skins *	Hand
Trillium grandiflorum	Large White Trillium	Ripe, yellowish-green capsules	Late Jul, Aug	Humid shade	Small, reddish-brown, oval with attached elaiosome	Macerate and float off pulp and skins *	Hand
Uvularia grandiflorum	Large-flowered Bellwort	Plump, greenish-yellow pods	Aug	Humid shade	Small, light brown, irregularly round with attached elaiosome	Seed pods split open and release seeds when ripe. *	Hand
Verbena hastata	Blue Vervain	Dry, brown seed heads	Late Sep, Oct	Room temp	Small, linear brown/silver	Crush dry heads or shake in a hard container	Window/large screen
Verbena stricta	Hoary Vervain	Dry, brown seed heads	Early Sep, early Oct	Room temp	Small, linear brown/silver	Crush dry heads or shake in a hard container	Window/large screen
Vernonia gigantea	Tall Ironweed	Dry seed heads with rust-colored fluff	Sep	Room temp	Medium, narrow triangular, gray to light brown	Thresh over 0.32–0.64 cm (0.125–0.25 in) screen	Winnow
Vernonia missurica	Missouri Ironweed	Dry seed heads with rust-colored fluff	Sep	Room temp	Medium, narrow triangular, gray to light brown	Thresh over 0.32–0.64 cm (0.125–0.25 in) screen	Winnow
Veronicastrum virginicum	Culver's Root	Dry, brown capsules	Early Sep, Oct	Room temp	Tiny, brown, ellipsoid	Shake/gently beat dry seed heads in a bag or hard container	Large screen/hand
Viola blanda	Sweet White Violet	Green, upright (not nodding) capsules	Late Jul, Aug, Sep	Room temp	Medium, creamy white-to-brown, obovoid with attached elaiosome	Allow capsules to dry in a paper bag until they release seeds *	Hand
Viola canadensis	Canada Violet	Green, upright (not nodding) capsules	Jun	Room temp	Medium, creamy white-to-brown, obovoid with attached elaiosome	Allow capsules to dry in a paper bag until they release seeds *	Hand
Viola pubescens	Downy Yellow Violet	Green, upright (not nodding) capsules	Jun	Room temp	Medium, creamy white-to-brown, obovoid with attached elaiosome	Allow capsules to dry in a paper bag until they release seeds *	Hand
Viola rostrata	Long-spurred Violet	Green, upright (not nodding) capsules	Jul	Room temp	Medium, creamy white-to-brown, obovoid with attached elaiosome	Allow capsules to dry in a paper bag until they release seeds *	Hand
Zizia aurea	Golden Alexander	Dry, mostly brown seeds	Early Aug to mid-Sep	Room temp	Small, dark brown, oblong with shallow ridges	Shake/gently beat dry seed heads in a bag or hard container	Large screen/hand

* Recommended that you cover ripening seed heads in a muslin or organza bag to facilitate seed collection

** Recommended that you use gloves when processing and cleaning the seeds

Appendix E: **Butterflies and Their Host Plants**

COMMON NAME	SCIENTIFIC NAME	HOST PLANT
American Lady	*Vanessa virginiensis*	*Anaphalis margaritacea* (pp. 48–49), *Antennaria neglecta* (pp. 58–59), *Antennaria parlinii* (pp. 60–61), *Vernonia missurica* (pp. 314–315)
Appalachian Azure	*Celastrina neglectamajor*	*Actaea racemosa* (pp. 32–33)
Baltimore Checkerspot	*Euphydryas phaeton*	*Chelone glabra* (pp. 102–103), *Mimulus ringens* (pp. 192–193), *Pedicularis lanceolata* (pp. 208–209), *Penstemon digitalis* (pp. 210–211), *Penstemon hirsutus* (pp. 212–213)
Black Swallowtail	*Papilio polyxenes*	*Zizia aurea* (pp. 326–327)
Columbine Duskywing	*Erynnis lucilius*	*Aquilegia canadensis* (pp. 62–63)
Common Buckeye	*Junonia coenia*	*Mimulus ringens* (pp. 192–193), *Verbena hastata* (pp. 308–309), *Verbena stricta* (pp. 310–311)
Crossline Skipper	*Polites origenes*	*Vernonia missurica* (pp. 314–315)
Fritillary spp.	various species	*Viola blanda* (pp. 318–319), *Viola canadensis* (pp. 320–321), *Viola pubescens* (pp. 322–323), *Viola rostrata* (pp. 324–325)
Frosted Elfin	*Callophrys irus*	*Lupinus perennis* (pp. 182–183)
Gorgone Checkerspot	*Chlosyne gorgone*	*Helianthus divaricatus* (pp. 148–149), *Helianthus giganteus* (pp. 150–151), *Helianthus tuberosus* (pp. 152–153), *Heliopsis helianthoides* (pp. 154–155), *Rudbeckia hirta* (pp. 232–233), *Rudbeckia laciniata* (pp. 234–235)
Gray Hairstreak	*Strymon melinus*	*Asclepias tuberosa* (pp. 80–81), *Fragaria vesca* (pp. 132–133), *Fragaria virginiana* (pp. 134–135)
Grizzled Skipper	*Pyrgus malvae*	*Fragaria vesca* (pp. 132–133), *Fragaria virginiana* (pp. 134–135)
Hairstreak spp.	various species	*Allium cernuum* (pp. 44–45)
Harris's Checkerspot	*Chlosyne harrisii*	*Doellingeria umbellata* (pp. 116–117)
Karner Blue	*Plebejus melissa samuelis*	*Lupinus perennis* (pp. 182–183)
Monarch	*Danaus plexippus*	*Asclepias incarnata* (pp. 74–75), *Asclepias purpurascens* (pp. 76–77), *Asclepias syriaca* (pp. 78–79), *Asclepias tuberosa* (pp. 80–81), *Asclepias verticillata* (pp. 82–83), *Asclepias viridiflora* (pp. 84–85)
Mustard White	*Pieris oleracea*	*Cardamine concatenata* (pp. 94–95)
Northern Crescent	*Phyciodes cocyta*	*Doellingeria umbellata* (pp. 116–117), *Symphyotrichum cordifolium* (pp. 270–271), *Symphyotrichum ericoides* (pp. 272–273), *Symphyotrichum laeve* (pp. 274–275), *Symphyotrichum lanceolatum* (pp. 276–277), *Symphyotrichum lateriflorum* (pp. 278–279), *Symphyotrichum novae-angliae* (pp. 280–281), *Symphyotrichum ontarionis* (pp. 282–283), *Symphyotrichum oolentangiensis* (pp. 284–285), *Symphyotrichum pilosum* (pp. 286–287), *Symphyotrichum puniceum* (pp. 288–289), *Symphyotrichum urophyllum* (pp. 290–291)
Painted Lady	*Vanessa cardui*	*Achillea millefolium* (pp. 28–29), *Anaphalis margaritacea* (pp. 48–49), *Antennaria neglecta* (pp. 58–59), *Antennaria parlinii* (pp. 60–61), *Doellingeria umbellata* (pp. 116–117), *Helianthus divaricatus* (pp. 148–149), *Helianthus giganteus* (pp. 150–151), *Helianthus tuberosus* (pp. 152–153), *Heliopsis helianthoides* (pp. 154–155), *Symphyotrichum cordifolium* (pp. 270–271), *Symphyotrichum ericoides* (pp. 272–273), *Symphyotrichum laeve* (pp. 274–275), *Symphyotrichum lanceolatum* (pp. 276–277), *Symphyotrichum lateriflorum* (pp. 278–279), *Symphyotrichum novae-angliae* (pp. 280–281), *Symphyotrichum ontarionis* (pp. 282–283), *Symphyotrichum oolentangiensis* (pp. 284–285), *Symphyotrichum pilosum* (pp. 286–287), *Symphyotrichum puniceum* (pp. 288–289), *Symphyotrichum urophyllum* (pp. 290–291), *Vernonia gigantea* (pp. 312–313), *Vernonia missurica* (pp. 314–315)
Pearl Crescent	*Phyciodes tharos*	*Doellingeria umbellata* (pp. 116–117), *Eurybia macrophylla* (pp. 124–125), *Rudbeckia hirta* (pp. 232–233), *Symphyotrichum cordifolium* (pp. 270–271), *Symphyotrichum ericoides* (pp. 272–273), *Symphyotrichum laeve* (pp. 274–275), *Symphyotrichum lanceolatum* (pp. 276–277), *Symphyotrichum lateriflorum* (pp. 278–279), *Symphyotrichum novae-angliae* (pp. 280–281), *Symphyotrichum ontarionis* (pp. 282–283), *Symphyotrichum oolentangiensis* (pp. 284–285), *Symphyotrichum pilosum* (pp. 286–287), *Symphyotrichum puniceum* (pp. 288–289), *Symphyotrichum urophyllum* (pp. 290–291)
Persius Duskywing	*Erynnis persius*	*Lupinus perennis* (pp. 182–183)
Pipevine Swallowtail	*Battus philenor*	*Asarum canadense* (pp. 70–71)
Queen	*Danaus gilippus*	*Asclepias tuberosa* (pp. 80–81)
Silvery Blue	*Glaucopsyche lygdamus*	*Lupinus perennis* (pp. 182–183)
Silvery Checkerspot	*Chlosyne nycteis*	*Doellingeria umbellata* (pp. 116–117), *Echinacea purpurea* (pp. 118–119), *Eurybia macrophylla* (pp. 124–125), *Helianthus divaricatus* (pp. 148–149), *Helianthus giganteus* (pp. 150–151), *Helianthus tuberosus* (pp. 152–153), *Heliopsis helianthoides* (pp. 154–155), *Ratibida pinnata* (pp. 228–229), *Rudbeckia fulgida* (pp. 230–231), *Rudbeckia hirta* (pp. 232–233), *Rudbeckia laciniata* (pp. 234–235), *Symphyotrichum cordifolium* (pp. 270–271), *Symphyotrichum ericoides* (pp. 272–273), *Symphyotrichum laeve* (pp. 274–275), *Symphyotrichum lanceolatum* (pp. 276–277), *Symphyotrichum lateriflorum* (pp. 278–279), *Symphyotrichum novae-angliae* (pp. 280–281), *Symphyotrichum ontarionis* (pp. 282–283), *Symphyotrichum oolentangiensis* (pp. 284–285), *Symphyotrichum pilosum* (pp. 286–287), *Symphyotrichum urophyllum* (pp. 290–291)
Southern Dogface	*Zerene cesonia*	*Coreopsis lanceolata* (pp. 108–109), *Coreopsis tripteris* (pp. 110–111)
Spring Azure	*Celastrina ladon*	*Actaea racemosa* (pp. 32–33)
Summer Azure	*Celastrina neglecta*	*Agastache nepetoides* (pp. 38–39)
Tawny Crescent	*Phyciodes batesii*	*Doellingeria umbellata* (pp. 116–117), *Symphyotrichum cordifolium* (pp. 270–271), *Symphyotrichum ericoides* (pp. 272–273), *Symphyotrichum laeve* (pp. 274–275), *Symphyotrichum lanceolatum* (pp. 276–277), *Symphyotrichum lateriflorum* (pp. 278–279), *Symphyotrichum novae-angliae* (pp. 280–281), *Symphyotrichum ontarionis* (pp. 282–283), *Symphyotrichum oolentangiensis* (pp. 284–285), *Symphyotrichum pilosum* (pp. 286–287), *Symphyotrichum puniceum* (pp. 288–289), *Symphyotrichum urophyllum* (pp. 290–291)
Variegated Fritillary	*Euptoieta claudia*	*Podophyllum peltatum* (pp. 218–219)
West Virginia White	*Pieris virginiensis*	*Cardamine concatenata* (pp. 94–95), *Cardamine diphylla* (pp. 96–97)
Wild Indigo Duskywing	*Erynnis baptisiae*	*Lupinus perennis* (pp. 182–183)

Appendix F: **SAR Codes and Definitions**

(explorer.natureserve.org/AboutTheData/Statuses)

SX – Presumed Extirpated: Species or ecosystem is believed to be extirpated from the jurisdiction (i.e., nation or state/province). Not located despite intensive searches of historical sites and other appropriate habitat, and virtually no likelihood that it will be rediscovered.

SH – Possibly Extirpated: Known from only historical records but still some hope of rediscovery. There is evidence that the species or ecosystem may no longer be present in the jurisdiction, but not enough to state this with certainty. Examples of such evidence include (1) that a species has not been documented in approximately 20 to 40 years despite some searching and/or some evidence of significant habitat loss or degradation; (2) that a species or ecosystem has been searched for unsuccessfully, but not thoroughly enough to presume that it is no longer present in the jurisdiction.

S1 – Critically Imperiled: At very high risk of extirpation in the jurisdiction due to very restricted range, very few populations or occurrences, very steep declines, severe threats, or other factors.

S2 – Imperiled: At high risk of extirpation in the jurisdiction due to restricted range, few populations or occurrences, steep declines, severe threats, or other factors.

S3 – Vulnerable: At moderate risk of extirpation in the jurisdiction due to a fairly restricted range, relatively few populations or occurrences, recent and widespread declines, threats, or other factors.

SU – Unrankable: Currently unrankable due to lack of information or due to substantially conflicting information about status or trends.

Those listed as S4 (Apparently Secure) and S5 (Secure) are not identified as such in this book.

ONTARIO
(*Endangered Species Act, 2007, S.O. 2007, c. 6*)

X – Extirpated: A species shall be classified as an extirpated species if it lives somewhere in the world, lived at one time in the wild in Ontario, but no longer lives in the wild in Ontario.

E – Endangered: A species shall be classified as an endangered species if it lives in the wild in Ontario but is facing imminent extinction or extirpation.

T – Threatened: A species shall be classified as a threatened species if it lives in the wild in Ontario, is not endangered, but is likely to become endangered if steps are not taken to address factors threatening to lead to its extinction or extirpation.

SC – Special Concern: A species shall be classified as a special concern species if it lives in the wild in Ontario, is not endangered or threatened, but may become threatened or endangered because of a combination of biological characteristics and identified threats.

INDIANA
(*Definitions were provided by the lead botanist, Indiana Department of Natural Resources [IDNR], as proposed updates to the definitions provided on their website*)

X – Extirpated: A plant species that is believed to be originally native to Indiana but currently without any known naturally occurring populations within the state.

E – Endangered: A plant species believed to be native to Indiana with five or fewer occurrences in Indiana or that is otherwise currently at the brink of extinction.

T – Threatened: A plant species believed to be native to Indiana with six to 20 occurrences in the state, or that is of conservation concern, or that is otherwise likely to become endangered within the foreseeable future.

WL – Watch List: A plant species believed to be native to Indiana that was once thought to have up to 20 occurrences in the state but that is now known to be more prevalent, or that is being considered for listing as endangered or threatened, or that is known to be harvested for medicinal purposes. A few species included as Watch List species are being researched to determine if they are truly present in Indiana. (Note that there is not currently a formal definition of Watch List within the IDNR.)

MICHIGAN
(Endangered Species Act of the State of Michigan [Part 365 of PA 451, 1994 Michigan Natural Resources and Environmental Protection Act])

X – Presumed Extirpated: A species assumed to have been annihilated.

E – Endangered: A species is in danger of extinction throughout all or a significant portion of its range.

T – Threatened: A species is likely to become endangered within the foreseeable future.

SC – Special Concern: A declining or relict species in the state. While not protected by law, these species need protection to prevent them from becoming threatened or endangered.

NEW YORK
(New York – Environmental Conservation Law [1990], section 9-1503)

E – Endangered: A plant species in danger of extirpation throughout all or a significant portion of its range within the state and requiring remedial action to prevent such extinction.

T – Threatened: A plant species that is likely to become endangered within the foreseeable future throughout all or a significant portion of its range within the state.

R – Rare: A native plant species that has from 20 to 35 extant sites or 3,000 to 5,000 individuals statewide.

EV – Exploitably Vulnerable: A native plant species that is likely to become threatened in the near future throughout all or a significant portion of its range within the state if causal factors continue unchecked.

OHIO
(Ohio Department of Natural Resources, Rare Native Ohio Plants, 2020–21 Status List)

X – Presumed Extirpated Species: A native Ohio plant species may be designated presumed extirpated when no natural populations of the species have been documented since 2000.

E – Endangered Species: A native Ohio plant species may be designated endangered if, based on its known status in Ohio, one or more of the following criteria apply:

1. The species is a federal endangered species extant in Ohio.
2. The natural populations of the species in Ohio are limited to three or fewer occurrences.
3. The distribution of the natural populations of the species in Ohio is limited to a geographic area delineated by three or fewer U.S. Geological Survey 7.5-minute quadrangle maps.
4. The number of plants in all the natural populations of the species in Ohio is limited to one hundred or fewer individual, physically unconnected plants.

T – Threatened Species: A native Ohio plant species may be designated threatened if, based on its known status in Ohio, one or more of the following criteria apply:

1. The species is a federal threatened species extant in Ohio but not on the state endangered species list.

2. The natural populations of the species in Ohio are limited to no less than four or more than 10 occurrences.
3. The distribution of the natural populations of the species in Ohio is limited to a geographic area delineated by no less than four or more than seven U.S. Geological Survey 7.5-minute quadrangle maps.

T(P) – Potentially Threatened Species: A native Ohio plant species may be designated potentially threatened if one or more of the following criteria apply:

1. The species is extant in Ohio and does not qualify as a state endangered or threatened species, but it is a proposed federal endangered or threatened species or a species listed in the Federal Register as under review for such proposal.
2. The natural populations of the species are imperiled to the extent that the species could conceivably become a threatened species in Ohio within the foreseeable future.
3. The natural populations of the species, even though they are not threatened in Ohio at the time of designation, are believed to be declining in abundance or vitality at a significant rate throughout all or large portions of the state.

PENNSYLVANIA

(Conservation of Pennsylvania Native Wild Plants [17 PA. Code § 45.1-91] – 2018)

X – Extirpated: A classification of plant species believed to be extinct in this Commonwealth.

E – Endangered: A classification of plant species that are in danger of extinction throughout most or all of their natural range in this Commonwealth if critical habitat is not maintained or if the species is greatly exploited by man.

T – Threatened: A classification of plant species that may become endangered throughout most or all of their natural range in this Commonwealth if critical habitat is not maintained to prevent their further decline, or if the species is greatly exploited by man.

V – Vulnerable: A classification of plant species that are in danger of population decline within this Commonwealth because of their beauty, economic value, use as a cultivar, or other factors that indicate that persons may seek to remove these species from their native habitats.

R – Rare: A classification of plant species that are uncommon in this Commonwealth because they have low population numbers or are only found in restricted geographic areas.

SC – Special Concern Population: A classification of plant species that the Department has determined to be a unique occurrence deserving protection. Among the factors used to classify a plant species as a Special Concern Population are the existence of unusual geographic locations, unisexual populations, or extraordinarily diverse plant populations.

TU – Tentatively Undetermined: A classification of plant species that are believed to be in danger of population decline but that cannot presently be included within another classification due to taxonomic uncertainties, limited evidence within historical records, or insufficient data.

Appendix G: **Reading List**

Armitage, J. (2016). *A Portable Latin for Gardeners: More than 1,500 Essential Plant Names and the Secrets They Contain.* University of Chicago Press, Chicago.

Art, H.W. (1987). *The Wildflower Gardener's Guide: Northeast, Mid-Atlantic, Great Lakes, and Eastern Canada Edition.* Storey Communications Inc., Pownal, VT.

Birdseye, C. and Birdseye, E. (1972). *Growing Woodland Plants.* Dover Publications Inc., New York.

Branhagan, A. (2016). *Native Plants of the Midwest: A Comprehensive Guide to the Best 500 Species for the Garden.* Timber Press, Portland, OR.

Branhagan, A. (2020). *The Midwest Native Plant Primer: 225 Plants for an Earth-Friendly Garden.* Timber Press, Portland, OR.

Brown, L. (2012). *Weeds and Wildflowers in Winter.* The Countryman Press, Woodstock, VT.

Burrell, C.C. (2006). *Native Alternatives to Invasive Plants.* Brooklyn Botanic Garden, Inc., Brooklyn.

Cullina, W. (2000). *The New England Wildflower Society Guide to Growing and Propagating Wildflowers of the United States and Canada.* Houghton Mifflin Co., New York.

Cutbirth, N. and Small, T. (2011). *Using Native Plants to Restore Community in Southwest Michigan and Beyond.* Cushing-Malloy, Inc., Ann Arbor, MI.

Daniels, J. (2020). *Native Plant Gardening for Birds, Bees & Butterflies: Upper Midwest.* Adventure Publications, Cambridge, MN.

Flanagan, J. (2005). *Native Plants for Prairie Gardens.* Fitzhenry & Whiteside Ltd, Markham, ON.

Goldberger, M. (2014). *Taming Wildflowers: Bringing the Beauty and Splendor of Nature's Blooms into Your Own Backyard.* St. Lynn's Press, Pittsburgh, PA.

Jennings, N.L. (2007). *Prairie Beauty: Wildflowers of the Canadian Prairies.* Rocky Mountain Books, Surrey, BC.

Johnson, L. (1998). *Grow Wild: Native Plant Gardening in Canada.* Random House of Canada, Toronto.

Johnson, L. (1999). *100 Easy-to-Grow Native Plants for Canadian Gardens.* Whitecap Books, Toronto.

Johnson, L. (2022). *A Garden for the Rusty-Patched Bumblebee: Creating Habitat for Native Pollinators.* Douglas & McIntyre, Madeira Park, BC.

Kershaw, L. (2002). *Ontario Wildflowers: 101 Wayside Flowers.* Lone Pine Publishing, Edmonton, AB.

Kimmerer, R.W. (2003). *Gathering Moss: A Natural and Cultural History of Mosses.* Oregon State University Press, Corvallis, OR.

Kimmerer, R.W. (2013). *Braiding Sweetgrass: Indigenous Wisdom, Scientific Knowledge, and the Teachings of Plants.* Milkweed Editions, Minneapolis, MN.

Ladd, D. and Oberle, F. (2005). *Tallgrass Prairie Wildflowers: A Field Guide to Common Wildflowers and Plants of the Prairie Midwest.* Morris Book Publishing, Guilford, CT.

Leopold, D.J. (2005). *Native Plants of the Northeast: A Guide for Gardening and Conservation.* Timber Press, Portland, OR.

Levine, C. (1995). *A Guide to Wildflowers in Winter: Herbaceous Plants of Northeastern North America.* Yale University Press, New Haven, CT.

Oldham, M.J. (2017). *List of the Vascular Plants of Ontario's Carolinian Zone (Ecoregion 7E).* Carolinian Canada and Ontario Ministry of Natural Resources and Forestry, Peterborough, ON. (Also available as a PDF at www.researchgate.net/profile/Michael-Oldham-4/publication /317731067_List_of_the_Vascular_Plants_of_Ontario)

Porter, F.W. (2013). *Back to Eden: Landscaping with Native Plants.* Orange Frazer Press, Wilmington, OH.

Richardson, M. and Jaffe, D. (2018). *Native Plants for New England Gardens.* Globe Pequot, Guilford, CT.

Small, N.C. and Small, T. (2011) *Using Native Plants to Restore Community in Southwest Michigan and Beyond.* Wild Ones, Kalamazoo, MI.

Steiner, L.M. (2006). *Landscaping with Native Plants of Michigan.* Voyageur Press, Minneapolis, MN.

Summers, C. (2010). *Designing Gardens with Flora of the American East.* Rutgers University Press, Piscataway, NJ.

Tallamy, D.W. (2016). *Bringing Nature Home: How You Can Sustain Wildlife with Native Plants.* Timber Press, Portland, OR.

Tallamy, D.W. (2020). *Nature's Best Hope: A New Approach to Conservation That Starts in Your Yard.* Timber Press, Portland, OR.

Vogt, B. (2017). *A New Garden Ethic: Cultivating Defiant Compassion for an Uncertain Future.* New Society Publishers, Gabriola Island, BC.

Williams, D. (2010). *The Tallgrass Prairie Center Guide to Seed and Seedling Identification in the Upper Midwest.* University of Iowa Press, Iowa City, IA.

Wilson, W.H.W. (1984). *Landscaping with Wildflowers and Native Plants.* Ortho Books, San Francisco.

Index of Common Names

Photo Credits